I0213928

American Seacoast Forts

A Directory with Period Military Maps 1890-1950

Volume 2

The Mid-Atlantic, South Atlantic, and Gulf Coasts: Delaware Bay to Galveston

Prepared by

Terrance C. McGovern
Mark A. Berhow
Glen M. Williford

Published by the Coast Defense Study Group Press
2025

All original compositions are © copyright 2025 by the individual authors
& the Coast Defense Study Group, Inc.

PHOTOGRAPHS ARE IDENTIFIED IN THE CAPTIONS AS TO SOURCE. ALL PHOTOGRAPHS ARE
COPYRIGHTED BY THE SOURCE, UNLESS THE SOURCE IS THE UNITED STATES GOVERNMENT,
IN WHICH CASE PHOTOGRAPHS ARE IN THE PUBLIC DOMAIN.
ALL RIGHTS RESERVED. NO PART OF THIS BOOK MAY BE USED OR REPRODUCED
WITHOUT WRITTEN PERMISSION OF THE AUTHORS.

PLEASE DIRECT ANY COMMENTS OR CORRECTIONS TO THE PUBLISHER - INFO@CDSG.ORG

IBSN 978-0-9748267-7-7 (Hardcover B&W)
LIBRARY OF CONGRESS CATALOG CARD NUMBER 0000000000

Library of Congress Cataloging-in-Publication data
American Seacoast Forts: A Directory / Terry McGovern, Mark Berhow and Glen Williford
p. cm.
Includes bibliographical references and index.
Library of Congress Control Number: 0000000000
ISBN 978-0-9748167-7-7 (h.c.)
1. Military History, 2. Artillery I. Terry McGovern, Mark Berhow and Glen Williford

First Edition: August 2025
Printed in the USA by Ingram Spark

Cover Photographs
Front cover: Fort Wool, Virginia (Terry McGovern)

Rear cover (clockwise from upper left): Map of the Entrance to Chesapeake Bay VA 1919 (National Archives);
Battery Cooper, Fort Pickens unit, Gulf Shores Natl Seashore, FL (Terry McGovern); Fort Saulsbury, DE 1930s
(National Archives); Battery Jasper and BCN 230, Fort Moultrie Natl. Hist. Site, SC (Terry McGovern)

THE COAST DEFENSE STUDY GROUP, INC.
CDSG.ORG

The Coast Defense Study Group, Inc. (CDSG) is a tax-exempt corporation dedicated to the study of seacoast fortifications. The purposes of the CDSG include educational research and documentation, preservation and interpretation of historic sites, and assistance to other organizations dedicated to the preservation and interpretation of coast defense sites. Membership is open to any person or organization interested in the study or history of coast defenses and fortifications. Membership in the CDSG will allow you to attend annual conferences, special tours, and receive quarterly newsletter and journal. To find our more about the CDSG, please visit the CDSG website at **cdsg.org.**

This work is copyrighted by the Coast Defenses Study Group ePress, all rights to original work are reserved by the CDSG, Inc. This book cannot be reproduced or distrubuted in anyway without the express written consent of the CDSG, Inc. All iamages credited to the US. Army and the National Archives are in the public domain, but all other text and photographs are property of the authjors and the CDSG, Inc. For additional information about this book, please inquire.

Acknowledgments

This book is dedicated to three key members of the CDSG who began research on this subject in the 1970s and provided much of the initial information for those that followed in this study. Robert Zink (member number 1) and Glen Williford (member number 2) were both dedicated in their studies and unfailing supportive to those that sought more information on American seacoast defenses. The third key CDSG member was Bolling W. Smith who did a large amount of research work in obtaining copies of documents from the National Archives of which many were used in this compilation.

Sighting a 10-inch gun at Fort Standish, Massachusetts
(Leslie Jones, Digital Commonwealth, Massachusetts Online Collection, Boston Public Library)

The CDSG Press

Coast Defense Study Group Press is a division of Coast Defense Study Group (cdsg.org), which publishes books of historical interest, especially concerning seacoast fortifications. The CDSG Press also offers an ever-expanding number of key reprints reports and manuals in electronic PDF format on compact disks. To order these books and other **CDSG Press** publications, please access the **CDSG Press** pages on the **CDSG web site** at **cdsg.org.**

CDSG Press is interested in new titles, especially those dealing with fortifications, please contact Terry McGovern at 703/538-5403 or at tcmcgovern@att. net if you have a title that you are seeking to have published. Visit www.cdsg.org/press.

Under the CDSG Press label, our organization has published:

Notes on Seacoast Fortification Construction by Col. Eben E. Winslow, 1920, 428 pp. 1994 reprint HC with bound drawings
Seacoast Artillery Weapons Technical Manual (TM) 9-210 by U.S. War Dept. 1944, 202 pp. 1995 reprint PB
The Service of Coast Artillery by F. Hines & F. Ward, 1910, 736 pp. 1997 reprint HC
Permanent Fortifications & Sea-Coast Defenses by U.S. Congress, 1862, 544 pp. 1998 reprint HC
American Coast Artillery Material Ordnance Dept. Doc#2042 by U.S. War Dept., 1922, 528 pp., 2001 Reprint HC
American Seacoast Defenses: A Reference Guide (3rd Edition) by Mark A. Berhow, (2015) 732 pp. HC
The Endicott & Taft Board Reports, reprint of original reports of 1886 and 1905 by U.S. Congress, 525 pp. 2007 HC
Artillerists and Engineers: The Beginnings of US Fortifications 1794-1815 by Col. Wade, U.S. Army, 226 pp. 2011 PB
World War II Harbor Defenses of San Diego by Commander (Ret.) Everett, U.S. Navy, 226 pp. 2020 HC

The CDSG Presss offers these Hole in the Head Press Books

Artillery at the Golden Gate by Brian B. Chin (Hole-in-the-Head Press), 176 pp. 2019 PB
Fort Baker Through the Years by Kristin L Baron and John A. Martini (Hole-in-the-Head Press), 99 pp. 2013 PB
Rings of Supersonic Steel (3rd Edition) by Mark Morgan and Mark Berhow (Hole-in-the-Head Press), 358 pp. 2010 PB
The Last Missile Site by Stephen Hailer and John A. Martini (Hole-in-the-Head Press), 158 pp. 2010 PB
To Defend and Deter by John Lonnquest and David Winker (Hole-in-the-Head Press), 432 pp. 2014 PB

The CDSG Press
1700 Oak Lane
McLean, VA 22101-3322 USA

AMERICAN SEACOAST DEFENSES
A DIRECTORY OF AMERICAN SEACOAST DEFENSES 1890-1950

Table of Contents

U.S. Coast Artillery 1890-1945 Harbor Defense locations

Directory to American Seacoast Forts 1885-1950 in 4 volumes
    (21 Continental US Harbor Defenses and Overseas Harbor Defenses)
• Volume 1 North Atlantic Coast: New England – Long Island Sound - New York
• Volume 2 Mid-Atlantic, South Atlantic, and Gulf Coasts: Cheasapeake Bay to Galveston
• Volume 3 Pacific Coast San Diego to Puget Sound
• Volume 4 Alaska and Overseas Bases: Hawaii - Philippines - Panama - Caribbean - Newfoundland
        - Bermuda

VOLUME 2: THE MID-ATLANTIC, SOUTH ATLANTIC, AND GULF COASTS: DELAWARE BAY TO GALVESTON

VOLUME 1: THE NORTH ATLANTIC COAST: PORTLAND TO NEW YORK

VOLUME 3: THE PACIFIC COAST: SAN DIEGO TO PUGET SOUND

VOLUME 4: ALASKA AND THE OVERSEAS BASES

The Harbor Defenses of Sitka, Alaska—WWII Program sites
The Harbor Defenses of Seward, Alaska—WWII Program sites
The Harbor Defenses of Kodiak, Alaska—WWII Program sites
The Harbor Defenses of Dutch Harbor, Alaska—WWII Program sites

The Harbor Defenses of Honolulu—Fort and Gun Battery Descriptions
The Harbor Defenses of Pearl Harbor—Fort and Gun Battery Descriptions
The Harbor Defenses of Kaneohe Bay and the North Shore of Oahu—Fort and Gun Battery Descriptions

The Harbor Defenses of Manila Bay—Fort and Gun Battery Descriptions
The Harbor Defenses of Subic Bay—Fort and Gun Battery Descriptions

The Harbor Defeses of Cristobal, Panama Canal Zone Atlantic side— Fort and Gun Descriptions
The Harbor Defenses of Balboa, Panama Canal Zone, Pacific side— Fort and Gun Battery Descriptions

The Harbor Defenses of Vieques Sound, Puerto Rico, Virgin Islands (Roosevelt Roads)
    —WWII Program sites
The Harbor Defenses of San Juan, Puerto Rico—WWII Program sites
Planned Defenses in Guantanamo Bay, Cuba
Planned Defenses in Jamaica
Planned defenses in Trinidad
Harbor Defenses in Newfoundland, Canada—WWII Program sites
Defenses in Bermuda—WWII Program sites

INTRODUCTION

The United States has long focused on defending its seacoasts against overseas enemies due to its geopolitical situation with its long coastlines and generally peaceful borders with Canada and Mexico. Earlier American fortification efforts resulted in the First System, the Second System, and the Third System of coastal defenses. The great brick and stone forts built or remodeled during 1820 to 1860 are well known from their military importance and use during the American Civil War. For many years after that great internal conflict most U.S. fortification efforts languished. After 1885, due to the great advances in military technology and America's increasing worldwide economic presence that the United States embarked on a new round of fortification building to protect its shores. The U.S. Army expended much of its limited manpower and resources to protecting America's coast from 1890 to 1950.

The technical development and tactical objectives of the coast defenses of the modern era (1890 to 1950) is a product of America's earlier policies and experiences. Until the advent of air power and the missile age, the defense of the United States has primarily been one of defending our shores from naval attack. Only during the nation's early years did the threat of land invasion exist. Accordingly, the United States relied on the ships of its Navy to provide the first line of defense. Its U.S. Army was called upon to provide the second line of defense by building forts at key points along its coastline to defend major harbors. This defense policy of denying an enemy fleet access to its major harbors and anchorages developed into the array of former coastal fortifications that remain today.

Based on concerns over external threats and on internal politics, the United States government has built coastal fortification in a series of construction programs. After inheriting the remains of the fortifications from the colonial era and the revolutionary war, the first in a series of national fortification construction programs began in 1794 and these programs continued into the late 1940s. For the ease of use in this book, these fortification periods have been organized into distinct groups and have been named the following: First System (1794-1801), Second System (1802-1815), Third System including the American Civil War (1816-1867), 1870s Period (1868-1879), Endicott Program (1885-1904), Taft Program (1905-1916), World War I & Interwar Period (1917-1939), and the 1940 Modernization Program including World War II (1940-1950).

The legacy of these seacoast defenses is a series of concrete structures scattered along America's coastline, many now in public shoreline parks. The nature of fortifications, the fact that they were designed to withstand the pounding of naval artillery, has allowed these massive structures to withstand the attack of both the natural elements and economic development. That these structures are still standing many years after their effective use ended draws our attention to them. They captivate us regardless of whether it's a large brick and stone multi-story structure surrounded by a dry ditch or an odd shaped, concrete structure covered with thick vines or bright graffiti and surrounded by worn fences sporting weathered warning signs. Visitors to our seashores are curious about the nature of these structures. Some of the questions that they ask are: What are these structures? Why are these structures here? When were these structures built? This directory provides answers to many of these questions. Many of these parks have visitor's centers and gift shops selling range of books and other items, but few have any books which explain the fortifications that once existed at that site. This directory will fill that void.

This directory is a companion work to the CDSG Press's *American Seacoast Defenses – A Reference Guide* by Mark Berhow. This directory is to aid students of American seacoast fortifications to locate and visit the key defenses of the "Modern Era" of coastal defense (1890 to 1950), which is defined by its use of concrete, steel, and breech-loading rifles. The following pages provide a brief review of the function and history of the development of coastal fortifications in the United States. This history reflects the politics of changing external threats to our nation and rapid advancement of military technology. The directory's focus is a guide to the "Modern Era" of fortifications along the Atlantic, Pacific, and Gulf coasts, as well as U.S. overseas bases by providing key maps, plans, photographs, and short description of their history.

The directory is organized into several sections. The first section is brief history of the modern era of American coastal defenses, including background on the U.S. coast artillery material, organization, armament, design of military reservations, and garrison life. The second section is a brief description of the history and its current status (ownership, public access, remaining assets, things to see, etc.) of the major military reservations that had seacoast artillery and a short history of each major concrete gun battery. This history includes each battery is described as to its rational, authorization, construction and transfer dates, engineering cost, naming citation, armament, service history, ultimate disarming, and current status. These battery histories do not include railway and mobile artillery sites, including those with Panama mounts. Also excluded are temporary batteries, especially those using loaned naval guns, and anti-aircraft batteries that did not also serve a seacoast role. The directory is organized by Harbor Defense around the United States clockwise from Portland Maine to the Puget Sound in Washington State, followed by the Alaskan defenses of World War II, and the defenses in Hawaii, the Philippines, Panama, the Caribbean, Newfoundland and Bermuda.

Supporting the directory of defense sites is a compilation of maps for all the harbor defense reservations utilized during the period of 1900 to 1946. The map collection includes general maps of the location of elements (sites) for each defended harbor and the individual location site maps showing buildings, gun emplacements, fire control stations and other elements. Each Harbor Defense section has an overall 1920s-30s period map of the defense sites and a selected set of Confidential Blueprint Maps for each military reservation.

The maps are arranged more or less in order from the south to the north. A series of map symbol and abbreviation keys from 1921 and 1945 are included in the introduction. Some harbors (Baltimore, Potomac River, Cape Fear, Port Royal, Savannah, Tampa, Mobile, and the Mississippi River) did not receive new defenses during World War II.

12-inch Rifle on a barbette mount, Battery Godfrey, Fort Winfield Scott, California,
(Golden Gate National Recreation Area Collection, NPS)

CHRONOLOGY

Key events during the Modern Era of American coast defenses

1875 – Funding for new construction of coast defenses is stopped by the U.S. Congress

1883 – The U.S. Navy begins the first new construction program since the Civil War

1885 – President Cleveland appoints a joint army, navy, and civilian board headed by the Secretary of War, William Endicott, to evaluate the threats and needs for U.S. coastal defenses (Endicott Board)

1886 – The Endicott Board reports on state of the U.S. defenses and recommends a $126 million construction program of breech-loading cannons and mortars, floating batteries, warships, and submarine mines in 29 locations around the nation

1888 – Congress creates the Board of Ordnance and Fortifications to test weapons and implement the Endicott Program
– Dynamite guns were developed to fire high explosive shells using compressed air

1890 – Congress approves funding for the construction of first Endicott Program batteries

1892 – The first Endicott Program battery is completed (Gun Lift Battery Potter)

1893 – First group of controlled mine casemates are completed

1894 – Buffington-Crozier disappearing carriage for 8-inch and 10-inch guns developed

1896 – Development of the Buffington-Crozier disappearing carriage for 12-inch guns

1898 – The Spanish-American War – 150 coast artillery pieces mounted
– U.S. adds the Philippines, Guam, Puerto Rico as colonies; establishes military bases in Cuba; annexes Hawaiian Is.

1899 – 288 heavy coast artillery guns, 154 rapid-fire guns, and 312 mortars have been mounted

1901 – Reorganization of U.S. Army artillery corps to 30 batteries of field artillery and 126 companies of coast artillery

1902 – Work begins on the fortifications of Corregidor, Philippines

1904 – Work begins on the Panama Canal
– First specially built mine planters constructed

1905 – President Roosevelt appointed a joint army, navy, and civilian board headed by the Secretary of War William Taft, to review the Endicott Program and to bring it up to date

1906 – The Taft Board reports on state of the U.S. defenses and recommends improvement in existing defenses by adding searchlights, electrification of defenses, and a modern system of fire control, as well as new defenses for newly acquired overseas bases

1907 – Establishment of the separate U.S. Army Coast Artillery Corps

1914 – World War I begins; Panama Canal opens

1915 – Report of the Board of Review on the coast defenses of the U.S., Panama Canal, and the Insular Possessions

1916 – First Coast Artillery anti-aircraft units formed

1917 – U.S. enters the World War II
– Construction begins on the first long-range barbette batteries using existing 12-inch gun barrels

1918 – End of the World War I

1920 – Construction of the first Panama mount for 155m GPF guns in the Canal Zone

1922 – Washington Naval Treaty limits naval construction and Pacific fortifications
– 16-inch gun and howitzer barbette batteries are constructed

1925 – Ten U.S. Harbor Defenses on active status and 15 are on caretaker status

1937 – Construction begins on first 16-inch casemated gun battery (Battery Davis)

1939 – Outbreak of World War II; U.S. Coast Artillery Corps has 4,200 troops

1940 – Congress approves the 1940 Modernization Program for 19 harbors in the U.S.
– U.S. draft begins, coast artillery units brought up to wartime strength, national guard units federalized

1941 – U.S. enters the World War II
– Establishment of the Harbor Entrance Control Posts (HECP)

1942 – U.S. Coast Artillery Corps has 70,000 troops

1945 – End of the World War II

1948 – All construction efforts cease, the coast defenses are abandoned, and armament salvaged

1950 – Disestablishment of the U.S. Coast Artillery Corps; remaining units reunited with Field Artillery

DESIGN AND FUNCTION OF AMERICAN SEACOAST DEFENSES IN THE MODERN ERA 1890-1950

Historical Development of American Coast Defenses during the Modern Era

The key development that led to the new American coast defense era was the development of new heavy rifled breech-loading guns that had a longer range, were more accurate and delivered a heavier projectile than the muzzle-loading smoothbore cannons of the Civil War. These new guns were made of high-quality steel that were lighter and stronger, which took advantage of new propellants that replaced gunpowder. Equally important was the development of effective breech mechanisms that could withstand the high pressures and temperatures generated by the new guns and allow for the gun to be loaded from the rear instead of the muzzle, which increased the rate of fire and allowed for improved protection of the gun crew. The new guns and mortars could accurately fire projectiles at effective ranges that were two to three times farther than the muzzle-loading smoothbore cannons used during the American Civil War. These developments coincided with the building of the new steel naval vessels that featured these new big guns starting in 1875. However, between 1875 and 1890 the U.S. Congress did not appropriate any funds for the construction of new coastal fortifications.

Seacoast Defenses built after the Endicott Board Report 1886-1904

As U.S. coastal fortifications were allowed to deteriorate in 1870s and early 1880s, new steam powered, ocean navigating iron warships were being built by foreign navies. As the U.S. Navy embarked on its new construction program, it required protected bases for its operations. The military began to lobby to overhaul the obsolete existing defenses. In 1885, a board was created by U.S. Congress to examine and report upon the state of U.S. coastal defenses. The board headed by the U.S. Secretary of War, William C. Endicott, was comprised of four officers from U.S. Army, two officers from the U.S. Navy, and two civilians. This joint board made an extensive study of fortifications, type of armament, and defense that would be needed, by evaluating current European developments. In 1886, the Endicott Board published its recommendations for new coastal fortifications to be built at 29 key harbors, along with floating batteries, torpedo boats, and submarine mines. The board's original plan called for over 1,300 guns and mortars of 8-inch or larger of the newest design to be installed. The costs of board's recommendations were estimated to be $126 million dollars (in 1886-dollar value). While the U.S. Congress took no immediate action on the board's report, the estimates provided in the report would be cited for the next 20 years as a measure of the construction progress of this new generation of U.S. coastal fortifications.

In 1888, Congress established an U.S. Army board for ordnance and fortifications who as charged with testing new weapons and to design new coastal fortifications. In 1890 Congress made the first appropriations for the first new construction of coastal fortifications in 16 years with an initial funding of $1.2 million dollars. This funding was for the first of new defenses: a 12-inch barbette battery at San Francisco; an 8-inch disappearing battery at New York Harbor; a 12-inch gun lift battery at Sandy Hook, New Jersey; and for 12-inch mortar batteries at Sandy Hook and at the Presidio, San Francisco. The design of these new coastal batteries would set the pattern of coastal defenses that would be duplicated at all of America's major harbors, and this was the beginning of what would become known as the "Endicott Program" of American seacoast defenses.

The designs used for the Endicott Program coastal fortifications demonstrated the shift in importance from the large multi-tiered multi-gunned "fortresses" to weapons emplaced in dispersed concrete "batteries" protected by earthen embankments. The "fort" became a defined reservation of land that contained guns of a range of calibers, along with the housing for the men required to man these defenses, and supply and maintenance buildings. The weapons were grouped into batteries containing from one to sixteen guns. The batteries were located along the shoreline to maximize their range and field of fire and were designed to blend into the landscape as not to be seen from the sea. The armament of these batteries ranged from weapons to engage enemy capital ships to small-caliber rapid fire guns to knock out fast moving torpedo boats, as well as to protect the fields of electrically controlled submarine mines from minesweepers. The dominance of the armament during this period is reflected by the dramatic increase in the time and cost in constructing a gun barrel and breech mechanism along with its carriage over that of the weapons of the earlier periods.

The primary weapons of the Endicott Program were 8-inch, 10-inch and 12-inch rifled breech-loading guns, a growth in size that reflected the need to match the increase in the size of opposing naval guns. These guns were mounted on both barbette and disappearing carriages that had a maximum elevation of 15 degrees and range of about seven to eight miles. The relatively unique American "disappearing" carriage allowed the gun to be raised over the parapet by using a counterweight to fire. The energy from the recoil caused the gun to drop back down behind the parapet into the emplacement to be reloaded while being protected from direct fire from attacking warships. These heavy weapons were mounted in large concrete emplacement with thick frontal walls that were in turn protected by many feet of earthen fill. Located below or adjacent to the firing platform were support areas that included the ammunition magazines containing projectiles and powder propellants. About three hundred of these heavy guns were installed around the United States during the Endicott Program in batteries of from one to six guns. It was a less expensive alternative to the armored turret mount favored by several European nations.

10-inch disappearing guns in Battery Hale, Fort Greble, Rhode Island (C.T. Gardner Collection)

The other large caliber weapon installed during the Endicott Program was the short-barreled 12-inch mortar. The mortar was designed to fire a shell in a high arc that descended down onto the lightly armored decks of warships of that era. To increase the opportunity of making a hit, these mortars were emplaced in groups of eight to sixteen mortars in square concrete pits that were protected by earthen hills. The use of these pits would give maximum protection to the mortars and their magazines from the flat trajectory of naval gunfire of the era. About four hundred of these mortars were installed around the United States during the Endicott Program.

The secondary smaller caliber weapons were installed to protect the controlled submarine mine fields from small craft that could sweep paths through the mines for larger warships, and to protect from attack by newly developed fast torpedo boats that could potentially penetrate the harbor and torpedo the shipping within. These threats called for guns that could be aimed, loaded and fired very rapidly. While not

12-inch mortar firing at Battery Alexander, Fort Barry, California (B.W. Smith Collection)

specified by the Endicott Report, several new gun and carriage systems were developed for this role ranging from 3-inch to 6-inch is size. These guns were generally mounted on either disappearing carriages or on pedestal carriages with simple steel shields. The concrete emplacements for these guns had low parapets and magazines below the guns. A rapid-fire battery had between two and six guns per battery. Over five hundred rapid-five weapons were installed during the Endicott Program.

While the use of submarine mines or "torpedoes," as well as channel obstructions or barriers, has a long history in defense of harbors, it was during the Endicott Program that a widespread and a structured use of submarine mines occurred. The U.S. Army developed a system of controlled submarine mines; stationary explosive devices located below the surface of water where ships were likely to pass. The submarine mines used from 1890 to 1930 were the buoyant type (floating but anchored to the sea bottom), though during World War II the buoyant mines were replaced with ground mines (stationed on the sea floor). The mines were only deployed during times of war or for practice, otherwise they were stored disassembled ashore—the mines and their control cables became defective after extensive exposure in the water. The controlled mines were connected to shore by undersea cables and could be exploded by electrical switches from a control board on shore by the soldiers manning the mine defenses when a warship passing over these mines or by direct contact. Controlled mines were usually laid in rows across the key shipping channels to create a group of mines, usually 19, which would cover a space of about 2,000 feet long in water up to 250 feet deep. Several groups of mines were to be deployed to create a field of mines. The U.S. Army Coast Artillery Corps had dedicated units to man the mine planting vessels, fire control stations, mine and cable storage facilities, mine casemates and switchboards, and loading wharves.

The Endicott Program roughly covered a period from 1885 to 1905, and the coast artillery function was a key mission of U.S. Army during this time (and made up a large percentage of total U.S. Army manpower). This also required a more technical trained soldiers to man them which led to the U.S. Army's Artillery branch to be reorganized in 1901 and 1907 to create the U.S. Coast Artillery Corps.

Planting a mine (Stillion Collection NPS, Gulf Shores Natl. Seashore)

Seacoast Defenses built after the Taft Board Report 1906-1916

In 1905, a new National Coast Defense Board headed by U.S. Secretary of War William Howard Taft was organized by President Theodore Roosevelt and charged with reviewing the progress of Endicott Program construction and update it. In the 20 years since the original Endicott Board report was presented, numerous technical and political developments had taken place. The Board, informally known as the "Taft Board" after its chairman, established new cost estimates and its recommendations were primarily concerned with modernization of existing coastal fortifications and adding coastal defenses to the overseas territories gained after the Spanish-American War including Hawaii and the Philippines and other locations.

The modernization of existing defenses included the electrification of lighting, communications, and ammunition handling equipment, both at the batteries and throughout the fort. The early emplacements had loading platforms widened and projectile hoists were installed to improve the rate of fire. The report recommended the use of searchlights for nighttime illumination of harbor entrances. During the Taft Program was the finalized development and implementation of a coordinated system of target information gathering and processing that greatly improved the target accuracy of the major caliber guns and mortars. Up to this time, the aiming of guns at a target had been generally done from each battery with basic sighting instruments and combination of luck and experience. The new system was based on triangulation using two observers with telescopic instruments at separate position finding stations or "base end" stations communicating with the newly developed telephones to a centralized battery plotting room that provided real-time tracking and firing coordinates on a moving target. The battery plotting room personnel would mathematically process this sighting information and other data into aiming instructions that would then be transmitted to each gun emplacement.

The 14-inch gun on a disappearing carriage the Taft-era Battary Osgood, Fort MacArthur, California
(Fort MacArthur Museum)

While the Taft Board's recommendations on the construction of new fortifications was largely limited to existing defenses at Eastern Long Island Sound, San Diego, Puget Sound, Columbia River, and Chesapeake Bay in the continental United States, major new construction projects were planned for the Philippines, Panama, Hawaii, Cuba, Puerto Rico, Alaska, and Guam. Plans for Cuba, Puerto Rico, Alaska, and Guam were not carried out. New defenses were added for the port of Los Angeles in 1909. These "Taft Program" defenses varied little from the overall designs used during the Endicott Program. Variations from the Endicott Program were the product of advancing naval armaments and the U.S. Army's twenty years of experience of operating coast defenses. To match the increased caliber of naval guns, a new disappearing gun of 14-inch caliber was developed. Another characteristic of the Taft Program batteries was the increased dispersion of batteries. The reduce density of weapons can be seen in the construction of several one-gun, 14-inch batteries and the reduction of mortar batteries from 8 mortars to 4 mortars (4 per pit to 2 per pit).

During the Taft Program, several one of kind of projects were undertaken. The Endicott Board Report called for 16-inch guns but work on the development stalled after the construction of one gun tube in 1895. Two unique 16-inch disappearing batteries were finally built in Panama and the Long Island Sound. In the Philippines army-designed armored turrets were custom built for a very small island in Manila Bay. Four 14-inch guns were mounted in two turrets at Fort Drum, which also became known as the "concrete battleship."

World War I and the Interwar Period (1917-1939)

The march of technical improvements in naval weapons continued through improvements in naval fire control and the ability for naval turrets to elevate their guns. By 1915 the newer battleships had guns that could out range the effective range of the coast artillery emplaced during the Endicott and Taft Programs. The increased angle of fire of the newer battleships also threatened the disappearing carriage batteries which were not protected from the plunging fire of these new battleships. In 1915, a National Board of Review on the coast defenses of the U.S., Panama Canal, and the Insular Possessions recommended the construction of new batteries mounting 12-inch and 16-inch guns on higher elevation, longer range barbette carriages. While efforts to introduce these coast artillery weapons had begun, the demands of World War I placed

Firing one of the 12-inch guns of the post-WWI Battery Kingman, Fort Hancock, New Jersey (NARA)

the modernization of coast defenses on hold. Many Coast Artillery units were transformed into field and heavy artillery units for service in France. As the United States was short on long range field artillery,12-inch mortars, 10-inch, 8-inch, and 6-inch gun barrels were removed from several coast artillery batteries. These existing gun barrels, ranging from 6-inch to 14-inch in caliber were quickly mounted on railway and tractor-drawn carriages. While the United States involvement in World War I was brief, it resulted in the Coast Artillery Corps mission to be divided into three specialized areas as compared to its single mission before the war. These missions, based on armament type, were fixed coast defense weapons (including controlled mines), mobile seacoast artillery, and anti-aircraft artillery.

The development of the airplane as a ground attack weapon during the World War I added the task of defending both the mobile ground army and the shores of United States from attacks by aircraft to the Coast Artillery Corps' mission. The U.S. Army developed fixed and mobile anti-aircraft weapons, as well as accessory equipment such as aircraft sound locators, rangefinders, searchlights, specialized fuses, and mechanical fire direction calculators. The primary weapon for the defense against aircraft was the 3-inch gun on a fixed carriage (in batteries of three or four guns) located at existing coast defense posts. This weapon was later supplemented with .50 caliber machine guns and mobile 3-inch AA guns. By 1938 larger caliber anti-aircraft guns were introduced including the 90 millimeters (mm), the 105 mm, and 120 mm guns.

The mobile coastal defense mission came about because of the lack of U.S. heavy artillery for the troops in Europe. Existing gun barrels, ranging from 6-inch to 14-inch in caliber were quickly mounted on railway and tractor-drawn carriages. The construction of the new mobile carriages for guns, such as the railway mounts, took months and most of these weapons never reached European theater before the war ended. The availability of this ordnance material, especially considering the economics of using existing weapons and increased desirability of weapon mobility in the interwar period, made mobile coast artillery an attractive alternative to building new fixed coast defenses. The primary railway guns selected for coast artillery use from this large stock of World War I material were 8-inch guns and the 12-inch mortars mounted on new railway carriages. Added later was an improved version of the wartime 14-inch railway gun of which only four were constructed by 1920. The surplus mobile field artillery mounted on carriages designed for road movement included the 155-mm GPF gun (derived from a 1917 French design) of which almost a thousand were available. This powerful gun became the standard tractor drawn weapon for coast defense use against secondary targets.

a 155 mm G.P.F. mobile mount in a field position at Long Point, California (Ruhlen Collection)

While mobile coast artillery had the advantage of being able to respond to coastal areas most threaten when enemy naval forces approached, both railway and tractor-drawn weapons lacked the accuracy and protection of fixed coast artillery. Without solid and steady firing platforms and the precision of pre-calibrated fire control networks, as well as the inability of the carriages of the mobile guns to quickly track horizontally moving targets made mobile artillery much less effective than weapons in fixed emplacements. Prepared locations with circular arcs of track were prepared at a few select locations. For the 155 mm GPF mobile artillery, simple circular concrete bases were designed. These circular bases improved stability during firing and provided for rapid azimuth adjustment for horizontal tracking. One of the most common base designs developed for the 155 mm GPF guns was a central pivot and a curved rail embedded in concrete, which the gun's split carriage would traverse. This design was first constructed in the Panama Canal Zone, so this design became known as the "Panama Mount". Given the limitations of mobile coast artillery, their use was primarily an augmentation of existing defenses or to provide protection during the construction of permanent fixed coast artillery. Due to the low level of military appropriations during the 1920s and 1930s, mobile coast artillery was the only available weapon to defend vital locations until new permanent defenses could be funded and constructed.

Given the low level of overall U.S. military funding during the 1920s and 1930s, the construction and development of new fixed coast defenses were limited. The need for economy and to allow for higher gun elevations led to the abandonment of the disappearing carriage and its complex two-level emplacements. Among the last disappearing carriages built were for two 16-inch single-gun batteries (one in Panama and the other in the Long Island Sound). A newly designed high angle barbette carriage for existing stocks of 12-inch Model 1895 guns allowed effectively doubled the range of the guns over the same 12-inch gun mounted on a disappearing carriage. Construction of fifteen long-range dual gun 12-inch batteries was started in 1917 and completed by the late 1920s. The emplacement design was a departure from those of the Endicott Program. The battery design had two guns located much further apart, each gun in the center of a large ground level concrete pad to allow for an all-around field of fire. Located between the two guns was an earth-covered reenforced concrete structure containing magazines for shells and powder, the power and plotting rooms, and storage rooms. Protection of the guns from naval fire was based on dispersion; the wide separation of the key elements of the battery. Other than camouflage and nearby anti-aircraft guns these batteries had no protection from air attack. The development of a new 16-inch gun and carriage with a range of nearly thirty miles which exceeded the range of all existing naval warships was completed in 1919. The 16-inch in emplacements that were very similar to those used for the long-range 12-inch barbette batteries, with an increased distance between the two guns of the battery and the dispersed location of magazines in simple storehouses connected by a rail system. Only a few of the U.S. Army designed barrels had been constructed when nearly sixty 16-inch barrels became available from the U.S. Navy. This windfall was due to the Washington Naval Treaty of 1922 that resulted in the cancellation of several U.S. battleships and battle cruisers then under construction. The naval 16-inch barrels were to be installed in modified U.S. Army barbette carriages after 1925. Six new twin-gun 16-inch batteries were built between 1922 and 1934. During the Interwar Period, the construction of new batteries including both long range 12-inch and 16-inch guns, amounted to little more than twenty new batteries. The coming of World War II would inject new life into building modern U.S. coast defenses.

The 1940 Modernization Program and World War II (1940 -1950)

During the 1930s the U.S. Army began discussing how to protect new coast artillery batteries from attack by aerial bombardment. The debate centered on the expense of designing and construction of turret mounts for 12-inch and 16-inch guns as compared to developing protective structures made of concrete and steel. It was practical economic and time frame requirements that resulted in the eventual selection of a concrete casemate structure design to protect the current type of barbette mounts.

The prototypes of this new type of major caliber battery were built at the San Francisco defenses, during 1937-1940. These emplacements were designed for two 16-inch guns located about six hundred feet apart with complete overhead cover. Located between the two guns along a service gallery were the ammunition magazines, power generators, and support areas. The 16-inch guns were enclosed in reinforced concrete casemates. The battery's structure was made up of eight to twelve feet of steel reinforced concrete which was topped by up to twenty feet of earth as additional protection. The entire battery structure was designed to withstand a direct hit from a naval projectile or an aerial bomb. When completed the southern San Francisco battery at Fort Funston emplacement looked like a small hill, especially when camouflage and natural ground cover was added to the structure. The only exposed portions of the battery were the casemates where the gun barrels projected out through armor shields and concrete canopies. A second casemated battery on a hilltop north of San Francisco was also undertaken. Four more casemated batteries were begun at Narragansett Bay, the Delaware River, and Chesapeake Bay in 1940-1941.

In 1940 the Harbor Defense Board was charged with developing a master plan to update the harbor defenses of the continental United States. Eighteen coastal areas in United States were selected for modernization due to their military and economic importance - Portland, Portsmouth, Boston, New Bedford, Narraganset Bay, Long Island Sound, New York, Delaware Bay, Chesapeake Bay, Charleston, Key West, Pensacola, Galveston, San Diego, Los Angeles, San Francisco, Columbia River, and Puget Sound. The Harbor Defense Board recommended the adoption existing stocks of 16-inch gun as the primary weapon and 6-inch gun as the secondary weapon for the modernization program. In all the board proposed building twenty-seven new 16-inch casemated batteries; the casemating of 23 existing primary batteries (both long-range 12-inch batteries and older 16-inch batteries; and building fifty new 6-inch two-gun barbette carriage batteries, which would provide long-range fire (15 miles maximum) against secondary warships. The new 6-inch batteries would be supported by 63 existing secondary batteries, mostly 6-inch and 3-inch barbette guns from the Endicott and Taft Programs, which would be retained. Upon completion of these new defenses 128 existing obsolete coastal batteries would be eliminated. The board estimated that the

One of the 16-inch guns of Battery Steele, Peaks Island M.R., Maine (Joel Eastman Collection)

A 6-inch gun of Battery Cravens, Peaks Island M.R., Maine, with a disguised SRC 296A radar behind
(Joel Eastman Collection)

whole program would require three years to complete and cost about $82 million during 1941-3. Formal approval of this modernization plan, which would become known as the "1940 Harbor Defense Modernization Program" or the "1940 Program," was approved in September 1940.

The 1940 Harbor Defense Modernization Program greatly simplified the task of Coast Artillery Corps by reducing the number of types of batteries as well as the overall number of batteries needed to carry out their coast defense mission. This allowed a reduction in personnel and the level of effort to maintain, training and supply the pre-1940 batteries. Some of coast artillery that was declared obsolete was shipped to Allied nations to supplement their defenses, but most were scrapped for the war effort. As the nation moved closer to war, additional coastal defense projects were added to the 1940 Program, especially at newly acquired overseas bases, such as Trinidad, Bermuda, Newfoundland, and in areas where the enemy threats seem greater, such as Alaska, Hawaii, Puerto Rico, and the Canal Zone. It also became apparent that planning, construction and emplacement of the many new batteries called for the 1940 Program was going to take a much longer time then original envisioned, especially as the program was competing with rapid expansion of the whole U.S. Army and U.S. Navy. By the middle of July 1941, only four 16-inch gun batteries were ready for action and construction work had been started on just five others. With pressure from the U.S. Army Air Corps, it was decided to limit active work to those batteries that could be completed by July 1944. As a result, all work on fourteen of the thirty-seven 16-inch batteries planned for the continental U.S. was discontinued. The expansion of overseas bases during 1941 impacted the construction of the new 6-inch gun batteries in the continental U.S. by priority assigned to the completion of twenty 6-inch batteries to guard these overseas bases.

The new batteries constructed under the 1940 Program were much more standardized that those of proceeding periods. The Army developed standardized designs for the 16-inch gun batteries and the 6-inch gun batteries which were used with only minor variations for local topography and soil conditions. Both the 16-inch and 12-inch guns, whether newly installed or retained from the Interwar Period, were emplaced within reinforced concrete casemates that limited their field of fire to about 180 degrees but gave them superior protection over the old open emplacements. The new 6-inch batteries were not casemated. A cast steel shield from four to six inches thick was placed around the gun and carriage. This shield would protect the gun and its crew from all but direct hits by heavy projectiles. Between the two 6-inch guns was an earth covered steel reinforced concrete structure contain the magazine, power generators, communications, air filtering equipment, storage, and plotting room. As these batteries were being built, they were assigned a "Battery Construction Number" for record keeping purposes. As many of the new batteries were never formally named, these construction numbers were the only designation they received. While the Army never referred to the 16-inch series of batteries as whole as the "100" series or the 6-inch series of batteries as whole as the "200" series, these terms are used by modern historians and are referred to as such in this work.

As the range of these new batteries was far greater than earlier batteries it was also necessary to update the fire control networks. The 16-inch batteries received new base end stations as far as twenty-five miles away from the gun's position to allow for gun's maximum range to be effectively used. These stations were built in wide variety of forms: houses, windmills, silos, water tanks, office buildings, or buried into hillsides. Radar was added as an early warning device and as a fire control instrument allowing the operation of coast artillery at maximum range during all weather conditions.

By the start of World War II, the Coast Artillery Corps' mobile coast artillery units had dwindled from the plans of the Interwar Period, especially the railway guns units. Several tractor-drawn 155mm GPF gun regiments were available in the continental U.S., but only part of one 8-inch railway regiment was on hand. The four 14-inch railway guns continued their role in Los Angeles and Canal Zone. The primary use of mobile coast artillery was to fill in for fixed coast artillery weapons until their completion or at secondary locations. The 155mm GPF gun units were reorganized into seventy-two 2-gun batteries along the Atlantic,

Gulf and Pacific coasts. Using 12-inch railway mortars and 8-inch railway guns from storage, several CAC units were formed and sent to both domestic and overseas locations to provide temporary harbor defenses until permanent works could be constructed.

As with earlier periods, an integral part of harbor defenses was the use of controlled mines across key ship channels. These mine defenses were supplement by U.S. Navy contact mines and the use of submarine nets and booms. As the primary threat during World War II turned out to be enemy submarines at most of these ports, the U.S. Navy added detection devices in outer harbor approaches and conducted offshore patrols. Because of the need for both the U.S. Army and U.S. Navy to coordinate their coast defense activities, a centralized harbor entrance command was created in 1941. The Harbor Entrance Control Post (HECP) used both army and navy personnel to provide a link between higher command and all subordinate elements of a harbor defense. These centers were responsible for monitoring all movement of shipping in and out of the harbor. To support this effort a secondary gun battery was on duty as commercial shipping traffic was examined upon entering the harbor. One of the concerns at this time was an attack by fast moving torpedo boats combined with the lack of modern rapid-fire guns. To fill this void, the 90mm anti-aircraft gun was selected to replace the existing 3-inch pedestal guns of the Endicott and Taft Programs. In late 1942, special anti-motor torpedo boat (AMTB) batteries were installed along the Pacific and Atlantic coastlines. These batteries usually consisted of two fixed mounted 90mm guns and two mobile mounted 90mm guns, and two mobile 37mm or 40mm anti-aircraft guns. These guns would be protected by earthen revetments with protected magazines. The active harbor defenses received two, three, four, or more of these AMTB batteries beginning in 1943.

Outside the continental United States, where the threat of attack and invasion was greater, new coast defense construction proceeded with greater speed and with the use of armament on hand rather than waiting for weapons sto be provided by 1940 Program. The coast defenses of Hawaii are a good example, as the Japanese attack on Pearl Harbor made new defenses the highest priority due to concerns of an invasion attempt. A series of batteries were constructed, using excess naval guns, ranging from the 14-inch turrets from the battleship USS *Arizona* to 8-inch gun mounts from the aircraft carriers USS *Lexington* and USS *Saratoga*. Throughout the Pacific Islands and Alaska, surplus U.S. Navy guns (5-inch, 6-inch, 7-inch & 8-inch) were mounted on shore to defend U.S. Navy installations.

With the tide of the war shifting toward the Allies after 1942 and the demands to produce war material for the mobile army, the navy, and the army air corps, the 1940 Program was pared back. While the construction of structures could keep pace with the original plan, the manufacture of weapons and their accessories could not. In response to these pressures, the 1940 Program was scaled back even further. By the war's end, the modernization program resulted in the completion of nearly 200 new batteries in the continental United States at a cost of $220 million, or about one-half the number of installations proposed in the 1940 Program, but still the most powerful collection of coastal defenses in America's military history.

The development in military tactics and technology during World War II brought about numerous changes to the concept of coast defense. It was no longer thought necessary to defend one's seacoast using just coast artillery and controlled mines. Air power and naval forces were to replace breech loading rifles and reinforced concrete. Already at the end of World War II, all except a few 90mm AMTB batteries were placed on caretaking status. During the transition years of 1946 to 1948, some new batteries started during the war were completed while many other batteries were being disposed of and guns scrapped. By 1949, the process was completed as the last of guns were scrapped. In 1950, the remaining harbor defense commands were disbanded, and the Coast Artillery Corps was abolished as a separate U.S. Army branch with its remaining units, all anti-aircraft artillery, recombined into the Field Artillery. After 150 years of being one of America's military prime missions, the building and manning of permanent coastal fortifications was over.

The U.S. Coast Defense Objective in the Modern Era

The objective of seacoast defense is to provide protection of the coastline from invasion by an enemy, and specifically the defense of important harbors, which includes securing the anchorages and bases needed for naval operations. Coast defense is not only protective in its strength but protects the nation's ability to carry war beyond its own coastline.

It is impractical to fortify the entire extent of any nation's long coastline in such a way that an enemy in command of the sea could not land upon some portion of it. The cost of such an undertaking would be excessive, as maintenance of these defenses and number of men required would make it prohibitive expensive. An example of this type of defense was the "Atlantic Wall" built by Germany in World War II (which stretched from Norway to Spain), which failed to prevent the Allies from landing in Europe in 1944. It was essential, however, that certain selected points be permanently fortified to make invasion more difficult and to protect key naval shore installations and fleet anchorages and important commercial harbors that support the nation's economy.

The resources to defending the coastline during this era were divided into two kinds of troops. The first was the Coast Artillery troops, made up the regular U.S. Coast Artillery Corps and the U.S. Coast Artillery Reserves. These technical troops manned both the fixed and mobile seacoast artillery and controlled mines defenses. The second resources were the supporting troops of the mobile ground forces of the U.S. Army which protected the both the coast defenses and unfortified coastline from enemy landings. The second would been the local National Guard troops (formally militia), while the U.S. Navy's role in coast defense was through both offensive and defenses operations against enemy warships.

To carry out this mission, seacoast weapons were divided into classifications according to their capabilities against enemy warships. Primary armament were those weapons that could theoretically destroy the primary or capital warships of enemy naval force. Throughout most of the Modern Era primary weapons were defined as seacoast artillery of initially 8-inch and larger caliber. Controlled submarine mines were also considered part of the primary armament. The second group of seacoast weapons was the secondary armament, which were designed to counter secondary or non-capital warships, such as cruisers, destroyers, and torpedo boats.

The selection of the numbers and type of seacoast weapons was determined by such factors as the importance of the coastal area, the hydrograph profile of the approaches, the topography of area, and effectiveness of seacoast weapons in defending the coastal area. The positioning of seacoast artillery was based on the attainment of effective fire and protective factors, such as concealment, other weapons, and local defense against ground or air attacks. Attainment of effective fire refers to a position which offers the widest field of fire and greatest range over navigable water. Also considered was the need to provide coverage to all areas in which an enemy warship may operate and the placement of a suitable concentration of fire on critical areas such as harbor entrances, approaches to mine fields, and narrow portions of the channel. Consistent with these requirements, batteries were sited to provide mutual support and defense against all forms of attack. The considerations for the location of primary armament included the ability to protect friendly naval forces while entering, within or leaving the harbor, and preventing hostile naval forces from approaching within effective range of the defended coastal areas. Submarine mine fields would be placed in the seaward area of the harbor entrance and within effective range of searchlight and rapid-fire secondary armaments. Both controlled and uncontrolled submarine mines are located to prevent entry into or close approach to the harbor of enemy surface warships or submarines at all times, including during night or during conditions of heavy fog or smoke. The secondary armament would be located to provide protection for mine fields, nets, booms, and other obstacles; and the attack of hostile secondary warships engaged in raids, reconnaissance, laying of mines, and torpedo fire. Since targets of the secondary armaments were within range of visual observation and assumed to move at high speed on rapidly changing courses, these batteries were sited in direct fire positions. Protective factors in site selection included protection for the power plant, plotting room, magazines, communications, exposure of the gun crew and ammunition during the service of piece, gas protection for command post and plotting room, distances between emplacements, and concealment.

U.S. Coast Defense Armament and Equipment in the Modern Era

Few weapons of the Modern Era of coastal fortification remain today. This is the result of the advancing technology that quickly made weapons obsolete and given the economic value of high-grade steel the military sold these obsolete weapons to salvage companies. The scrapping of coast artillery material also holds true for most its supporting equipment, machinery, and instruments. As a result, today we mainly only have period images of these armaments or supporting equipment.

The development of new armament and equipment over this era usually went through cycles where the level of perceived external threats to the United States generated appropriations from Congress to allow the funding of new weapon systems. The development process for new weapons required several steps. First, was the design stage which led to the prototype and testing period and then to production and installation phase. Finally, while the weapon was in service it received modifications and improvements until it was declared obsolete. The life cycle of seacoast artillery varied from a few years to as long as fifty years.

The construction of the Endicott and Taft Programs defenses relied on the growth of heavy industry in the United States. Many of items used in coast artillery forts were invented specially for that purpose and represented the cutting edge of that technology. Early defense works relied on steam, coal, and manual energy to make things work. The use of oil and the advances in electricity brought motor driven equipment, telephones, radar, computers, and electric lighting to become key ingredients in U.S. coastal defenses.

It is also important to note that different U.S. Army branches had specialized functions that need to work together to complete a weapon system. A seacoast weapon would be designed and constructed by Ordnance Department while the emplacement was designed and constructed by the Corps of Engineers. These activities were all support by the Quartermaster Corps, Signal Corps, and so forth. The final product was then turned over to the Coast Artillery Corps for use. As you may imagine sometimes the priorities of these various organizations were not always in agreement, so delays or undesired weapons systems did occur.

A 10-inch rifle on a disappearing carriage
Battery Benson, Fort Worden, Washington (Puget Sound Coast Artillery Museum Collection)

6-inch rifles on dissapearing carriages
Possibly Battery Tolles, Fort Worden, Washington (Puget Sound Coast Artillery Museum)

For coast artillery material, the U.S. Army insured that all items were assigned a "type" and a "model". For seacoast artillery, the type for the gun or barrel refers to the size of bore (diameter) in inches while type for carriage or mount refers to the style of operation. Associated with the type is the model which refers to year of development and any subsequent modifications until 1930 when use of the year was dropped. This nomenclature extends to projectiles, fire control instruments, searchlights, submarine mines, ammunition hoists, power generators, radar, etc.

Carriage or mount types were either fixed or mobile, they allowed the guns to elevate and provide for some horizontal movement while taking up the recoil of the discharge and return the piece to the loading position. The major caliber fixed carriages were classified as Barbette (BC) carriage, which allowed the gun to remain above the parapet for loading and firing; Mortar (MC) carriage, which allowed a short-barrel gun to fire in a high arc; the Barbette long-range (BCLR) which allowed for greater firing elevations and ranges; and the Turret (TM) mount which was a barbette carriage protected by an armored housing with ammunition supplied from below. Guns of 7-inch or lesser caliber were mounted on the Pedestal (PM) mount, which had a fixed cylindrical base on which rotated a yoke that held the gun in a cradle equipped with recoil absorbing cylinders; the Anti-aircraft (AA) mount, a pedestal mount that allowed fire at high attitude. The Fixed retractable carriages included the Gun-Lift (GLC) carriage which was a BC on an elevator platform; the Disappearing (DC) carriage where the gun is raised above the parapet for firing and retracts behind the parapet for loading. The earlier smaller caliber guns had the Balanced Pillar (BPM) mounts and the Masking Parapet (MPM) mounts, which enabled the gun to be lowered below the parapet to protect it from view. Guns on mobile carriages were used in the Interwar Period as the Railway (RY) mount cars and Tractor-drawn (TD) mounts. Other temporary coast defenses made use of available weapons with a range of carriage types, primarily former naval models.

Controlled mines were anchored to the bottom of a harbor, either sitting on the bottom itself (ground mines) or floating (buoyant mines) at depths which could vary widely, from about 20 to 250 feet. These mines were fired electrically through a vast network of underwater electrical cables at each protected harbor. Mines could be set to explode on contact or be triggered by the operator, based on reports of the position of enemy ships. The networks of cables terminated on shore in concrete bunkers called mine casemates, that were usually partly buried beneath protective coverings of earth. The mine casemate housed

A 6-inch gun on a pedestal mount, Battery Carpenter, Fort McKinley, Maine
(Joel Eastman Collection)

3-inch guns on pedestal mounts in Battery O'Rorke, Fort Barry, California (NPS, GGNRA)

12-inch mortars in pit A, Battery Worth, Fort Pickens, Florida (Stillions Collection, NPS)

On the deck of mine planter (Stillions Collection, NPS)

electrical generators, batteries, control panels, and troops that were used to test the readiness of the mines and to fire them when needed. Each protected harbor also maintained a small fleet of mine planters and tenders that were used to plant the mines in precise patterns, haul them back up periodically to check their condition (or to remove them back to the shore for maintenance), and then plant them again. Each of these harbors also had onshore facilities to store the mines and the TNT used to fill them, rail systems to load and transport the mines (which often weighed over 750 lbs.) each when loaded), and to test and repair the electrical cables. Fire control structures were also built that were used first to observe the mine-planting process and fix location of each mine and second to track attacking ships, reporting when specific mines should be detonated. The preferred method of using the mines was to set them to detonate a set period of time after they had been touched or tipped, avoiding the need for observers to spot each target ship.

Key to the successful use of coast artillery was fire control and position finding as if the guns, mortars, and controlled mines failed to strike their intended targets their mission was incomplete. Early aiming efforts relied on the skill of the gunner to hit the target, but as the weapon's range increased so did the need for specialized fire control. Using geometry, optical instruments, telephones, timing interval bells, and mechanical devices a system was devised to point weapons successfully at their targets. Key equipment included the Depression Position Finder (DPF), the Azimuth Instrument (AI), Coincidence Range Finder (CRF), Plotting Board, Range Correction Board, Fire Adjustment Board, Deflection Board, Spotting Board, Range Percentage Corrector, Data Transmission Devices, Telephone Sets, and Timing Interval Bells. Many of the devices were replaced or supplemented by the development of radar (for both surveillance and fire control duties) and gun computers (combining many of plotting room devices) during the 1940 Program.

Until the advent of radar, the use of searchlights (plus star shells and airplane flares) was used to illuminate naval targets at night. Both mobile and fixed searchlights were used for both harbor defenses and anti-aircraft defense. At first 36-inch and 60-inch searchlight were used, but the 60-inch became the standard. Searchlights were located as close as possible to the water-edge to maximize their effective range of between 8,000 and 15,000 yards. Fixed searchlights were provided with a shelter to protect the searchlight from elements, to house the electrical generator, and provide concealment of the light when it was not in use. Some positions placed the searchlight on small rail cars that allowed the searchlight to move a short distance to a more exposed operating site. Searchlights were also housed in towers, pits, and even tower that "disappeared" by pivoting. After 1940 all new searchlights assigned to the defenses were the mobile type.

By 1943, two technological advances significantly changed coast artillery fire control. The most striking was the development of radar, which, as noted, could function in any weather or visibility. The use of radar greatly reduced the need for searchlights and for fire control stations as spotting enemy war-ships and aircraft could now be undertaken by radar units. In addition, after decades of experimentation and development, largely stymied by inadequate funding, the coast artillery adopted gun data computers, primarily for the last generation of batteries. These replaced the plotting boards and, coupled with direct-reading observation instruments, substantially automated the fire control process, reducing the human error that had always plagued the system.

Searchlight and shelter/powerhouse, Fort Flagler, Washington (Puget Sound C.A. Musuem)

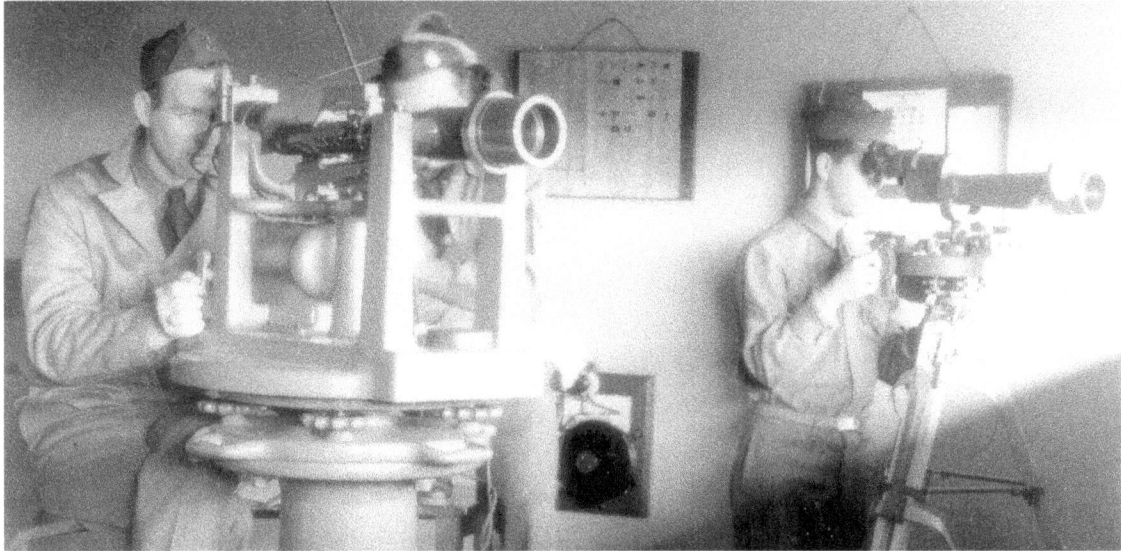

A Depression Range Finder (left) and a azimuth scope (right) in a base end station
(Al Scroeder Collection)

U.S. Coast Defense Fire Control Structures in the Modern Era

The development and changes in the optical instrument fire control system from 1900 to 1945 was a long and complicated process that changed equipment, operating procedures, and designations frequently. The reader is encouraged to consult *American Seacoast Defenses: A Reference Guide* and articles in the *Coast Defense Journal* for more detail and references to U.S. Army manuals and reports. The maps included in this guide have an extensive set of symbols indicating the locations of the various fire control structures.

By 1909, each battery was under the immediate command of the officer stationed at the battery commander's station (BC). Each battery may have had one or more additional base end stations (B) with optical spotting instruments. Small caliber batteries usually had a coincidence range finder station (CRF) nearby. Mine commanders manned their posts at the mine primary (M') station. In the defended harbor areas, called the Coast Defense Command, batteries were grouped into Fire Commands, each under the overall command of the fire commander stationed at the fire command primary station (F'). The Fire Commands were then grouped together by geographical areas under the command of the officer in command of that entire sector of the coast defense. This command was initially called the Battle Command but later was changed to the Fort Command. This officer was stationed at the primary fort command station (C'). In 1925, this chain of command was changed slightly. All forts and/or groups were under the Harbor Defense command (H). Forts (F) were also used as tactical commands. Individual gun batteries were assigned to a gun group (G). Later an additional tactical organization, the groupment (C), was added below the Harbor Defense command composed of two or more groups.

In general, batteries in each harbor defense were assigned tactical number designations, generally in numerical sequence from the south (Tactical Battery #1) to the north (Tactical Battery #2, 3, 4, etc.) on the Atlantic coast; from the north to the south along the Pacific coast; and from the east to west on the Gulf coast and along the Puget Sound during the 1940s. Note that by 1940 base end stations (B) and spotting stations (S) were often combined. This is useful in deciphering the symbols for designating the fire control observation stations on the maps: $B^1_1S^1_1$, $B^2_1S^2_1$, etc. The lower number is the tactical battery number to which the station is assigned, the upper number (or "prime" mark) is the station designation number in the series of stations assigned to that tactical battery. The number of base end stations assigned to each battery ranges from a single station to as many as 14 stations. Each station had at least one azimuth scope and/or depression range finder (DPF) scope as well as connected telephone communication equipment. 3-inch small caliber batteries had one base station with a coincidence range finder (CRF) located close to the battery.

During the period 1905 to 1940 the fire control structures were generally located on existing military reservations. The location and identities of these stations can be found on the confidential blueprint maps; in the reports of completed batteries; in the reports of completed works; and in the harbor defense engineer notebooks that are part of the CDSG ePress harbor defense document collection. After 1940 the ranges for the new guns were longer and the fire control stations were more dispersed, which resulted in the acquisition of a number of new small reservations along the coastline of each active harbor defense.

Battery Plotting Room circa 1944 (NARA)

The U.S. Coast Defense Organization Before World War II

The following organization structure of the administration and tactical command of the U.S. Coast Artillery Corps is for the 1930s period. The organization prior to 1924 was on a company basis and after 1942 on a separate battalion basis and is not discussed here.

Earlier organizations had similar purposes but used different terminology. Earlier tactical structures had Artillery Districts that were divided into Battle Commands, Fire Commands, Mine Commands, and Battery Commands, each with their own commanders, while for administrative and training purposes the CAC was divided into companies which in turn were assigned to coast artillery forts or posts.

A Harbor Defense Command is a subdivision of a Defense Command, which would cover an entire region. All elements, including materiel and personnel of a Harbor Defense Command, were located at one or more coast artillery forts. These forts consisted of defined land areas within a harbor defense in which the harbor defense elements were assigned. The forts were organized primarily to provide a centralized control over administrative and technical components of the harbor defense. The materiel provided for a harbor defense may have included various types of seacoast artillery guns, anti-aircraft guns, searchlights, controlled submarine mines, underwater listening posts, radar, observation and fire control systems, and harbor patrol boats. Harbor defenses were designated by the name the harbor or coastal area which they were defending, or by the name of the largest city in their immediate area. Examples are "The Harbor Defenses of San Francisco" or the "Harbor Defenses of Chesapeake Bay."

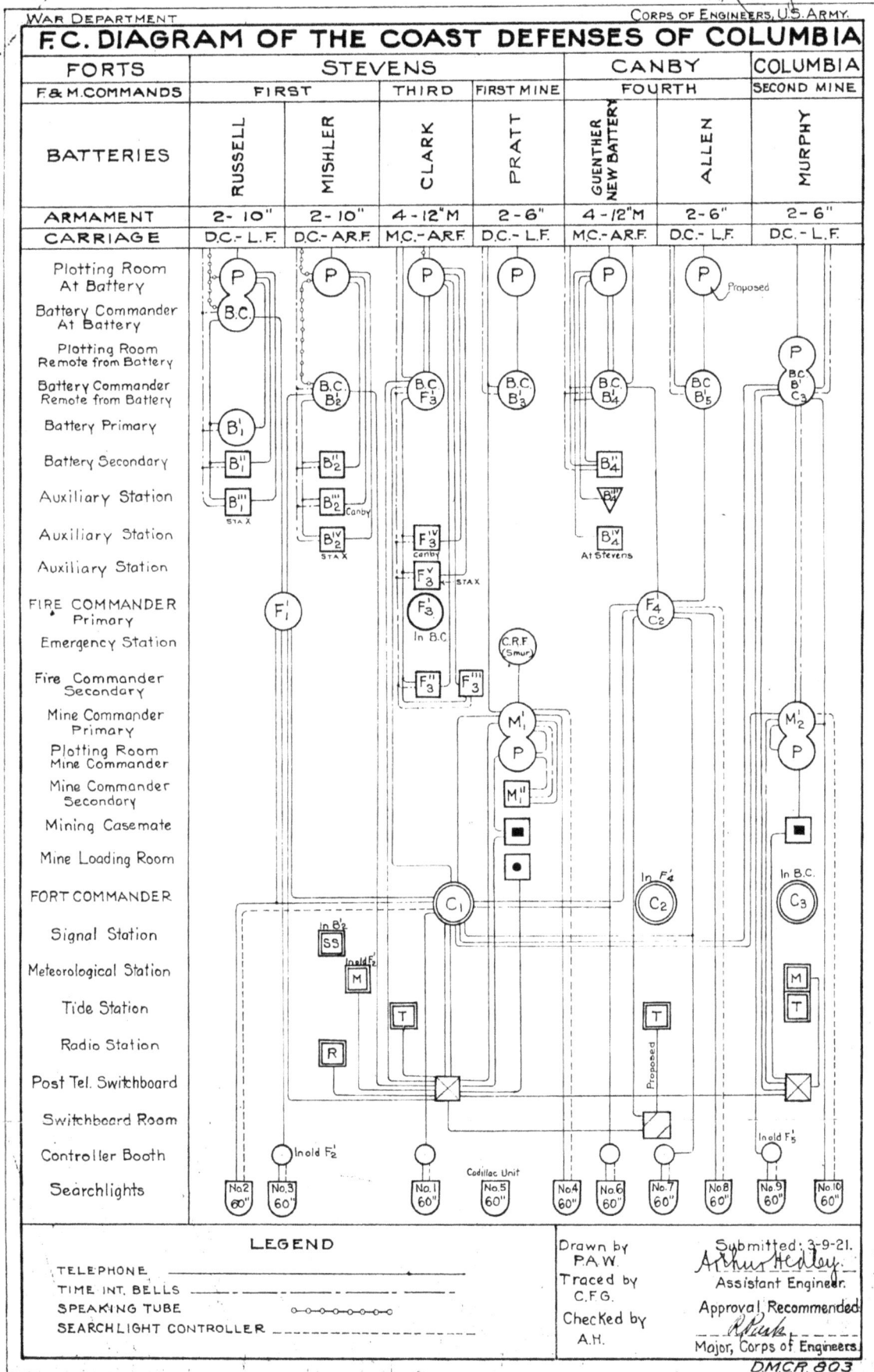

A fire control diagram showing the communications lines between the various stations.

A senior U.S. Coast Artillery Corps officer was usually designated the harbor defense commander responsible for both the administration and tactical commands. He was supported by a harbor defense headquarters staff and service units from the Quartermaster, Ordnance, Medical, Signal, Engineers, and Military Police organizations. The service units usually staff the administrative headquarters and the Coast Artillery Corps the tactical headquarters. Each fort was organized with its own headquarters and fort commander, who was responsible for the administration of the post. While the fort commander was not included in the tactical chain of command, he was responsible for the training and supervision of damage control to all the fort's structures and the activities of the service units.

The basic units of the coast defense tactical command were the battery, battalion, and group. The battery was the basic combat unit of the harbor defense and contained enough men required to man one primary battery. Batteries were classified by the type according to the material with which they were equipped. The gun battery consisted of one or more fixed or mobile guns of the same caliber and characteristics to be employed against a single target and of being commanded by a single individual. It included all structures, equipment, and personnel necessary for emplacement (or mobile weapons), the conduct of fire, and the performance of service. The strength and organization of a battery depended upon the type, number, and caliber of the guns of the battery. It was divided into a battery headquarters section, a range section (containing a battery commander's detail, an observing detail, and a range detail), a maintenance section, and a gun section for each gun or mortar. Special gun batteries were the anti-motor torpedo boat (AMTB) battery and the fixed anti-aircraft battery. The mine battery consisted of the personnel, structures, and equipment other than mine planters necessary for the installation, operation, and maintenance of all or part of the controlled mine fields. It was divided into a battery headquarters section, an operations section (containing a command post detail and range detail), a casemate section, a loading and property section (consisting of loading, cable, explosive, and maintenance details), a planting section (consisting of mine planter, distribution box boat, and small boat (yawls) details) and a maintenance section. The searchlight battery consisted of the personnel, material, equipment, and structures necessary for the operation and maintenance of seacoast and anti-aircraft searchlights.

These batteries were normally administerial combined into battalions with each battery commander reporting to the battalion commander. The battalion was organized to provide administrative, training, and tactical functions. Gun battalions were composed of from two to five-gun batteries, while a mine battalion consisted of the personnel, submarine mine material, structures and vessels necessary to plant, operate, and maintain part or all the controlled mine fields. The primary purpose of the coast defense battalion was providing effective fire direction through the coordination of various types of batteries. When a harbor defense command was large, battalions will be organized into groups. A group was a tactical command containing from two to five battalions or independent batteries. As with battalions, the primary mission for the group and the group commander was to provide effective fire direction. The use of groups occurred when the number of units is greater than can be controlled by the harbor defense commander. The basis for battalions or groups was to organize batteries that covered same field of fire or water area. When large number of batteries covered the same water area then the organization was based on target selection, such as primary and secondary armaments.

For administrative and training purposes, battalions were organized into regiments up to 1942. The garrison of a harbor defense consisted of part or all of one or more regiments, and the organization of different regiments varied to conform to the special requirements of the different harbor defenses. Generally, a coast artillery regiment assigned to fixed armament consisted of a headquarters battery, a searchlight battery, a band, and three battalions. The forts were assigned to Coast Artillery Districts. The district commander commands all coast artillery troops stationed within the territorial limits of the district, including the coast artillery units of the Organized Reserves and those of the National Guard when in the service of the U.S. At the start of World War II, the headquarters for Coast Artillery Districts were in Boston, MA (1st CAC

District), New York, NY (2nd CAC District), Fort Monroe, VA (3rd CAC District), Fort MacPherson, GA (4th CAC District), and Presidio, CA (9th CAC District). Overseas coast artillery units were assigned to local U.S. Army Departments, such as the Hawaii Department, etc.

View of Fort Flagler, Washington (D. Kirchner Collection)

A Typical U.S. Coast Artillery Fort in the Modern Era

While each coast artillery fort has its own unique design, it is possible to provide a general blueprint of the type and purpose of structures that you would find at a U.S. coast artillery fort built during the Modern Era. It is important to remember that each fort was like small self-contained city. All the services that were required to support the daily needs of its garrison and to operate the fort's weapon systems were included within the military reservation.

The reservation was typically surrounded by a fence. There was a main entrance gate with a guard house. While very few coast artillery forts had any land defenses, the use of security fencing was widespread. Recognizing this fencing is usually the first indication of a former U.S. military reservation. The main cantonment area contained a variety of buildings spread over a large area, not much different in appearance of a rural college campus. This support area was subdivided into functional sections surrounding by a large parade ground area. While the overall fort was under the Coast Artillery Corps, each of the support services (Quartermaster, Engineer, Medical, etc.) had their own buildings or reservations within the fort.

The main parade ground is the focal point of the post. The fort headquarters, officer's quarters, non-commission officer housing, service clubs, and enlisted barracks usually surround it. Most of the non-tactical structures at the forts constructed during the Endicott-Taft Programs were designed to be permanent structures. These wood-frame buildings were built on stone foundations with slate roofs, sided with local brick, clapboard, or stucco. The Quartermaster Corps architect's office created standard plans for all types of buildings. Those designed at the turn-of-the-century—when most Coast Artillery forts were constructed—were of Colonial Revival style with elements of Queen Anne style in the officers' quarters. As the century progressed, new styles were adopted, such as Italianate and Spanish Revival, and these styles were used when additional buildings were constructed. Store houses and pumping plants used more practical industrial or utilitarian styles.

Officer's quarters varied in size and elaborateness depending upon the rank of officer for whom the building was intended. The Commanding Officer's Quarters was usually the largest and most elaborate of the officer's quarters, and it was placed, if possible, on the highest and most prominent location on the parade ground. Other senior officers were assigned single quarters, while many of the quarters were double quarters for two families. Large forts had a Bachelor Officer's Quarters with its own mess. Non-Commissioned Officer's quarters were usually double sets.

Parade ground and officer's quarters, Fort Casewell (BW Smith Collection)

The interiors of buildings were finished with wood floors, plaster walls with wood trim, and pressed metal ceilings. All structures where officers and men lived or worked had electricity, running water and flush toilets. Each barracks was designed to house a company or battery of 100 men and was self-contained with its own kitchen, dining room, day room, barber shop, and tailor shop. Sleeping quarters were on the second floor, while the lavatory and latrine were located in the basement in northern climates. In the south, separate lavatory and latrine buildings were sometimes built. Large forts had double barracks–two 100-man barracks-built end-to-end–which functioned as two separate barracks. Forts which served as the headquarters post for a harbor defense usually had a band barracks.

Although the parade ground was used as a general athletic field, tennis and handball courts, and baseball fields were also built in open areas of the fort. A system of permanent roads served the entire fort, and the streets were usually named. Railroads and tramways were built during the construction of the forts, and these lines often continued to be used. These forts eventually had their own water, sewer, telephone, and electrical systems. If municipal water and commercial power services were available, the army used them, but at many sites the engineers built their own water and electrical plants and distribution systems. Sewer pipes ran into the ocean. Ice houses, and in northern areas, ice ponds, were also built to provide refrigeration for food in the years before electrical cooling became available. Systems for the disposal of garbage and rubbish were also created. Garbage and combustible waste were burned in crematoria, while non-combustible materials were disposed of in landfills or dumped into the ocean. The major fuel at forts was coal, and a system of unloading, transporting, and storing the fuel was developed, usually relying on mule-drawn wagons.

A large portion of fort's reservation would be devoted to the Quartermaster Corps. The Quartermaster was tasked with providing housing, supplies, and transportation for all the troops assigned to the fort. The Quartermaster oversaw the construction of most of fort's support buildings, as well as the installation of its own quartermaster wharf and tramway to transport supplies within the fort. Storehouses, commissary, workshops, and stables were usually centered near the quartermaster wharf.

The Corps of Engineers were responsible for construction the actual fortifications known as the tactical structures (emplacements, fire control stations, casemates, power houses, etc.); the Ordnance Department provided the weapons, machinery, and instruments that went into these structures; and the

Signal Corps provided the technical equipment as new technology was developed. Near the shoreline were located the fire control stations along with protected telephone exchanges, command posts, meteorological stations, seacoast searchlights positions, and reserve magazines that support the fort's weapons systems.

Fort Terry buildings (BW Smith Collection)

The fort's two main coast defense weapon systems were controlled mine fields and seacoast artillery. Controlled mine fields required an extensive infrastructure within the fort. Principal structures for the mine defense included the mining casemates from which the mines were operated; the conduits connecting the casemates with the shore; the cable terminals on the shore; the cable tanks in which the mine cables were stored when not in use; the mine storehouses in which were kept the mine cases; the loading rooms in which the mines were loaded; the magazines in which the dynamite was stored; the range stations, plotting rooms, and dormitories, the mine wharves at which the mine planter used to land and receive the loaded mines; and the tramway connecting the wharves with the cable tanks, storehouses and loading rooms.

Closer to the shoreline are the emplacements of the fort's other main weapon system the large caliber gun batteries. These gun batteries consisted of both of large caliber breach loading rifles mounted on disappearing or barbette carriages and smaller rapid-fire guns on pedestal mounts during the Endicott and Taft Programs. The purpose of these gun emplacements was to provide a stable base for these guns and carriage and a convenient platform for the personnel serving the gun. The emplacement also designed to provide the armament and the personnel the maximum protection as possible, as well as providing a safe storage place of the ammunition. These thick concrete structures were covered with earthen fill on the seaward side, while they were partially buried, these batteries are easily accessible from the rear due their open back design.

The gun emplacements from the Interwar Period and 1940 Program are quite different from these earlier designs as American coast artillery responded to the progression of larger caliber naval weapons with longer firing ranges and the advent of military aviation with aerial bombs. These emplacements are usually one-story high but usually completely buried. The gun position consists of a gun well surrounded by a circular concrete pavement. Later many of these emplacements were completely rebuilt with thick reinforced concrete casemates to protect their weapons from aerial attack and naval bombardment.

Another primary seacoast weapon of the Endicott and Taft Programs was the seacoast mortar, actually a short barrel breech loading rifle. These batteries by definition did not require direct fire, so they were often located away from the shoreline. They were located within or behind the fort's cantonment area. A

Fort Mott 1936 (NARA)

typical mortar battery had a high reinforced concrete parapet with traverses that formed a series of pits. These pits were usually open to the rear, but early designs were completely surrounded with access through a tunnel. A battery had one to four pits with two or four mortars in each. Between the pits or around their sidewall were ammunition magazines, power generator rooms, shot truck areas, storerooms, and a plotting room.

The secondary gun batteries mounted rapid fire guns for the defense of the controlled mine fields from minesweepers and to repulse fast moving naval vessels and were installed after the first round of primary gun batteries. These are simple emplacements that basically provided a stable firing platform for the weapons and a protected magazine for their ammunition. Also located around some forts were groups of three or four concrete gun blocks for anti-aircraft guns that were added in later years.

A key feature of all coast artillery forts are the fire control stations which provided the target information for the mine and gun defenses. These stations come in all shapes and size. They range from a single below-grade room with observing slots to large multi-level, multi-room towers. Constructed of both wood and concrete, these stations have been disguised as non-military structures ranging from summer cottages to grain silos. Associated with World War II fire control stations were radar stations that by 1944 replaced their function. These radar stations had antennas which were mainly located on steel towers but could be mount on other structures. These antennas sent their signals to operating rooms where measurements provide location data to plotting rooms. Support these stations were power rooms and dorms for troops manning the stations.

Several to many of the structures at most of remaining U.S. coast artillery forts. However, you may only view piles of rubble and mounds of dirt as their status and condition are constantly changing. Nearly all the seacoast armament and equipment were scrapped after World War II which accounts for the lack of actual coast artillery at the forts.

Soldiers in barracks (Stillions Collection, NPS GSNS)

Garrison Life at a U.S. Coast Artillery Fort in the Modern Era

The soldiers assigned to the defenses experienced a great change in quality of life during the years from 1890 to 1950. The early years were certainly the roughest. In general military service in the U.S. armed forces was not well compensated or widely respected in some quarters. As the permanent posts were being established, physical living conditions were sometimes poor, and relationship with the local, civilian community at times strained. Officers could afford higher standards of living for themselves and their families as well as greater social involvement with the local community.

By the end of the early modernization programs in the 1910s, the living and work conditions had greatly improved. In particular the Coast Artillery was an elite assignment, with considerable prestige. The Coast Artillery Corps was relatively well funded and equipped, had a strong technical and professional dedicated career officer contingent, and was based on teamwork activity that encouraged close camaraderie. Opportunities for duty at oversea bases in exotic tropical locations like Hawaii, the Philippines, and the Panama Canal had its advantages, especially as many of the tropical diseases had been conquered. Training was emphasized, but in all the workload was reasonable. Pay was not extravagant for the enlisted man– but decent food, recreation, and athletic events were provided on post. Soldiers tended to stay in this branch of service, often re-enlisting, and were quite good at what they were taught and with what equipment they practiced on.

The daily schedule of the Coast Artillery troops focused on drill, inspections, maintenance, meals, and recreation. The center of activity for enlisted men was their barracks which was designed to house one or two companies or batteries of 100 men each with its own kitchen, lavatories, dining room, day (recreation) room, barber shop, and tailor shop. The barrack along usually arrayed around a parade ground. The officer quarters were usually located on the opposite side of the parade ground. The day would begin with meal, roll call, and assignment of duties. This usually was training/drill in the mornings with maintenance tasks or recreation events in the afternoons. Recreation was considered important by the U.S. Army after

1900 as it was believed that it not only maintained physical fitness but promoted competitiveness which made the men more effective in combat. Most large forts were provided with a gymnasium and bowling alley, as well as athletic fields, handball and tennis courts. Other recreation activities included visiting the post theater, service clubs, libraries, chapel, and the Post Exchange, as well as leave to visit the local cities and towns. Another aspect of garrison life were weekly inspections and parades, and soldiers who failed these inspections would end up spending their weekend cleaning barracks and latrines, rather than having a weekend pass to visit the local communities.

63rd Coast Artillery Company on parade gound at Fort Worden, Washington in 1908
(Puget Sound C.A. Museum)

Mess hall set for Christmas Dinner 1911, 126th Coast Artillery Company, Fort Worden Washington
(Puget Sound CA Museum)

MODERN ERA SEACOAST FORTS TODAY

In 1950, the remaining harbor defense commands were disbanded, and the U.S. Coast Artillery Corps was abolished as a separate branch with its remaining units, all anti-aircraft artillery, moved into the Field Artillery. Meanwhile, the responsibility for limited harbor defense, primarily underwater defenses, was transferred to the U.S. Navy. The U.S. Army retained several of the old coast artillery forts for other missions, while the Navy acquired several reservations for thier use including for its new role in harbor defense. Other federal agencies had an opportunity to claim all or portions of the former coast artillery sites. Those not transferred were turned over for disposition to the U.S. General Services Administration (GSA), who offered them to state, county, and local governments, and finally to private citizens. Many of the smaller, independent plots of land which had been leased or purchased for fire control and searchlight positions were returned to original owners or sold to private owners, before selling or transferring these former forts, the U.S. Army either returned to its depots all usable equipment or auctioned items in lots to the public.

Several coast defense sites had been abandoned by the U.S. Army as active defenses by 1928 including those at the Mississippi River, Mobile Bay, Tampa Bay, Savannah, Port Royal Sound, Cape Fear River, Baltimore, and the Potomac River, and the smaller inner harbor defenses at East New York, San Francisco, and Puget Sound. Many of these reservations were reclaimed for use during World War II. The next large-scale transfer of harbor defense properties from the U.S. Army began in 1947 and continued through the mid-1950s. In the early 1970s a general series of military base closures occurred throughout the U.S. Department of Defense to reduce basing costs. Several large former harbor defense sites, including military reservations around San Francisco, New York, and Pensacola, were included. Given the large size and value of these properties, Congress passed several laws that directed the ownership of these former forts to be transferred to the U.S. National Park Service (NPS). Base closure commissions in the 1990s, 2000s, and 2010s recommended the closure and transfer of other former harbor defense sites, which included the Presidio of San Francisco, Fort Wadsworth on Staten Island, Fort Monroe in Hampton, and Fort Trumbull in New London. Only a handful of old coast defense reservations remain in military hands in 2025 — Fort Story, VA, Fort Hamilton, NY, a large part of Fort Rosecrans, CA, Fort Kamehameha, HI, Fort Hase, HI and a few other sites. Other national agencies, state agencies, and local governments acquired numerous coast artillery sites for parks and recreation areas, since they inevitably had scenic river or ocean views. Depending on how diligently the GSA protected the sites, and the length of time it took to dispose of them, some sites and structures survived in excellent condition, while others suffered at the hands of salvagers and vandals.

While many of the Modern Era forts and batteries are now located within parks, they have not been accorded the same level of protection or care as the remaining brick and stone forts. Most of the old coast defenses structures are considered to be, at the worst, a legal liability or at best, an eyesore to the park. Remaining structures have been built on, fenced in, buried, or destroyed. They have been removed as interfering with the park's primary mission of providing recreation space. Vandalism has caused considerable damage over the years. Abandoned and neglected coast defense structures have suffered from freeze-thaw cycles cracking and spalling the concrete and brick, rusting metal rebar and materials has hastened deterioration. Unchecked vegetation growth has caused some structures to collapse. And rising sea levels and increasingly violent storm surges are eroding away shoreline and destroying major structures. While most gun emplacements have been constructed in such a way to resist these attacks, many other tactical structures have collapsed, and even brick structures have been damaged or destroyed by vandals and neglect. Non-tactical structures, particularly officers' quarters, have survived at many parks and government-owned sites through adaptive reuse, but at some former posts such structures have been completely removed.

However, public interest in the history of American coast defenses has grown since the publication of *Seacoast Fortifications of the United States: An Introductory History*, by E.R. Lewis in 1970. The book publication was a pivotal event, giving the public and park personnel a well-documented interpretive history of American coast defenses. A group of coast defense history enthusiasts gathered at a meeting in 1978 and organized the Coast Defense Study Group (CDSG) in 1985. The CDSG's annual conferences, Journal, Newsletter, web site, and reprints of key coast defense books have played important roles in fostering interest in the history of American coast defense and assisting both the public and park staffs in understanding the fascinating history of these defenses and to interpret their surviving elements. These massive seacoast batteries have been able to withstand both the natural climate and economic development longer than other military features from the same periods. These structures incorporated the leading edge of technology of their time and that draws interest in studying them and interpreting purpose and history. Hopefully this will translate into efforts to preserve and restore these sites for current and future generations.

Battery Winchester, Fort Armistead, Baltimore, Maryland (Terry McGovern)

PERIOD MILITARY MAPS

This book contains a compilation of maps for all the harbor defense reservations utilized during the period 1900 to 1950. The harbor defense projects show a general map of the location of elements (sites) for each harbor and the individual site maps showing the fire control elements. A series of map symbol and abbreviation keys from 1921 and 1945 are included.

The directory is organized by Harbor Defense around the United States clockwise from Portland Maine, down the Atlantic coast to Key West Florida, across to Gulf coast to Galveston Texas, then up the Pacific Coast from San Diego California to the Puget Sound in Washington, then up to the Alaskan defenses of WWII, and followed by the defenses in Hawaii, the Philippines, Panama, the Caribbean, Bermuda, and Newfoundland, Canada. While the status information is fairly comprehensive of the larger fort and military reservations, the status of many of smaller WWII-era fire control stations is not. The authors would appreciate receiving any updated information to correct or add to what has been presented here.

Notes on Coast Defense Maps

Site maps; site plans; exhibits from project plans, supplements and annexes; confidential blueprints; D-series maps—these are all terms that have been used to describe various maps which depict sites used by the U.S. Army, at one time or another, in connection with harbor defense fortifications and fire control. These maps have been keys to ferreting out the identification of the various remaining structures during site visits, yet there is some confusion over where these maps come from, what their cryptic symbols mean, and even what they are called.

Most maps of harbor defense installations are located in the Cartographic Branch of the National Archives. Many of the more frequently seen maps have come from a variety of National Archives holdings. The two concrete-era (1890-1950) map formats most frequently seen are the Confidential Blueprint map series (1900-1935 and 1940-1948) and the exhibits from the annexes/supplements to the harbor defense projects (1940s), which cover the 1940 Modernization Program (WWII-era) construction.

Confidential Blueprint Series Maps (1915-37)

As new construction finished, maps were created, revised, and updated by the Corps of Engineers. A series of maps was reproduced as negatives from a master positive in blueprint style, which meant maps were composed of white lines on a blue or dark background. As they were classified "confidential" by the War Department, they became known as "confidential blueprints."

A number of these confidential blueprints have been found in various cartographic and textual Corps of Engineers records in the National Archives. The confidential blueprint series of maps have general maps of each defended harbor, and general maps of each of the forts and military reservations in the harbor defense. If it was warranted, larger scale maps of parts of some forts were also included. These were labeled "D" for "detail" and followed in series, D-1, D-2, D-3, etc., as required. These maps show the location of batteries, various components of the fire control and communication system, mine facilities, and all the post buildings. Identification of each structure was shown by name, symbol, abbreviation, or number.

After 1900 an optical system for fire control based on trigonometric principles was developed for more precisely aiming coast artillery guns. The structures that were built to house the optical and communication elements of this system were often numerous and small in relation to the other major buildings on a military reservation, and many required a detailed description making it complicated to label them on a map, so a set of map symbols was developed to indicate the fire control structures. As these fire control structures were built in the years following 1905, they were incorporated into the maps on which the

Corps of Engineers recorded the location of all the structural elements of the fortifications in the seacoast defenses.

Keys to the fire control map symbols began appearing in coast artillery manuals, such as drill regulations, training regulations, and later field manuals. A complete update of these maps was performed during the years 1920-1922, just after the major construction projects of the Endicott and Taft programs were completed and before some of the smaller harbor defense areas were eliminated. These maps were kept as part of the records of the various Corps of Engineer district offices around the country. Copies were turned over to qualified parties in the army, such as the Coast Artillery Corps, the Quartermaster Corps, etc. On July 12, 1922, the Coast Artillery Board at Fort Monroe requested a complete list and set of these maps for their records, which were provided in August 1922. The 1922 collection contains about 290 maps of 29 harbor areas. Other versions of these maps were found in the notebooks kept by the engineer assigned to each harbor defense. In due course, the records of the Corps of Engineers and the other branches of the army have been turned over to National Archives. The map collections have been scanned and digitally "cleaned up" to remove extraneous lines and smirches from the scanning process.

WW II-Era Harbor Defense Project Maps

The 1940 "Modernization Program" brought a new set of harbor defenses, some on existing reservations, some on entirely new reservations. The fire control system was much more widespread and frequently located on newly obtained smaller reservations located around the harbor defense shoreline. Maps for these works in this guide come from the 1944-46 supplements to the various harbor defense projects published by the army.

A Harbor Defense Project was a written document which described all existing and projected harbor defense elements, including structures, first prepared in 1932-33. Supplements to the Harbor Defense Projects were prepared 1943-44 and updated during 1945-46. The supplements detailed the progress on the construction of the new 1940s modernization program defenses with descriptions and a set of maps that showed where these new structures were located, the field of fire of the guns, radar coverage, etc. The supplements provide extensive detailed information on all tactical and physical aspects of the harbor defenses on the date of the annex, both existing and proposed, and a number of exhibits detailing the locations of elements. The supplements are generally composed of 7 annexes:

A- Armament
B- Fire Control (including optical instruments and radar installations)
C- Seacoast Searchlights
D- Underwater Defenses (mines)
E- Antiaircraft Artillery
F- Gas Defense
G- Equipment (usually detailing what was on hand and what was needed)
H- Real Estate Requirements (usually detailing sites not yet obtained

These supplements and other forms of the Harbor Defense Projects have been scanned from the National Archives and are available from the CDSG ePress as electronic PDFs. These supplements contain a very comprehensive listings and exhibits of everything that was to be in place at the completion of new rearmament program and are the key references to consult for information on the final state of the American seacoast fortifications in 1945.

A few comments on the items that appear on the confidential blueprint maps and the Harbor Defense Project maps—

A **Harbor Defense** (called a "Coast Defense" before 1925) consisted of a series of land reserves (named as "Forts" and in some cases "Camps" and "Military Reservations") on which the various components of the seacoast defense fortifications were built to guard a major commercial and naval seaport. When the harbor defenses of the United States were modernized in 1890-1910, a new system of defensive works were created. The modern forts consisted of tactical and non-tactical structures spread over hundreds of acres of land. The U.S. Army Corps of Engineers selected the locations, purchased additional land, sited, designed, and constructed the tactical structures—gun batteries, mine facilities, observation stations, plotting rooms, power plants, switchboard rooms, and searchlight shelters.

Gun Batteries: The modern seacoast artillery consisted of guns, mortars and antiaircraft weapons mounted in concrete support structures varying from the simple to the quite complex. Guns were mounted on barbette, pedestal and "disappearing" carriages. Mortars were emplaced in protected pits. Antiaircraft weapons, usually the 3-inch guns, were mounted in simple concrete platforms. The term "battery" was used to describe a set of guns under a single commander together with the entire structure erected for the emplacement, protection, and service of those guns.

Fire Control Structures: The target range and azimuth for seacoast artillery guns were determined using command and equipment systems collectively referred to as fire control and position finding. The standard systems of position finding used by seacoast artillery were based on trigonometry. Components of the system included widely spaced base end stations, command stations, plotting rooms, tide stations, meteorological stations, and cable linked telephone communication systems with protected switchboards. Radar installations were deployed for the major gun batteries and as general surveillance after 1942. The radar installations included power/control buildings and antenna towers.

Searchlights: Most searchlights installed during the period 1901-1920 were fixed, located in a structure for concealment and protection during the day, with their electrical power generator. Over the years after WWI, the mobile searchlights became more reliable, durable, and rugged. By the late 1930s, the Coast Artillery switched to using mobile searchlights and replaced fixed searchlights where at all possible so after 1940 the US seacoast defenses used mostly mobile searchlights.

Controlled Mine Facilities: Throughout the modern or "concrete" era of American harbor defenses (1890-1950), mines were considered to be one of the primary harbor defense weapons. Mines were only deployed during times of war or during limited training expertises. The mines and cables were stored ashore between use. The mine shore facilities included torpedo storerooms, loading rooms, mine wharfs, explosives storage, tramway systems, cable tanks, mine casemates, and cable vaults.

Electrical Generator Power Plants: By the turn of the century, electricity had become a vital necessity for the Coast Artillery. It was used to traverse and elevate some of the large guns, to light emplacements, to operate ammunition hoists, to power searchlights, to control submarine mines, and for communications, in addition to standard garrison uses. Most large forts had a central power plant with electrical generators. The requirement that coast defenses be self-contained resulted in power rooms being included in most batteries and mining casemates, and separate searchlight powerhouses were constructed.

Protected Switchboard Rooms: As seacoast defense artillery covered increasing distances, a need for remote accurate and instant communications was required. Telephones connected by phone lines were integrated into the fire control system utilizing protected switchboard rooms after 1906. As radio communication developed in the 1930s, fixed radio sets were often integrated with the telephone communication system in their protected switchboard rooms or housed in separate protected structures.

Garrison Buildings: These are shown in the Confidential Blueprint series maps but not on Supplement series maps.

The system of numbering for buildings was the same for all Confidential Blueprint maps in the period 1915 to 1937. All buildings of the same "type" were given the same number on all the maps. For example all barracks buildings were numbered "7."

1.	Administration Building
2.	Commanding Officer's Quarters
3.	Officer's Quarters
4.	Hospital
5.	Hospital Steward's Quarters
6.	Non-commissioned Officer's Quarters
7.	Barracks
8.	Guard House
9.	Post Exchange

10 to 19 and 100 to 199	Post Buildings
20 to 29 and 200 to 299	Quartermaster Buildings
30 to 39 and 300 to 399	Ordnance Buildings
40 to 49 and 400 to 499	Engineer Department Buildings
50 to 59 and 500 to 599	Signal Corps Buildings
60 to 69 and 600 to 699	Reserved for future requirements
70 to 79 and 700 to 799	Religious and Social Buildings
80 to 89 and 800 to 899	Government Buildings not under War Dept. Control
90 to 99 and 900 to 999	All Private Buildings (Private dwellings, stores, contractor's buildings and buildings purchased with the land but not assigned to public use.)

Fort Columbia, Washington 1913 (NARA)
From left to right is the Post Exchange, a Company Barracks, the Administration Building,
a Double Officer's Quarters and the Commanding Officer's Quarters.
Just visible behind the front row of buldings is the Quartermaster's Storehouse and the Post Hospital

Symbols and Abbreviations—1921 Confidential Blueprints

Name	Abbr.	symbol	Sta. w/o roof
Fort Commander's Station	C		
Primary Station, Fire Command	F'		
Secondary Station, Fire Command	F''		
Supplementary Station, Fire Command	F'''		
Primary Station of a Battery	B'		
Secondary Station of Battery	B''		
Supplementary Station of a Battery	B'''		
Battery Commander's Station	BC		
Primary Station, Mine Command	M'		
Secondary Station, Mine Command	M''		
Supplementary Station, Mine Command	M'''		
Double Primary Station, Mine Command	M'-M'		
Double Secondary Station Station, Mine Command	M''-M''		
Separate Plotting Room	P		
Separate Observing Room	O		
Self-contained Horizontal Base	C.R.F.		
Emergency Station	E		
Spotting Station	Sp		
Meteorological Station	Met		
Tide Station	T		
Searchlight (30, 60, etc., relates to the size of the lights)	S		
Controller Booth	C.B.		
Watchers Booth	W		
Signal Station	S.S.		
Radio Station	R		
Cable Terminal	C.Ter.		
Post Telephone Switchboard	P.S.B.		
Mining Casemate	M.C.		

Name	Abbr.	symbol
Loading Room	L.R.	 ▣
Switchboard Room	S.W.B.	 ◩
Central Powerhouse	C.P.H.	 ◙⊢
Powerhouse (and Searchlight Powerhouse)	P.H.	▢⊢
Combined Stations, in same room		 (B.C)
Combined Stations, in communicating rooms		  (F')(B) (B.C)(P)
Combined C and F' Station in same room		 (C.F')
Differentiation of auxiliary plants		 a⊢ b⊢ c⊢ etc.

Abbreviations used on maps

Cable Gallery	C.Gal.
Cable Tank	C.T.
Cable Hut (commercial cable)	C.H.
Coast Guard Station	C.G.S.
Engineer Wharf	Engr. Whf.
Gasoline Tank	G.Tk.
Guard House	G.H.
Latrine	L.
Lighthouse	L.H.
Lighthouse Wharf	L.H.Whf.
Magazine	Mg.
Mining Boathouse	M.B.H.
Mining Derrick	M.D.
Mining Tramway	M.T.
Ordnance Machine Shop	O.M.S.
Mine Wharf	M.Whf.
Private Wharf	Pvt.Whf.
Radio (commercial station)	Rad.
Railway Wharf	Ry.Whf.
Saluting Battery	Sl.B.
Searchlight Shelter	S.Sh.
Service Dynamite Room	S.D.R.
Steamship Wharf	S.S.Whf.
Sunset Gun	S.G.
Tide Gauge (not a Tide Station)	T.G.
Torpedo Storehouse	T.S.
Tower	Tw.
Water Tank	W.Tk.
Weather Bureau	W.B.

Additional Symbols and Abbreviations

Name	Abbr.	symbol
Pumping Plant	P.P.	
Radio Powerhouse	R.P.H.	
Searchlight Powerhouse	S.P.H.	
60 inch Searchlight No. 7	$S.^{60}_{7}$	
Coincidence Rangefinder	C.R.F.	
Quartermaster Wharf	Q.M.Whf.	

Subscripts for use in both Legend and on Face of Plat are—

Imp.   Improvised. $\quad$ B″ imp.
$\qquad$ (for temporary fire control structures only.)

p.   Portable. $\quad$ S^{36}_{p2}
$\qquad$ (Principally used for portable searchlights etc.)

s.   Superseded. $\quad$ 24s.
$\qquad$ (for abandoned buildings, etc.)

t.   Temporary. $\quad$ 19t.
$\qquad$ (For all uses except fire control structures.)

Datum Point—location indicated by intersection of lines or by dot at end of arrow.

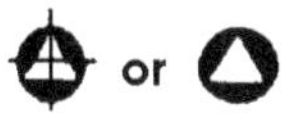

Triangulation Station.

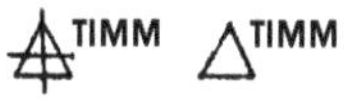

Intersection Point.

Benchmark.

Lighthouse.

Such other topographic signs as were necessary were taken from the *Engineer Field Manual* (Professional Papers, Corps of Engineers, No. 29) pages 74 to 97.

Note: Maneuver buildings were classed as post buildings.

SYMBOLS and ABBREVIATIONS 1940
FM 4-155, Reference Data (Seacoast Artillery and Antiaircraft Artillery) 1940
TABLE C.-Symbols for seacoast artillery fire-control maps, diagrams, and structures

Part 1.—Basic symbols

Name	Abbreviation	Symbol
Harbor defense command post	H D C P	
Groupment command post	Gpmt C P	
Fort command post	Ft C P	
Gun group command post	G C P	
Mine group command post	M C P	
Seacoast battery command post	B C P	
Harbor defense observation station	H D O P	
Groupment observation station	Gpmt O P	
Fort observation station	Ft O P	
Gun group observation station	G O P	
Mine group observation station	M O P	
Battery observation station	B O P	
Emergency observation station	E O P	
Antiaircraft observation post	A A O P	
Battery spotting station	S O P	
Separate observation station	O P	

Name	Abbreviation	Symbol
Operations and plotting room	O P R	 ⊚ (O)
Plotting room	P	 (P)
Self-contained base range-finder station	R F	 (RF)
Magazine	Mg	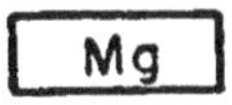 [Mg]
Shellroom	S Rm	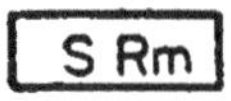 [S Rm]
Temporary or improvised fire-control structures	Imp	 ☐ Imp
Mine casemate	M C	 ■ MC
Mine loading room	L R	 [LR]
Searchlight, 60-inch seacoast	S L	
Searchlight, seacoast, other than 60-inch	S L	
Antiaircraft searchlight	A A S L	 AA
Searchlight shelter	S Sh	 [S Sh]
Searchlight powerhouse	S P H	 ⊣[S]⊢
Searchlight controller booth	C B	 ◯
Data booth	Data B	■
Watchers booth	W Bth	 ⊕
Meteorological station	M E T	 [M]

Name	Abbreviation	Symbol
Tide station	Td	
Signal station	S S	
Fire Control switchboard room	F S B	
Post telephone switchboard room	P S B	
Combined fire-control & post telephone S B room	F S B P S B	
Cable terminal	C Ter	
Powerhouse	P H	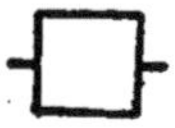
Radio powerhouse	R P H	
Central powerhouse	C P H	
Pumping plant	P P	
Datum point		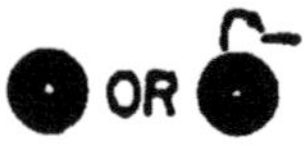
Triangulation station		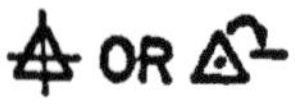
Intersection point		
Benchmark	B M	
Lighthouse	L H	

Other abbreviations used in this guide

BS - base end station & spotting station

BC - battery commander's station

HECP - harbor entrance command post

C - fort commanders station

HDOP - harbor defense command observation post

G- group command station

HDCP- harbor defense command post

M- mine station

SBR -telephone system switchboard or radio room

SCR - signal corps radar

AMTB- Anti-motor torpedo boat BC station

SL - searchlight

Part 2.-Numbers for harbor defense installations.—a. In harbor defense, seacoast artillery installations of each type are numbered consecutively from right to left, facing the center of the field of fire of the harbor defense. Antiaircraft installations pertaining to the harbor defense may be numbered in any convenient sequence.

b. Groupments, gun groups, mine groups, batteries, and all installations functioning directly under the harbor defense commander, such as harbor defense observation stations, searchlights, and underwater listening posts, are numbered consecutively, each type in a separate series, beginning with number 1. These numbers normally are shown as subscripts to the letter included in the appropriate symbol. Exceptions are included among the examples that follow.

Name	Abbreviation	Symbol
Harbor defense observation station	$HDOP_3$	
Fort observation station	$FtOP_3$	
Antiaircraft observation post	$AAOP2$	
Magazine or shell room	$Mg\ 2$ or $S\ Rm\ 2$	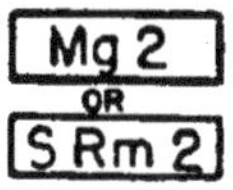

c.Groupment, group, and battery observation and spotting stations assigned to a unit are numbered consecutively within the unit, each type in a separate series, beginning with number 1. These numbers are shown as superscripts to the letter included in the appropriate symbol, the unit number remaining as the subscript.

Name	Abbreviation	Symbol
Groupment observation station	$Gpmt_2\ OP_2$	
Gun group observation station	$G_2\ OP_1$	
Mine group observation station	$M_2\ OP_1$	
Battery observation station	$B_1^1\ OP$	
Spotting station	$S_3^1\ OP$	
Emergency observation station	$E_2^1\ OP$	
Temporary or improvised fire control structures	$B_3^2\ Imp.$	

d. In certain cases it is desirable to show additional information regarding an installation, such as its size and whether fixed, portable, or mobile. Such information is placed either in the symbol or to the right thereof.

Name	Abbreviation	Symbol
60-inch seacoast searchlight; fixed, portable or mobile.	SL 2F (P or M)	2F (P or M)
Seacoast searchlight other than 60-inch	SL$^{36}_{3P}$	36'
Antiaircraft gun battery or composite battery, fixed or mobile.	A A No. 2 (F or M)	AA 2 (F or M)

e. Where two stations are combined in one room, the symbols are superimposed one upon the other, and the letters representing each station are inclosed in the combined symbol.

Name	Abbreviation	Symbol
Combined groupment command post and fort command post.	Gpmt Ft Cp	CF
Combined battery observation and spotting station.	$B^2_1 S^2_1$ O P	$B^2_1 S^2_1$
Combined group command post and battery command post.	$G_1 B_2$ C P	G_1 BC_2
Combined battery command post and battery observation station.	B_2C P B_2^2O P	B^2_2 BC_2

f. Where stations are adjacent in the same structure, the symbols are tangent to each other and are arranged to show the relative location, as:

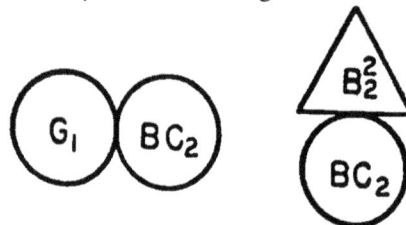

G_1 BC_2

B^2_2 BC_2

g. Where communication may be had by voice through a passage, door, window, or voice tube, the symbols are left open at the point of contact, as:

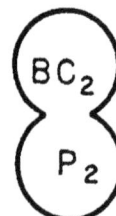

BC_2 P_2

Part 3.—Communications symbols for use on harbor defense fire-control charts and diagrams.

Telephone cable (numerals indicate number of pairs and gage)	 26-19
Speaking tube	
Mechanical data transmission line	
Electrical data transmission line	
Searchlight controller line	
Zone signal and magazine telephone line	
Firing signal line	
Time interval bell line	
Submarine cable (numerals indicate number of pairs and gage)	 50-19

Part 4.-Abbreviations

Cable gallery	C Gal
Cable tank	C T
Cable hut (commercial cable)	C H
Coast Guard station	C G S
Engineer wharf	Engr Whf
Gasoline tank	G Tk
Guardhouse	G H
Latrine	L
Lighthouse wharf	L H Whf
Mine boathouse	M B H
Mine derrick	M Drk
Mine tramway	M Tmy
Mine wharf	M Whf
Ordnance machine shop	O M S
Private wharf	Pvt Whf
Radio (commercial station)	Rad
Railway' wharf	Ry Whf
Saluting battery	Sl B
Service dynamite room	S D R
Steamship wharf	S S Whf
Quartermaster wharf	Q M Whf
Superseded (for abandoned buildings, etc.)	24 s
Temporary (for all uses except fire-control structures)	19 t
Sunset gun	S G
Tide gage	T G
Torpedo storehouse	T S
Tower	Tw
Water tank	W Tk
Weather bureau	W B

A DIRECTORY OF AMERICAN SEACOAST DEFENSES 1890-1950

This directory is a comprehensive guide to all the major locations and sites used for harbor defense, with maps showing what was at each site and comments on the current status of each site (extant, in ruins, destroyed, privately owned, current U.S. military use, federal, state, county, city parks, etc.) as far as information is known to the authors. While the status information is fairly comprehensive for the larger forts and military reservations, the status of many of smaller World War II-era fire control sites is not. The authors would appreciate receiving any updated information to correct or add to what has been presented here. Terms used in this reference work to describe the various periods of construction such as "Endicott-Era," "Taft-Era," "Post-World War I-era," "World War II-Era," the "100-Series" and "200-Series" batteries, etc. are terms used by modern historians and were not used by the Army to describe these programs in progress. Note that several of the planned batteries in the 1940 program were cancelled before any work was done as denoted by their *battery # in italics* and as (planned).

This directory does not cover the following artillery used for seacoast defense at various times between 1898 and 1945—the Rodman guns emplaced or re-emplaced during the Spanish American War; the Navy guns and mounts installed during the World War II years, mostly in the Pacific theater; Hawaii's World War II temporary and provisional defenses; the fixed antiaircraft gun batteries emplaced in the defenses from 1920; mobile artillery which had prepared positions including those for 12-inch railway mounted mortars and 8-inch railway guns; the Panama mount positions for the tractor drawn 155 GPF guns; and the positions on Oahu for the 240 mm howitzers.

The directory is organized by Harbor Defense around the United States clockwise from Portland Maine to the Puget Sound in Washington State, followed by the Alaskan defenses of World War II, and the defenses in Hawaii, the Philippines, Panama, the Caribbean, Newfoundland, and Bermuda.

This directory includes detailed brief histories of modern-era American coast artillery concrete gun batteries. Glen Williford created this as a personal reference guide over many years of research and study of U.S. Coast Artillery history. This battery listing includes the histories of all modern (post-1886) "fixed" or permanent concrete seacoast gun batteries emplaced by the U.S. in the country and outlying territories. The emplacements were mostly built by the Corps of Engineers, and manned by the Coast Artillery. Each battery description includes the following information where possible. The battery name in capital letters if an officially conferred name, in lower case if just an informal, local, or construction designation. The description then briefly covers the purpose for construction and the general location on the reservation, particularly in relation to other elements. In most cases the act or source of original funding (which does not include the cost of coast artillery) and date of plan submission follows. Major design features or significant variations from Mimeograph Type plans are discussed. The general dates of construction, transfer date to the Coast Artillery and engineering costs may also be included. This is followed by a description of the armament, including gun and carriage models and specific serial numbers and date of mounting if known. In general, the manufacturer of the guns and carriages are only designated only when there are multiple producers and thus duplicate serial number runs. The general order and date, that names each battery is included with a brief description of person honored. Subsequent service events, including any major alterations, accidents, armament, or name changes follows. The date of gun dismounting or at least the date of authorization for deletion is covered. A brief statement on whether the battery still exists, or when destroyed, and park or status if on public property concludes the description.

The major sources consulted were: Reports of Completed Batteries, and Reports of Completed Works, Engineer Letters of Submission, surviving Fort Record Books, Seacoast Gun Record cards and earlier Ordnance Department Seacoast Gun Ledger Books, Annexes to Seacoast Projects, General Orders naming citations, supplemented by various records in archive primary engineer correspondence files, annual reports of the Chief of Engineers, private Williford studies on emplacement accidents, temporary defenses, defenses of the Spanish American War.

THE HARBOR DEFENSES OF THE DELAWARE BAY –
DELAWARE AND NEW JERSEY

Delaware Bay is a large estuary fed by the Delaware River. Dutch and Swedish settlements along the river began in 1623. The Pennsylvania colony was established by charter in 1681, and Philadelphia grew as a port due to its deep-water navigation. The defenses directly below Philadelphia were begun by the British and saw action during the Revolutionary War. The Philadelphia defenses were upgraded during the First System of fortification construction. A large Third System for was built on Pea Patch Island near Delaware City and the entrance to the Chesapeake and Delaware Canal which opened in 1829. New batteries were initiated in 1870 but not completed across from Fort Delaware. The Endicott Program built 16 new concrete batteries at three locations across the river at Delaware City. These defenses were augmented by two long-range 12-inch batteries built closer to the entrance of the bay at Slaughter Beach during the 1920s. Finally, the 1940 Program batteries at Cape Henlopen and Cape May covered the wide entrance by 1945. All the defenses were inactivated after 1947, but parts of Cape Henlopen remained under military control until the 1990s.

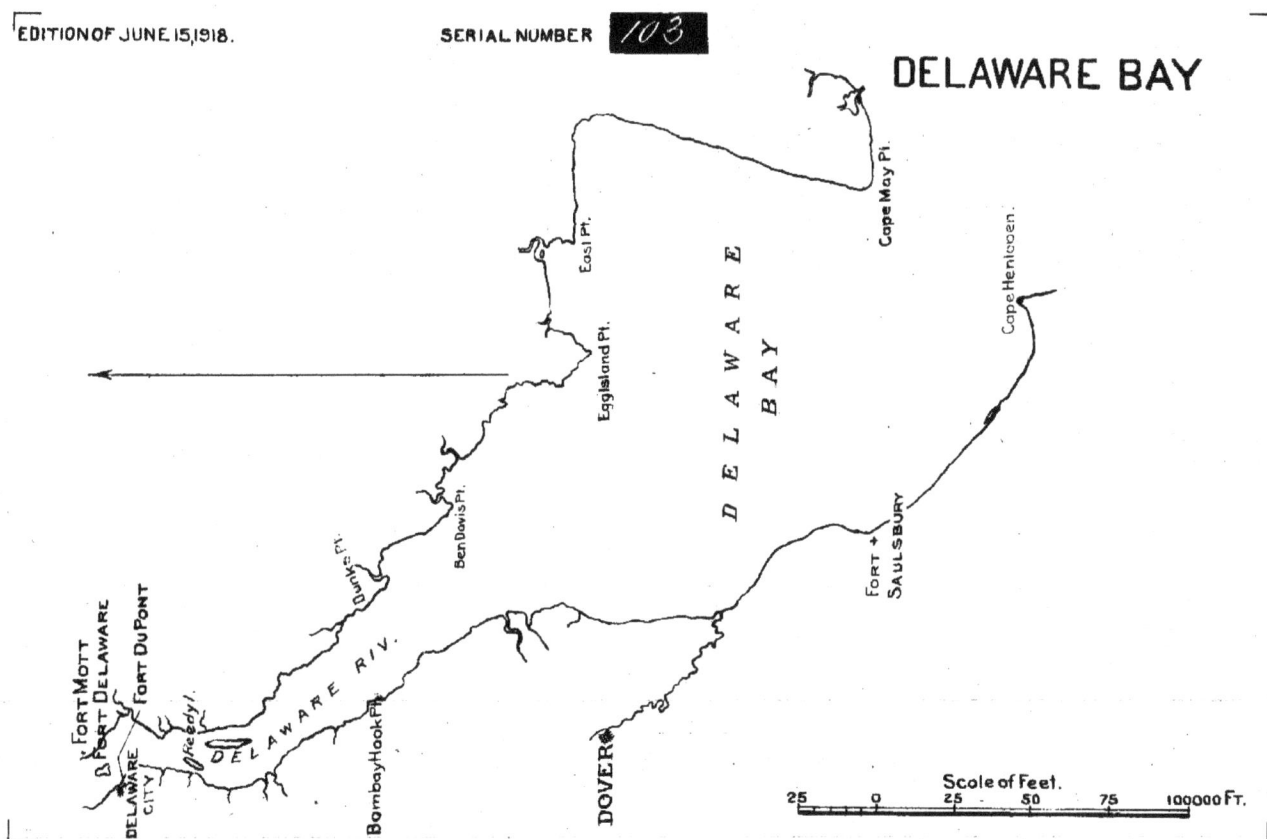

Fort Mott (1875-1943) is located on Finns Point on the Delaware River, about seven miles from the town of Salem, New Jersey. The first seacoast batteries built here in 1872 but only two emplacements were completed by 1876. The Endicott Program saw the construction of five batteries between 1896 and 1903. It was named in General Orders 72 of 1897 for Maj. Gen. Gershom Mott, USV of Civil War service. Of note, was the use of two of the 1870s Period magazines to create a casemated 3-inch pedestal battery and the protective parados and moat behind the main gun line. Fort Mott was placed on caretaker status for most of its service life. Several of the fort's guns were transferred to other defenses in Hawaii and Canada. In 1943, the last battery was removed from service and the reservation was sold to State of New Jersey in 1947 as a state park. An extensive renovation program has been carried out by New Jersey State Parks to stabilize and restore historic fabric of the remaining defensive fortifications at Fort Mott. Restoration and interpretive efforts include Battery Harker, a restored fire control tower, a restored administration building, and a restored peace magazine. Open daily, some restorations are only open for special events. Fort Mott is connected with Fort Delaware and Delaware City by a seasonal passenger ferry operated by the Delaware River and Bay Authority.

Fort Mott Gun Batteries

- **ARNOLD**: A battery for three 12-inch disappearing guns emplaced in the center of the old 1870s earthen battery of Finns's Point, New Jersey, firing to the south. It was part of, and contiguous to the three guns of Battery Harker, together being one six-gun sequence. Plans were submitted on July 21, 1896, using funds from the Fortification act of June 6, 1896. The approved plans closely followed type designs, with lower-level magazines on the right flank and ammunition service with hoists. All three emplacements were interior type, although they each had an additional block of concrete within the western corner of the loading platform for added protection from gunfire from the river. With the bend in the Delaware River behind Fort Mott, there was a danger of reverse fire into the emplacement from only 3.5-miles away. On February 4, 1897 a plan was submitted for a full parados of the same height as the battery and protective moat behind it for construction behind all six main gun line emplacements. This was built at the same time as the emplacements. Most of the concrete work was done in the year of 1897, though completion was delayed by the late arrival of the ordnance and installation of lifts and lights that took until 1898. It was transferred, including adjacent Battery Harker on January 6, 1899 for a cost of $268,351.93. It was named on General Orders No. 105 of October 9, 1902 for Lt. Colonel Lewis G. Arnold of Seminole and Civil War service. It was armed with three 12-inch Model 1888 Watervliet guns on LF Model 1896 disappearing carriages (#12/#1, #8/#2 and Model 1888MII tube #33/#4). The loading platforms were extended in 1907-1910. New hoists were installed in 1907. The carriages were provided with 5-degrees of increased elevation around 1916-1917. Arnold was the site of a fatal accident during practice loading on the night of September 25, 1904. Two crew members were seriously injured (one subsequently dying) when crushed by the sudden release of the chain hoist mechanism on the No. 2 emplacement. The battery served through the 1920s, being placed in a readiness class of "C" during the 1932 Review. Battery Arnold was recommended for immediate abandonment in the July 1940 Program, but the actual disarmament was not implemented until granted by authority on August 31, 1943. The emplacement still exists at the Fort Mott State Park. The battery is open to the public.

- **HARKER**: A battery for three 10-inch disappearing guns emplaced on the left flank of the old 1870s earthen Rodman battery, firing to the south. It was part of a six-gun sequence, with the three guns of Battery Arnold on its immediate right flank. The project was approved on May 10, 1896 a couple of months ahead of Battery Arnold. Submission of plans was made on May 12, 1896. Construction began on just the first two emplacements, but a similar third emplacement was soon added. Plans followed the type mimeographs, with in-line, separate interior platforms, also with the added protection on the river flank exposure. Ammunition service was by lifts, later modified to new chain hoists in 1910. The traverses between the platforms were deeper and narrower than in the type plans. Work was done from the spring of 1896 to mid-1897. Only the mounting of armament and lifts being the outstanding tasks. All of this was done by mid-1898. The battery was transferred along with Battery Arnold on January 6, 1899 for a total of $268,351.93. It was named on General Orders No. 105 on October 9, 1902 for Brigadier General Charles G. Harker who was

SERIAL NUMBER 124

DELAWARE RIVER
FORT MOTT
Finns Pt. N.J.

EDITION OF APR.23,1915.
REVISIONS: DEC.7,1915; APR.10,1916.
MAR.31,1920;MAR.29,1921.

500' 0 500' 1000'

U.S. Boundary

GREGG
HARKER
ARNOLD
KRAYENBUHL
EDWARDS
SEAWALL
Q.M.Whf.
Engr.Reservation
PARADE
FLAG
MOAT
True Meridian Var.1912, 7°15'W.
U.S. Boundary
U.S. Boundary
U.S.Boundary
ROAD TO NATIONAL CEMETERY
NATIONAL CEMETERY
TIDE BANK
Target Butt

LEGEND

1. ADMINISTRATION BLD'G.
2. COMMANDING OFF. QRS.
3. OFFICER'S QRS.
4. HOSPITAL.
5. HOSPITAL STWD'S.QRS.
6. N.C.OFFICER'S QRS.
7. BARRCKS.
8. GUARD HOUSE.
9. POST EXCHANGE.
10. CONCRETE RESERVOIR.
11. HOSE HOUSE.
12. ARTESIAN WELL.
13. PUMPING STATION
14. ENGINEER'S QUARTERS.
14. ARTILLRY ENGINEERS STORE-ROOM.
15. AMBULANCE SHED.
16. BAKERY.
17. OIL HOUSE.
18. COAL SHED.
19. CARPENTER SHOP.
100. WAGON SHED.
101. BATH HOUSE.
21. Q.M.STOREHOUSE.
22. C.S.STORES.
23. Q.M.STABLE
24. TEMP.Q.M.ST.HO.
25. Q.M.CONTONMENT BLDG.
31. ORDNANCE ST.HO.
41. ENGINEERS ST.HO.
70. READING ROOM. AND LIBRARY.
71. Y.M.C.A.BUILDING:
102. WAGON SHED.

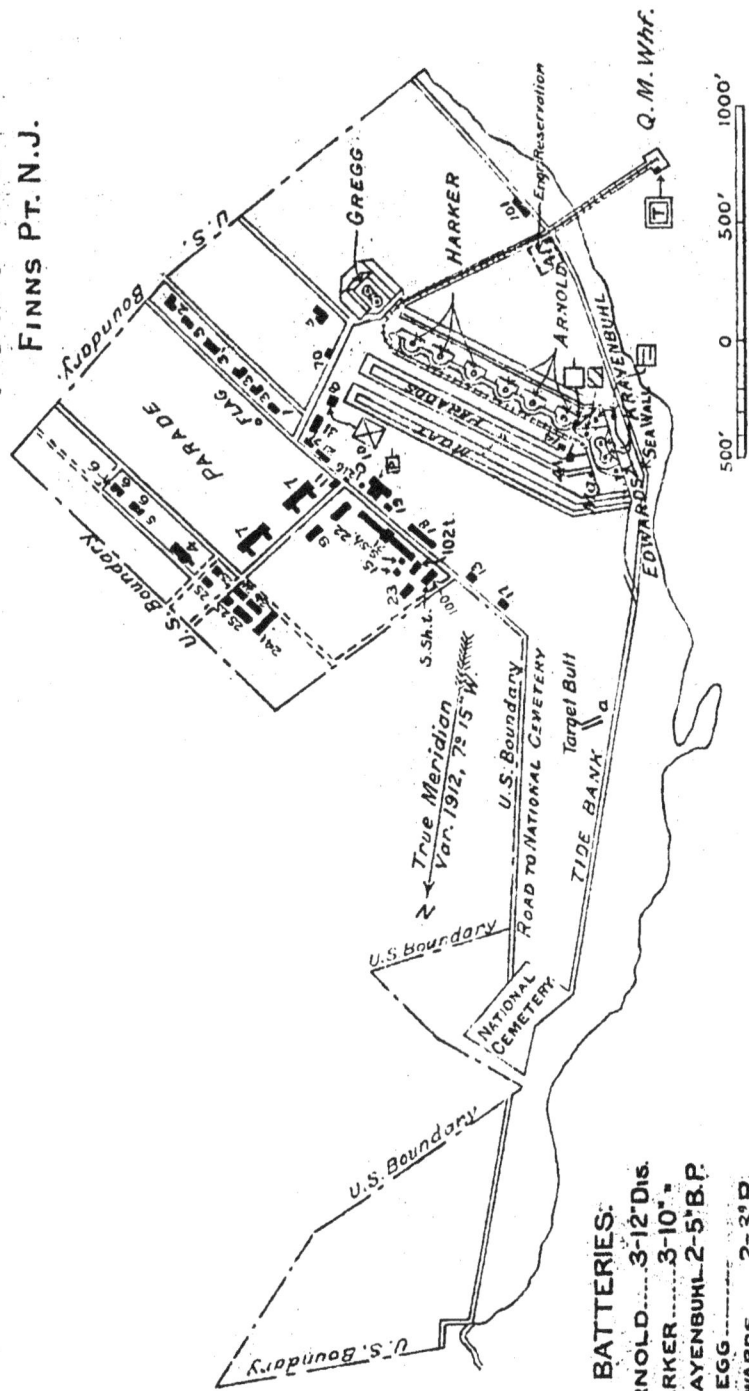

BATTERIES:

ARNOLD....3-12"Dis.
HARKER.....3-10".
*KRAYENBUHL 2-5"B.P.
*GREGG........
*EDWARDS....2-3"P.
*No guns nor Carriages.

Fort Mott 1940 (NARA)

Fort Mott 1936 (NARA)

mortally wounded at Kennesaw Mountain in 1864. It was armed with three 10-inch Model 1888 Watervliet guns on disappearing carriages Model LF 1894M1 (#3/#15, #7/#22 and Model 1888MII #41/#14). The usual modifications to loading platforms, new hoists and an enlarged battery commander's station were done around 1910. The battery served through the 1930s. It was slated for deletion by the 1932 Review. In 1940 the armament (both guns and carriages) were transferred to Canada and used to arm new emplacements at Fort Prevel and Cape Spear. The emplacement still exists at Fort Mott State Park. The battery site is open to the public.

- **KRAYENBUHL:** A battery for two 5-inch balanced pillar guns emplaced on the western end of the emplacement line, at the bend of the old earthen battery, firing to the southwest. Plans were submitted on March 27, 1897. The two emplacements were situated on top of an old 1870s emplacement at the turn of the platforms line. The design used the old magazines below the platforms and built new access passageways to the rear of the entries. Hoists were included to bring shells to the passage between the platforms that were set level to the surface all-around. Excavation started on June 30, 1896 and the concrete work was done by August 15, 1897. Transfer was made on December 19, 1900 for a cost of $19,341.80. It was named on General Orders No. 46 of April 6, 1901 for Captain Maurice G. Krayenbuhl, killed during the Philippine Insurrection in 1899. It was armed with two 5-inch Model 1897 Watervliet guns on Model 1896 balanced pillar mounts (#6/#24 and #8/#25). These guns were removed about 1918. The emplacement still exists at the Fort Mott State Park. The battery site is open to the public.

- **GREGG:** A battery for two 5-inch pedestal guns emplaced on a separate section of the old 1870s earthen battery to the left flank (eastern) end of the emplacement line. It fired to the south. The original plan for rapid-fire guns at Fort Mott called for a 5-inch battery on each flank of the main gun line—this was the second one, on the eastern flank. It was submitted on September 20, 1899, intended for the Brown wire-wound type of gun that was never acquired. Care was taken to move it enough to the rear so it would not be in the field of fire of Battery Harker not far from it on its right flank. The battery was placed atop an old magazine mound so its height would be a 35-foot crest, and it could actually fire over the parados behind the gun line in case a troublesome enemy ship reached those waters. It was generally of the mimeograph design type, but the increased height allowed the magazines to be beneath the platforms instead of alongside. That increased protection but necessitated the use of bigger cranes with triplex blocks for hoisting ammunition. Work was done in the summer of 1900. Transfer was made on June 7, 1901 for a cost of $17,500. It was named on General Orders No. 46 of April 6, 1901 for Captain John C. Gregg who was killed during the Philippine Insurrection in 1899. With the substitution of a new type of armament for the cancelled wire-wound guns, armament was delayed until 1907. It was then armed with two 5-inch Model 1900 Watervliet guns on Model 1903 pedestal mounts (#18/#20 and #20/#21). The battery did not serve long, the armament was removed on August 30, 1913 and shipped to Hawaii to be eventually emplaced in Battery S.C. Mills at Fort Ruger. Subsequently the emplacement was used as the location for a CRF station and heavily modified in the early 1920s. It still exists at the Fort Mott State Park. The battery site is open to the public.

- **EDWARDS:** A battery for two 3-inch, rapid-fire guns emplaced in a unique casemated work close to the river on the right flank of the emplacement line. It fired to the southwest. A proposal of February 4, 1897 advocated using some of the existing old 1870s magazines on the right (western) flank of the old earthen battery as emplacements for light, 3-inch guns. Originally it was planned to convert four magazines by removing the earth in front and breaking through the end of the magazine to place an embrasure for the light gun. In September 1900 plans for the ordnance were prepared, the gun to be a standard Model 1898 3-inch barrel on a special embrasure pedestal made either by the American Ordnance Company or of an Ordnance Department design made by Bethlehem Steel. It was to have a 60-degree field of fire out the embrasure opening. Eventually it was decided to convert just two adjacent magazines as a single battery for two guns. They utilized the two old magazines on either side and below 5-inch battery Gregg building directly above. Work was done from March 25 to July 27, 1901. Transfer was made on January 16, 1902 for a cost of only $5,841.11. It was named on General Orders No. 78 of May 25, 1903 for Captain Robert Edwards, killed in action at Frenchtown, MI in 1813. The armament was a unique combination. It consisted of two standard Model 1898 Driggs-Seabury gun tubes (#62 and #63) on modified Model 1902 pedestals, now designated as Casemate Mounts #1 and #2. This was declared obsolete on July 2, 1920 and removed and scrapped. The two embrasure emplacements still exist at the Fort Mott State Park. The battery site is open to the public.

Fort Delaware (1814-1943) is located on Pea Patch Island in the middle of the Delaware River, about a mile down river from Delaware City, Delaware. The first earthwork batteries were built on the island during the War of 1812 and then a timber fort was constructed in 1828, but in 1831 most of the fort burned down. As the 178-acre island is more a mud bank than a true island, a series of dikes and elaborate foundations were needed to provide a suitable site for fortifications. It was officially named in General Orders 32 of 1833. In 1847, construction begun on the Third System fort, a large, pentagon-shaped, granite fort covering six-acres with gun positions for 169 cannons. The three-level fort was completed in 1859 and served as a military prison throughout the Civil War. At its peak, the island prison held 12,500 Confederate prisoners-of-war and over 2,700 died of disease. The Endicott Program between 1892 and 1901 saw the construction of five batteries and a mine casemate. The main battery was three 12-inch guns which were to be barbette carriages on a gun lift platform, but this was changed to use disappearing carriages. Fort Delaware was placed on caretaker status until 1942 when a detachment manned the fort's remaining 3-inch guns, until the guns were transferred to Fort Miles. The fort's 12-inch guns were used to arm a casemated 12-inch battery in San Juan, Puerto Rico during World War II. The site became a Delaware State Park in 1951. Over the years the fort has been cleaned up and stabilized. Today, Fort Delaware State Park can be visited by a passenger ferry from Delaware City during the summer season. The park has a visitors' center and gift shop and an active Civil War re-enactment program. The site hosts a number of special events during the year.

Fort Delaware Gun Batteries

- **TORBERT:** An emplacement for three 12-inch guns emplaced across the parade on the interior of the old masonry Fort Delaware on Pea Patch Island. This was one of the original sites allocated a gun lift battery by the original Endicott Report. The Fortification Act of August 1, 1894 authorized an expenditure of $260,000 for a battery of three 12-inch lift guns. After almost a year of testing whether the ground on the island could support such weight, a plan was submitted on July 30, 1895. It would have been a three-story work, taking the crest to the top level of the old work, and requiring new pilings placed underneath the mass of the new concrete structure. Then on July 19, 1898 a new plan was submitted changing the battery to three conventional 12-inch disappearing guns on Model M1896 carriages. Work had already been done with placing new foundation piles and concrete pouring for the lower levels. The new plans were also carried to the three-story level, but did require cutting down some of the old ramparts in order to get a 3-degree depression for the carriages. Trunnions were at a 50-feet height. The magazines were kept one level lower than the gun platforms, store and support rooms being on the lowest levels. The space for three platforms in the available space was necessarily tight. Gun centers were only 78-feet apart. They were shaped

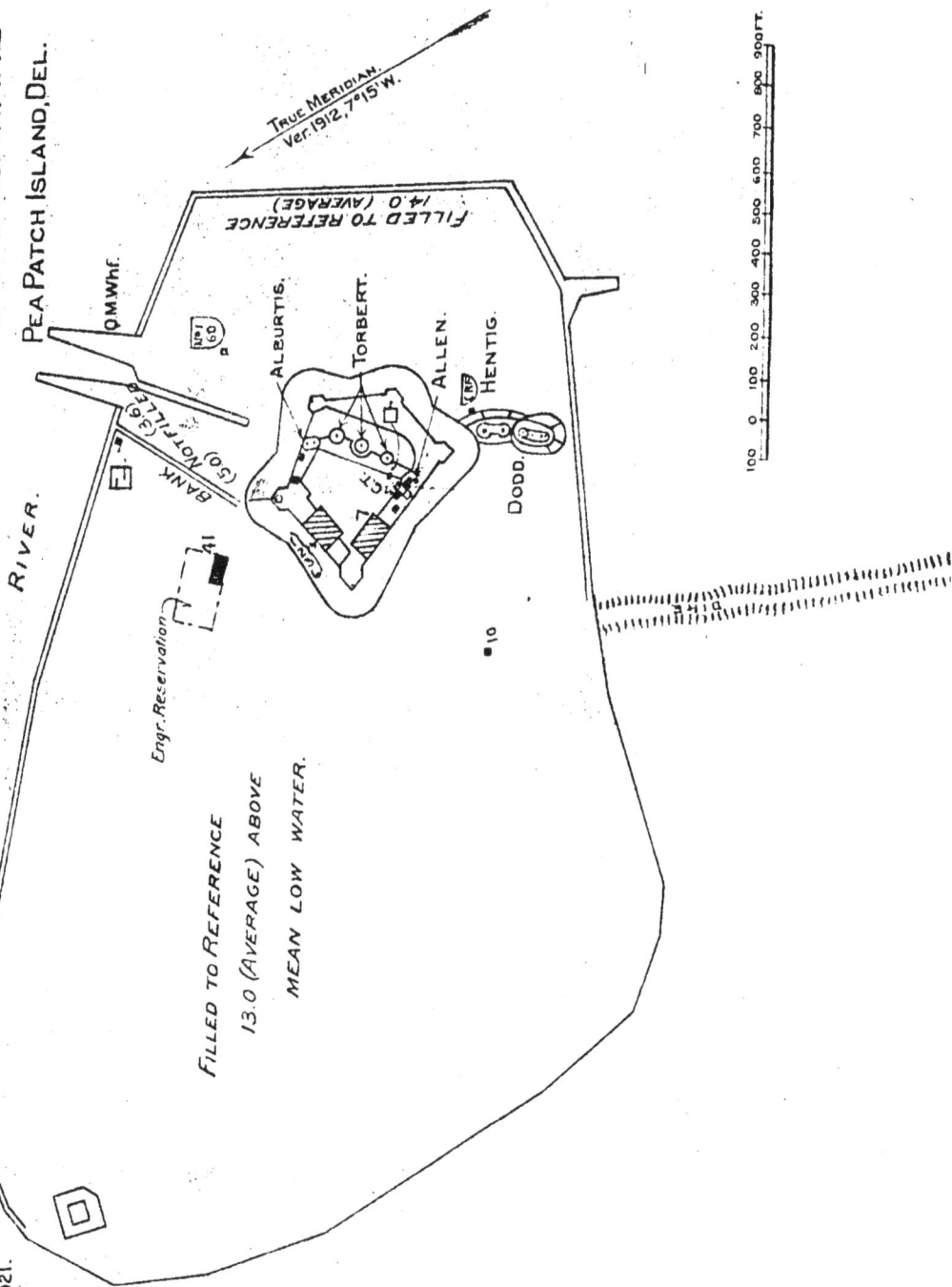

Fort Delaware 1940 (NARA)

Battery Torbert Fort Delaware (Mark Berhow)

as one interior with a right and left flank emplacement respectively on the sides. Work on the new design was done from December 11, 1899 to October 19, 1900. Transfer finally came on April 18, 1901 for a total cost of $350,000. It was named on General Orders No. 78 of May 25, 1903 for Captain Alfred T. A. Torbert of Civil War service. The armament was delayed when the carriages were wrecked in a transportation train accident in late 1899, and many parts needed to be repaired or replaced. It was eventually armed with three 12-inch Model 1895 Watervliet guns on Model LF 1896 disappearing carriages (#15/#13, #18/#14, and #20/#15). Numerous changes occurred in the battery in its early years. Hoists were replaced with new models and then modified for long-point shells. A new BC station was added. Carriages were modified in 1916-1917 for extra elevation to provide longer range. The battery was not disarmed during World War I and continued to be one of the most powerful battery of the defenses. It was finally scheduled for deletion in 1940, the guns being taken to be modified and two were mounted in a new battery in Puerto Rico, the old carriages were scrapped in place in March 1943. The emplacement still exists at the Fort Delaware State Park. The battery site is closed to the public.

- **DODD**: An emergency Spanish American War emplacement for two purchased 4.7-inch guns emplaced on Pea Patch Island. The battery plan was submitted on April 13, 1898 for a pair of the rapid-fire guns purchased in England at the start of the war. Temporarily they were mounted on the barbette tier of the old fort, where work being started on April 14 and was completed by April 30th. The guns arrived on April 23rd and were mounted on May 5, 1898. A considerable discussion ensued about the permanent site and type of battery to be built. Local engineers wanted it to be on the southeast corner of the island where it could cover all the mine fields. In such an exposed position they wanted to build a fully casemated battery. But with that type of construction and the piles or grillage necessary to sustain the weight, it was estimated to cost $27,000. Washington overruled this choice and instructed that a conventional emplacement be built to the southwest of the old fort. The exposure to enemy fire and the limits to coverage due to the mass of the work behind it being accepted. Resubmission was made on June 8, 1898. Work Began immediately and was completed by June 14, 1899. Transfer was made on January 28, 1899 for $19,429.85. It was named on General Orders No. 78 on May 25, 1903 for Captain Albert Dodd killed in action at Gaines Mill, VA in 1862. It was armed with two 4.7-inch Armstrong guns and pedestals (#9723/#10839 and #9724/#10845). There was apparently some problem with the construction of the work, as the 1910 Report of Completed Batteries lists it as in poor condition and even the gun breechblocks removed and in storage. In any event it was disarmed during World War I and finally abandoned by authorization of June 23, 1919. The emplacement still exists at the Fort Delaware State Park. The battery site is open to the public.

- **HENTIG**: A battery for two 3-inch pedestal guns emplaced outside of old fort Delaware near and just to the west of Battery Dodd. Plans were submitted on June 23, 1900 for the site on the left flank of the completing 4.7-inch battery. It was planned as a sunken battery, with the capability of all-round fire, except obviously beyond the mass of the masonry fort behind. The platforms were complete circles, the two magazines were in front of the side passageways into the emplacement. The platforms were placed 52-feet apart, and the trunnion height was 21-feet. Like the other batteries on the muddy foundation of Pea Patch Island, it required expensive grillage and piles. Work was done from September 1900 to April 1901. Transfer was made on September 16, 1901 at a cost of $15,758.67. While it was designed and built for masking parapet carriages (with barrel niches and space for the well of the pillar) it was converted during completion for pedestal mounts. The armament was not mounted until November 1910. It had two 3-inch Model 1903 guns and pedestals

(#54/#52 and #87/#95). It was named on General Orders No. 78 of May 25, 1903 for Captain Edmund C. Hentig killed in action against hostile Indians in Arizona Territory in 1881. A new CRF station was added in 1919 of the left flank of the battery. The guns served until removed and sent to Fort Miles to be emplaced in the Examination Battery in May 1942. The emplacement still exists at the Fort Delaware State Park. The battery site is open to the public.

- **ALBURTIS**: A battery for two 3-inch masking parapet guns built atop the Fort Delaware curtain wall on the left flank of Battery Torbert. Plans were submitted on January 20, 1899. The site, obviously exposed to direct fire against the masonry below, did have the benefit of an excellent field of fire to the east over the channel and mine field. Also, the emplacement was inexpensive, not needing work on the foundation and using protected magazines directly below the gun in the casemates of the old work—essentially only the new gun platforms were needed. Chain hoists were used to move the ammunition to the platforms. Work was done during 1899. It was transferred on April 18, 1901, and cost $3800. The battery was named on General Orders No. 78 of May 25, 1903 for 1st Lieutenant William Alburtis killed in 1847 at Vera Cruz, Mexico. Alburtis was armed with two 15-pounder Model 1898 Driggs-Seabury guns on masking parapet mounts (#22/#22 and #24/#24). A new CRF was built on the flank in 1919. However, in June 1920 the gun type was declared obsolete and removed. The emplacement still exists at the Fort Delaware State Park. The battery is closed to the public.

- **ALLEN**: A battery for two 3-inch masking parapet guns built atop the Fort Delaware curtain wall on the right flank of Battery Torbert. Plans were submitted on January 17, 1899. The site, obviously exposed to direct fire against the masonry below, did have the benefit of an excellent field of fire to the west over the main channel and mine field (and firing over Batteries Dodd and Hentig mounted below). Like its sister Battery Alburtis, the emplacement was inexpensive, not needing work on the foundation and using protected magazines directly below the gun in the casemates of the old work—essentially only the new gun platforms were needed. Chain hoists were used to move the ammunition to the platforms. Work was done during 1899. It was transferred on April 18, 1901, and cost $4000. The battery was named on General Orders No. 78 of May 25, 1903 for 1st Lieutenant Robert Allen Jr., killed in action in 1862 at Gaines Mill, VA. Allen was armed with two 15-pounder Model 1898 Driggs-Seabury guns on masking parapet mounts (#18/#18 and #21/#21). A new CRF was built on the flank in 1919. However, in June 1920 the gun type was declared obsolete and removed. The emplacement still exists at the Fort Delaware State Park. The battery is closed to the public.

Fort DuPont (1863-1943) is located on the Delaware River, just south of Delaware City, Delaware. The military reservation was first used as a supporting battery to Fort Delaware during the Civil War when a ten-gun battery was constructed. During the 1870s Period, a twenty-gun earthen barbette battery was installed and mortar battery begun, along with a controlled mine casemate. Due to the importance of the Chesapeake and Delaware Canal, the 320-acre reservation received six Endicott Program batteries and a controlled submarine mine complex between 1899 and 1904. It was named in General Orders 134 of 1899 for Rear Adm. Samual F. DuPont, USN. The main armament was 12-inch mortars that covered the river from Fort DuPont to Fort Delaware to Fort Mott. The Fort DuPont was placed on caretaker status in 1922 and then it became a U.S. Army Engineer post. During World War II, the post served as the headquarters for the HD of the Delaware and a mobilization center. Later, it became a German prisoner of war camp until 1945 when it was declared surplus. The State of Delaware converted part of the fort into the Governor Bacon Health Center in 1948, which used many of the post garrison buildings. This mental health facility was phased out in 1984. In 1992, 322 acres of former fort along the river was turned into the Fort DuPont

FORT DUPONT

DELAWARE RIVER

DELAWARE CITY.

EDITION OF APR. 23,1915
REVISIONS: DEC.7, 1915, APR.10,1916
MARCH 31, 1920.
MAR. 29,1921.

SERIAL NUMBER 124

LEGEND.

1. ADMINISTRATION BLDG.
2. COMMANDANTS QUARTERS.
3. OFFICERS QRS.
3a. FIELD OFFICER'S QRS.
3b. BACHELOR QRS.
4. HOSPITAL.
5 HOSPITAL STWD'S. QRS.
6. N.C. OFFICERS' QRS.

7. BARRACKS.
7a. BAND QUARTERS.
8. GUARD HOUSE.
9. POST EXCHANGE.
10. ARTY. ENGR. OFFICE.
11. ST. HO.
12. FIRE STATION.
13. FIREMAN'S QUARTERS.
14. RESERVOIR (CONCRETE
    30,000 GALLONS)
15. FIRE APP. HOUSE.
16. ARTESIAN WELLS.
17. BAKERY.
18. GARAGE.
19. GREENHOUSE.
100. STOREHOUSE (TEMP.)
101. WOOD SHED.
102. OIL HOUSE.
103. COAL SHED.
104. CREMATORY.

105. WAGON SHED.
106. SCALE.
21. Q.M. STOREHOUSE.
22. Q.M. SUBSISTENCE BLDG.
23. Q.M. WORKSHOP.
24. Q.M. STABLE.
31. ORDNANCE ST. HOUSE
41. ENGINEER ST. HOUSE.
71. SCHOOLHOUSE.
72. Y.M.C.A.

LEGEND.

200. GARAGE.
201. TOOLHOUSE.

25 Q.M. CANTONMENT BLDG.
26 ISOLATION WARD AND
   BOILER ROOM. TEMP.

27. PIGEON HOUSE.
28. POST CARPENTER. TEMP.
29. CEMENT SHED.
107. ENGINE HOUSE (SEWER
     PLANT.
108. MOTOR CONTROL PLANT.
109. STABLEMEN'S QRS.
110. GARAGE.
111. PAINT SHOP.
112. TARGET BUTTS.
113. CONCRETE RESERVOIR
40. OFFICE.
42. CARPENTER SHOP.
43. BLACKSMITH SHOP.

BATTERIES.

BEST......4-12" M.
RODNEY...4-12" M.
*READ......2-12"N.DIS
*GIBSON...2-8" DIS.
*RITCHIE..2-5" P.
ELDER....2-3" P.

* NO GUNS NOR CARRIAGES.

Location #10 2 AA GUNS.

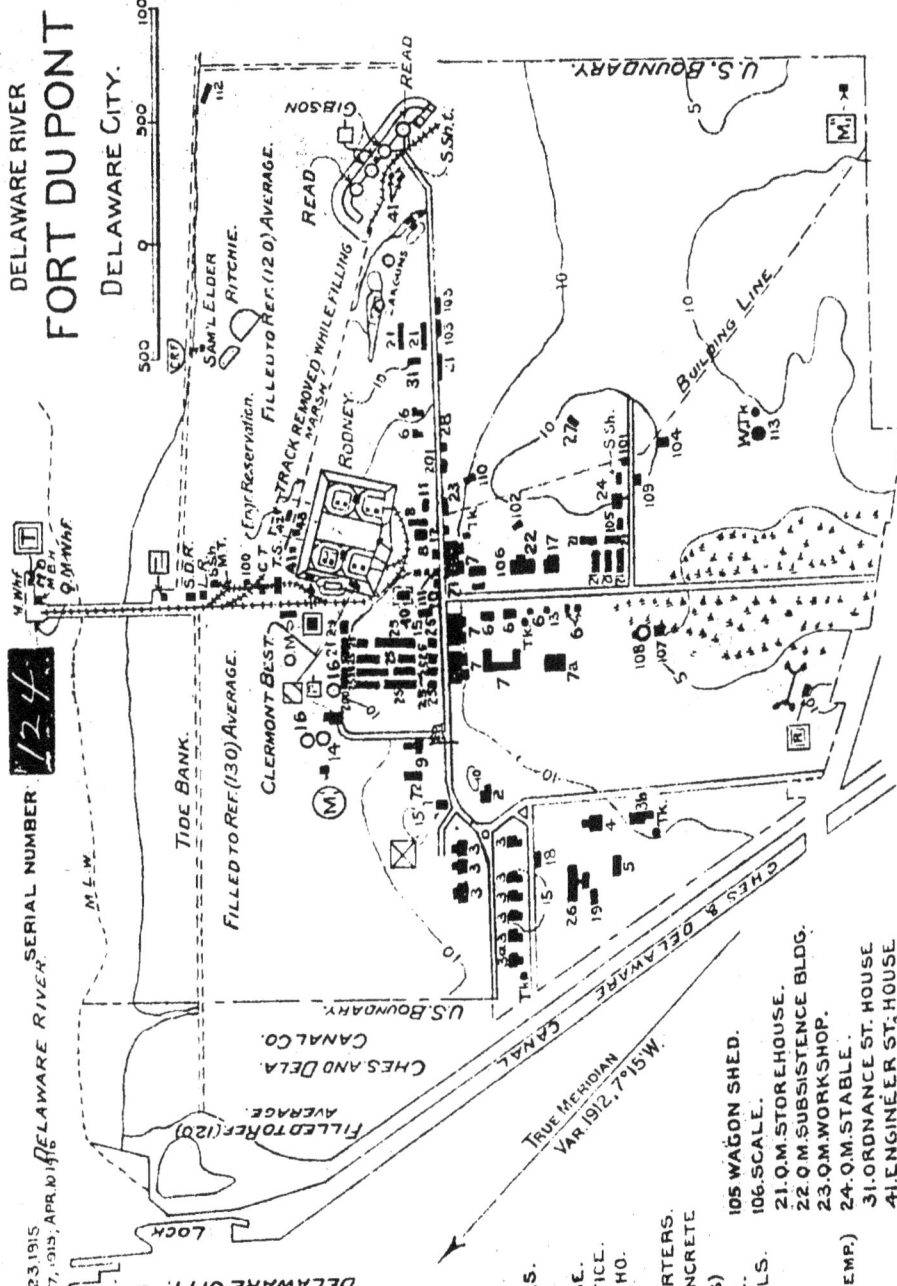

Fort DuPont 1940 (NARA)

Battery Harker Fort Mott (Mark Berhow)

Fort Delware State Park (Terry McGovern)

Fort DuPont State Park (Terry McGovern)

Fort Mott State Park (Terry McGovern)

Fort Mott State Park (Terry McGovern)

State Park. In 2014, the Fort DuPont Redevelopment and Preservation Corporation assumed responsibility for about 400 acres of the former fort (including the state park) to preserve the fort's crumbling structures and to provide for economic development. In 2017, new homes were built along the Branch Canal, site of Army's officers' quarters, with plans for a townhome development along the Delaware River. The former battery area is now considered park land, and the structures are currently abandoned and closed to the public.

Fort DuPont Gun Batteries

- **BEST – RODNEY**: The mortar battery for the Delaware defenses erected at the Delaware City defenses at Fort DuPont. Plans were submitted on April 28, 1897. Due to the battery's exposure over an arc of 200-degrees the engineers utilized the more protected, Abbott quadrangular design with the pits arranged two-by-two square. There were practical modifications to what was rapidly becoming an obsolete design. The less-exposed western side had reduced protection thickness. This western flank also received an electric plant and firing room. The design had been criticized as being too cramped, so the powder magazines were enlarged, consequently being fitted to carry 200 rounds and charges per mortar. Most significantly the rear and front set of pits were closer together by some 40-feet over the type plan, the overall perimeter of the pits being 120 by 245-feet. The battery was sited in the center of the reservation, inland from the wharf, pointing downriver to the southeast. It was a low elevation site, trunnion height being just under 16-feet. Work was delayed until the new wharf and the new plant for construction was built. Work was done from late 1897 through mid-1898. All concrete work was done by June 30, 1898, including mounting the first eight mortars. Transfer was made on May 2, 1900 for a construction cost of $217,721.13. It was named on General Orders No. 16 of February 14,1900 for Cesar Rodney of the Continental Congress. It was armed with sixteen Model 1890M1 mortars on Model 1896 carriages (Builders tubes: #3/#76, #4/#87, #6/#89, #9/#56, #10/#88, Watervliet tubes: #13/#58, #62/#106, #75/#105, #92/#103, #94/#90, #96/#59, #98/#107, #118/#108, #119/#118, and Bethlehem tubes: #27/#83 and #30/#84). In 1906 the battery was administratively split, the two southern pits of eight mortars being named in General Orders No. 10 of January 5, 1906 for Major Clemont L. Best, Artillery Corps, of Spanish American War service. On June 8, 1912 a plan was submitted to entirely rebuild the battery, similar to what was being done at Fort Banks in Boston. However, the estimated cost of $78,710 for what was admitted to be a less-than-ideal location (thoughts were already contemplating new defenses further downriver) quickly ended the project. In 1914 eight mortars and carriages were removed, two from each pit (guns No. 2 and No. 4 in each). These mortars were installed in new batteries at San Diego and Hawaii (Battery Rodney supplied Battery White at Fort Rosecrans and Battery Best supplied Battery Birkhimer at Fort Ruger). The remaining armament continued to serve with eight guns until the 1940s. Final disarmament was in 1940-41. The emplacement was used afterwards for a number of purposes, including the State of Delaware's civil defense center. In recent years the batteries have been abandoned with modern day renovations removed. It still exists at the Fort DuPont Preservation and Redevelopment site and Fort DuPont State Park. The battery is open, but the interior is closed to the public.

- **READ**: A battery for two 12-inch barbette guns emplaced on the Fort DuPont reservation, south of the mortar battery, firing to the southwest down the river approaches. Plans were submitted on May 17, 1898 using funding from the National Defense Act of March 9, 1898. It was one of the batteries slated to use a special order of Model 1892 barbette carriages that could be produced much more quickly than disappearing carriages to match up with a surplus of barrels available. It was an integral part of a four-gun battery with Battery Gibson's two 8-inch disappearing gun emplacements started with the same appropriation slightly earlier. Gibson's emplacements were in the center, with Read's 12-inch barbette guns on each flank. The plans were of the type design, with lower-level magazines and ammunition service with lifts. Work was done from May 1, 1898 to December 31, 1898. Transfer was made on January 12, 1899 for a cost of $80,829.68. The guns arrived on July 23, 1898 and the carriages in April and May 1899—all were mounted by August 1899. The armament was two 12-inch Model 1888M1-1/2 Watervliet guns on Model 1892 barbette carriages (#37/#21 and #38/#22). The battery was named on General Orders No. 16 of February 14, 1902 for George Read, a signer of the Declaration of Independence. The guns were removed in 1918 for use on railway carriages, and the carriages either scrapped or sent away for use elsewhere. The emplacement still exists at the Fort DuPont Preservation and Redevelopment site and Fort DuPont State Park. The battery is open, but the interior is closed to the public.

- **GIBSON**: A pair of 8-inch disappearing guns emplaced near the center of the Fort DuPont reservation, south of the mortar battery. Like Battery Read, it was funded with the March 1898 National Defense Act at the start of the Spanish American War. Plans were submitted on March 31, 1898. They were of typical design type, closely following the proscribed mimeographs. Internal type, separate platforms were separated with a traverse. Magazines were to the lower left; ammunition service was by hoists. Work was done from April 1, 1888 until the guns and carriages were mounted in May 1899. Transfer came on January 12, 1899 at a cost of $67,955.04. It was named on General Orders No. 78 of May 25, 1903 for Colonel James Gibson, killed in action in 1814 during the War of 1812. It was armed with two 8-inch Model 1888MII Bethlehem guns on Model LF 1896 disappearing carriages (#19/#8 and #20/#13). Modifications were made in the early 1900s to the loading platforms and BC station. By 1915 the upper Delaware defenses were being downgraded, and the battery was not in active service. In late 1917 the guns were removed for use on railway carriages, and in 1918 the carriages were ordered scrapped. The emplacement still exists at the Fort DuPont Preservation and Redevelopment site and Fort DuPont State Park. The battery is open, but the interior is closed to the public.

- **RITCHIE**: A battery for two 5-inch pedestal guns built near the shore between the river and the mortar battery on the eastern side of the reservation. Plans were submitted on July 27, 1899 for a battery to carry two 5-inch Brown-pattern wire-wound guns. It was located some 700-feet from the 12-inch Battery Read and designed to fire across the rivet to protect the mine field in both the main and secondary channels. It followed the type plan recommendation. However, the only available land was low and marshy, calling for extra pilings and concrete increasing the cost considerably. Funds were allocated on October 5, 1899 and came from the July 7, 1898 Fortification Act. Work was done from April 28 to September 30, 1900. However, the original armament was never produced, and the substitute 5-inch Model 1900 guns on Model 1903 pedestal carriages only became available after several years' delay, postponing the battery's eventual service. It was transferred on December 19. 1900 for an emplacement cost of $16,702.71. It was named on General Orders No. 78 of May 25, 1903 for Captain John Ritchie, killed in action in 1814 at Lundy's Lane. It finally received two 5-inch Model 1900 guns and Model 1903 pedestals (#10/#18 and #11/#19). This was not mounted until mid-1905. In turn this armament was removed in May of 1917 to arm a temporary battery at Fisherman Island in the Chesapeake defenses. The emplacement itself was not subsequently re-armed and was eventually destroyed in 1930s. No traces of it remain today.

- **ELDER**: A battery for two 3-inch pedestal guns mounted near and just to the north of Ritchie on the river front. It was funded with the June 6, 1902 Fortification Act. Submission was made on February 19, 1903. The plan followed the type plans, with a flank position on the left and an interior position on the right. Platforms were at a 42-foot distance, the traverse covering two separate magazine rooms. The low and marshy ground required the use of pilings for the foundation. The battery was electrified from the beginning. Construction was done from March 23 to October 1903. Final transfer came on March 4, 1904 for a total cost of $18,383.36. It was named on General Orders No. 194 of December 27, 1904 for Brevet Lieutenant Colonel Samuel S. Elder, of Civil War service. The battery was not armed until March of 1910, when it received two 3-inch Model 1903 guns and pedestal mounts (#77/#61 and #80/#62), A new CRF station was added nearby and transferred on August 19, 1922 for $2,303.65. The guns served until World War II, when with authority of April 9, 1942 they were removed and placed in a new temporary battery on Reedy Island. The emplacement still exists at the Fort DuPont Preservation and Redevelopment site and Fort DuPont State Park. The battery is open to the public.

Liston Front Range Light Military Reservation (1942) is located adjacent to the lighthouse at Bayview Beach, Delaware on the Delaware River. Built in 1942 as a new battery (aka Battery Liston), with guns relocated from Fort DuPont in order to better protect the U.S. Navy's AMTB boom defense at Reedy Island. The gun battery was active only between May to December 1942 before the guns were removed in March 1943 when the Navy removed their river boom defense. The concrete emplacements and magazines still exist. Private property.

Liston Front Range Light M.R. Gun Battery

- **Liston:** The two 3-inch Model 1903 guns from Battery Elder at Fort DuPont (#77/#61 and #80/#62) were removed from that battery on May 11, 1942 and reported to be relocated to a new site at the Liston Front Range Lighthouse to protect the boom located on Reedy Island. The site is variously reported as "Reedy Island" and "Liston Light." From maps these appear to be different locations, and it is not clear if there were two locations or only one. In any event the guns are reported as having been removed early, taken out on December 12, 1942 and going into storage locally. The two concrete emplacements and magazine structure remain next to the Liston Front Range Lighthouse at Bayview Beach, while the status of any remains near Reedy Island is not known. The battery site is closed to the public.

Fort Saulsbury (1917-1946) is located near Slaughter Beach, Delaware about a mile inland from the Delaware River. Off State Highway 36, six mile east of Milford, Delaware, this 162-acre military reservation was acquired during World War I for two long-range 12-inch barbette batteries of the Interwar Period. These two batteries covered the lower Delaware Bay and made Fort Mott, Fort Delaware, and Fort DuPont redundant. Completed in 1924, these batteries remained in use until World War II. It was named in General Orders 130 of 1917 for the Hon. Willard Saulsbury, a US Senator from Delaware. Plans were made to build protective gun houses over the four guns, but instead, the guns and carriages of one battery was relocated to a new casemated battery at Fort Miles, at the mouth of the Delaware Bay, while the other was retained until 1946 without being casemated. The fort was used as prisoner-of-war camp after 1943. Fort Saulsbury was declared surplus in 1946, and the reservation was sold to private interest. The emplacements for the two batteries remain today on privately owned land.

Fort Saulsbury Gun Batteries

- **HALL:** A battery for two 12-inch long-range barbette guns emplaced as part of the 1915 Board of Review Program at the new reservation of Fort Saulsbury. It was the northern of the two dual batteries, firing to the east. It followed closely the type plans for these long-range batteries, except it

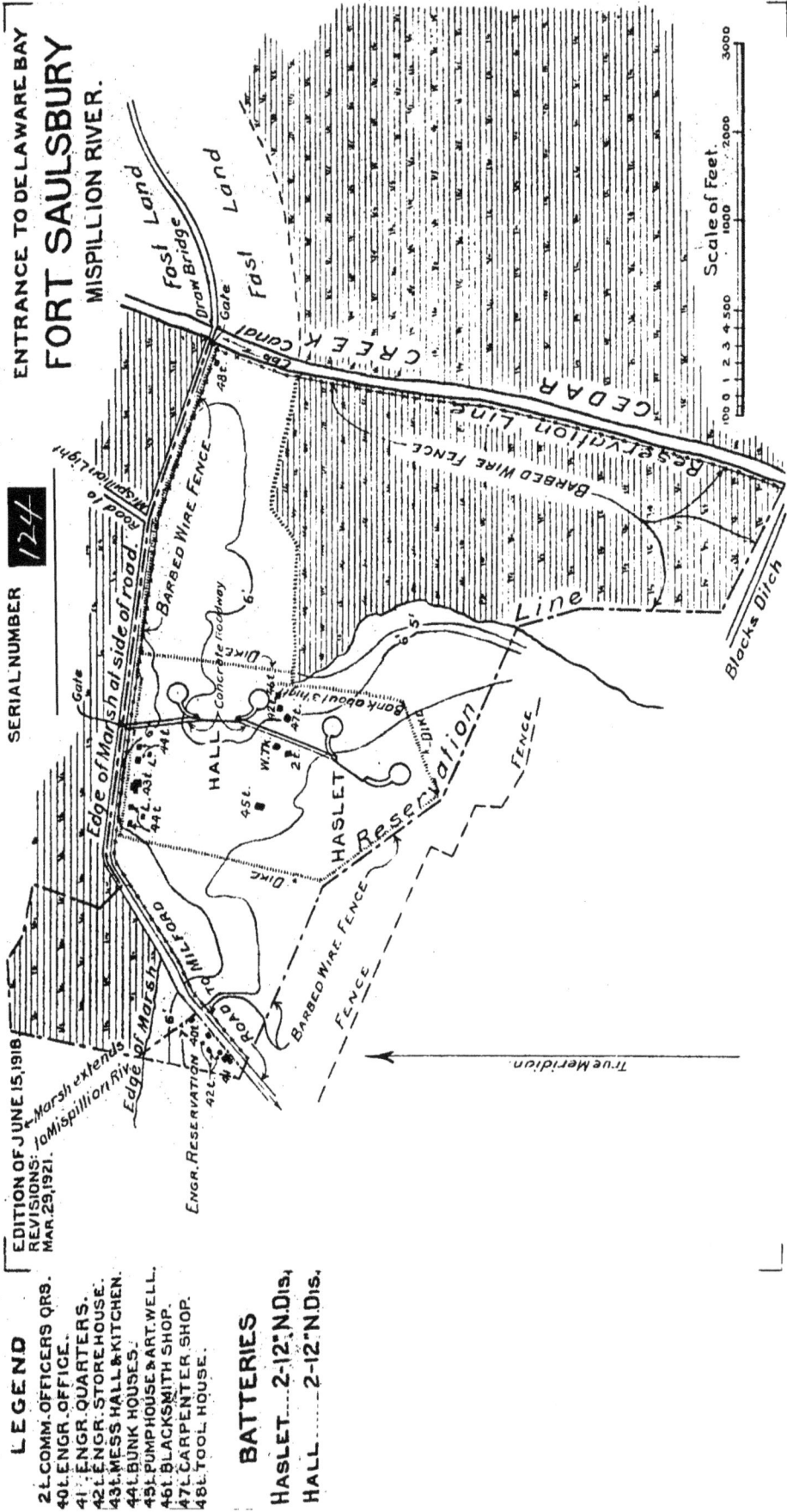

ENTRANCE TO DELAWARE BAY
FORT SAULSBURY
MISPILLION RIVER.

SERIAL NUMBER 124

EDITION OF JUNE 15, 1918.
REVISIONS: MAR. 29, 1921.

Scale of Feet.

LEGEND

2t. COMM. OFFICERS QRS.
40t. ENGR. OFFICE.
41t. ENGR. QUARTERS.
42t. ENGR. STOREHOUSE.
43t. MESS HALL & KITCHEN.
44t. BUNK HOUSES.
45t. PUMPHOUSE & ART. WELL.
46t. BLACKSMITH SHOP.
47t. CARPENTER SHOP.
48t. TOOL HOUSE.

BATTERIES

HASLET ---- 2-12" N.Dis.
HALL ------ 2-12" N.Dis.

CEDAR CREEK

Draw Bridge

Gate

Ebb Canal

Fast Land

Fast Land

Reservation Line

Barbed Wire Fence

Blocks Ditch

Road to Mispillion Light.

Edge of Marsh at side of road.

Gate

Barbed Wire Fence

HALL

HASLET

Reservation

Concrete Footway

Dike

Bank about 3' high

Dike

Dike

Barbed Wire Fence

FENCE

FENCE

ROAD TO MILFORD

Marsh extends

To Mispillion Riv.

Edge of Marsh

Engr. Reservation

True Meridian

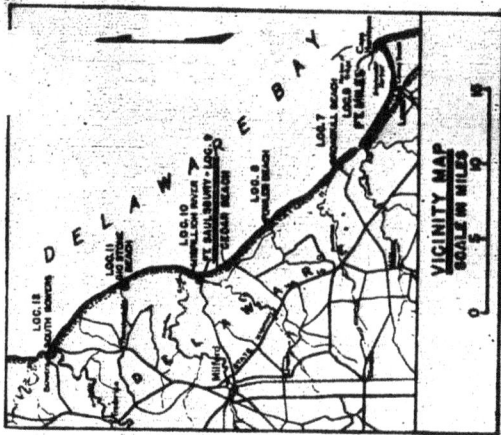

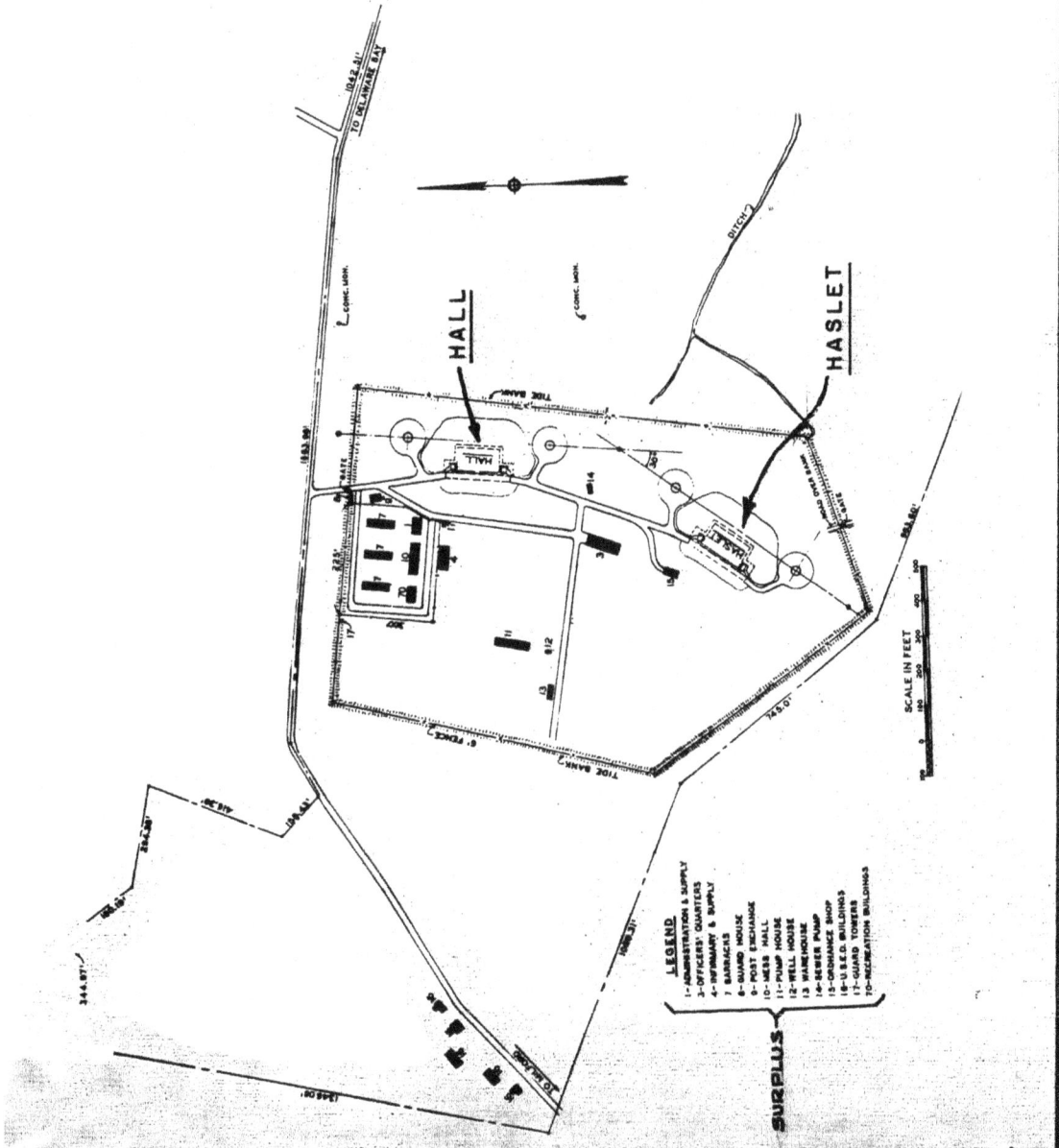

Fort Saulsbury 1940 (NARA)

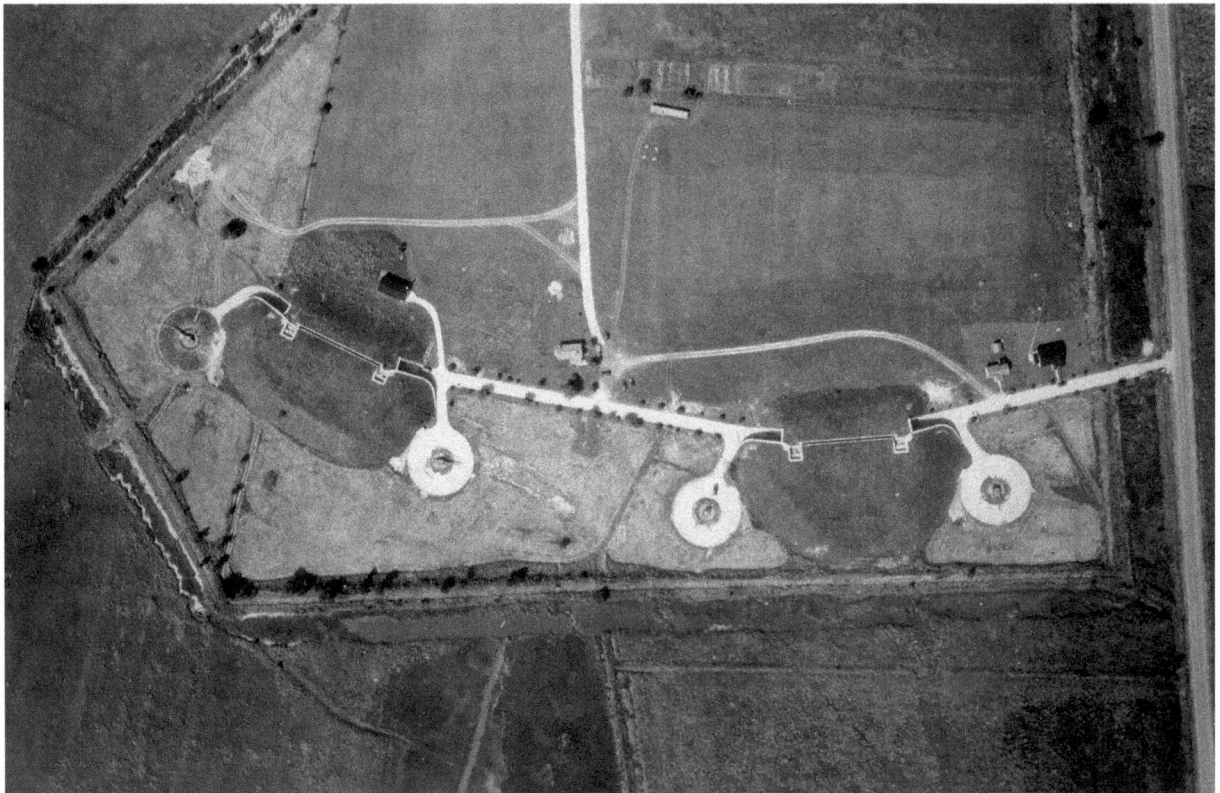

Fort Saulsbury 1940 (NARA)

was of the variety with a closed back, using the additional gallery created for barracks and kitchen for the gun crew (a feature that was used at more geographically isolated locations). Work was done begun in August 1917. Transfer was made on December 27, 1920 for $835,585. It was named on General Orders No. 130 of 1917 for Colonel David Hall, of Revolutionary War service. It was armed with two 12-inch Model 1895M1A4 guns on Model 1917 barbette carriages (#52/#6 and #57/#7). The battery was retained in the postwar reviews, and in the 1940 Program was to have been provided with increased overhead protection. However, the casemating of the battery was never implemented. Replaced in function by the more modern batteries at Fort Miles, the battery retained its original armament and protection until reduced to reserve status on July 15, 1944. The armament was finally removed in late 1945. The emplacement still exists on private property. The battery site is closed to the public.

- **HASLET**: A second battery for two 12-inch long-range barbette guns emplaced as part of the 1915 Board of Review Program at the Fort Saulsbury reservation. It was the southern of the two dual batteries, firing to the east, southeast. It followed closely the type plans for these long-range batteries, except it was of the variety with a closed back, using the additional gallery created for barracks and kitchen for the gun crew that was used at more geographically isolated locations. Work was begun in August 1917. Transfer was made on December 27, 1920 for $839,585. It was named on General Orders No. 130 of 1917 for Colonel John Haslet, Continental Army of Revolutionary War service. It was armed with two 12-inch Model 1895M1A4 guns on Model 1917 barbette carriages (#51/#23 and #72/#24). The battery was retained in the postwar reviews, and in the 1940 Program was to have been provided with increased overhead protection. However, the casemating of the battery was never implemented. Eventually, with the deletion of the second 16-inch battery (#119) at the Delaware Capes, it was planned to move one gun each from Battery Hall and Haslet to a new 12-inch battery at Fort Miles. Eventually both guns at Haslet were selected for this move, dismounted on January 8, 1943. They were emplaced at Battery #519 at Fort Miles shortly after the move. The emplacement still exists on private property. The battery site is closed to the public.

Fort Miles (1917-1946) is located on Cape Henlopen at the entrance of Delaware Bay. Off State Highway 404, two mile east of Lewes, Delaware. Fort Miles military reservation was primarily developed during World War II, though a temporary World War I examination battery was located on the military reservation. It was officially named in 1941 for Lt. Gen. Nelson A. Miles, U.S. Army. The 1940 Program resulted in the construction of a #100 Series 16-inch casemated battery and two #200 Series 6-inch shielded batteries. Additionally, a 12-inch casemated battery using two guns and carriages from Fort Saulsbury was built on the site of a planned 16-inch casemated battery. Supporting these main batteries were submarine mine facilities including a mine casemate, eight 8-inch gun railway guns, several large concrete fire control towers, a four gun 155mm GPF battery, and several AMTB batteries mounting 90mm and 3-inch guns. After World War II, Fort Miles was used as a training center for anti-aircraft units and amphibious landings. By 1950 the U.S. Army's coast defense role had been transferred to the U.S. Navy. The U.S. Army continued to use portions and in 1962 the U.S. Navy established Naval Facility (NAVFAC) Lewes, a Sound Surveillance System (SOSUS) shore terminal. The NAVFAC was in commission 1 May 1962 to 30 September 1981. Fort Miles was declared surplus by the U.S. Army in 1958 and most of the reservation became the Cape Henlopen State Park over the next 30 years. The Fort Miles Museum is housed in the renovated 1940s casemated battery #519. The battery has been partially restored with a 12-inch barrel in one casemate and includes museum displays and event space. The location also features an artillery park with displays of a number of artillery pieces, an accessible fire control tower, along with a number of other coast artillery fortifications. The park is open daily, admission fee required, the battery museum is open for regularly scheduled tours. Today, Fort Miles is an excellent example of the 1940 Program and a World War II coast artillery fort.

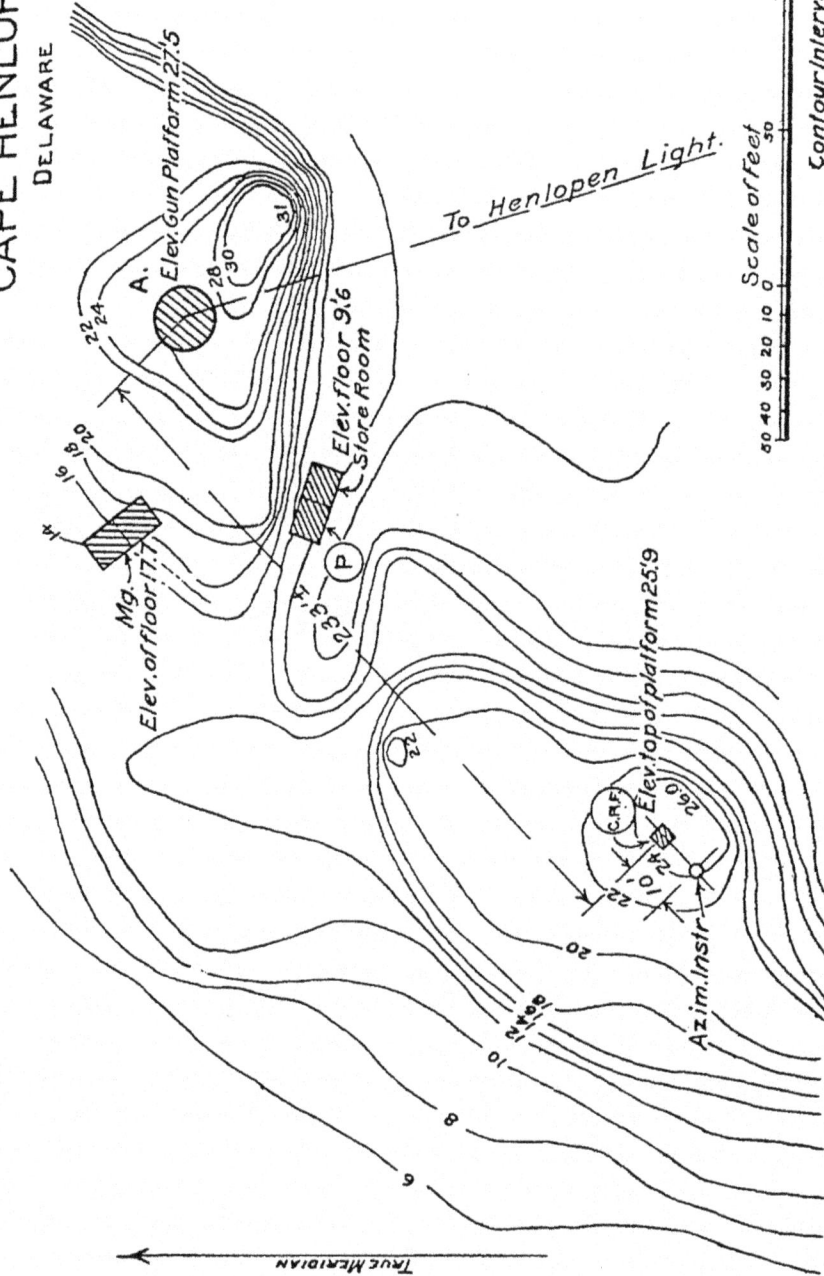

ENTRANCE TO DELAWARE BAY
BATTERIES
A............1-6" N.DIS.

CAPE HENLOPEN
DELAWARE

SERIAL NUMBER

EDITION OF JUNE 15, 1918.

A. Elev. Gun Platform 27.5

To Henlopen Light.

Elev. floor 9.6
Store Room

Mg.
Elev. of floor 17.7

Elev. top of platform 25.9

C.R.F.

Azim. Instr.

TRUE MERIDIAN

Scale of Feet
50 40 30 20 10 0   50   100

Contour Interval 2 Feet.

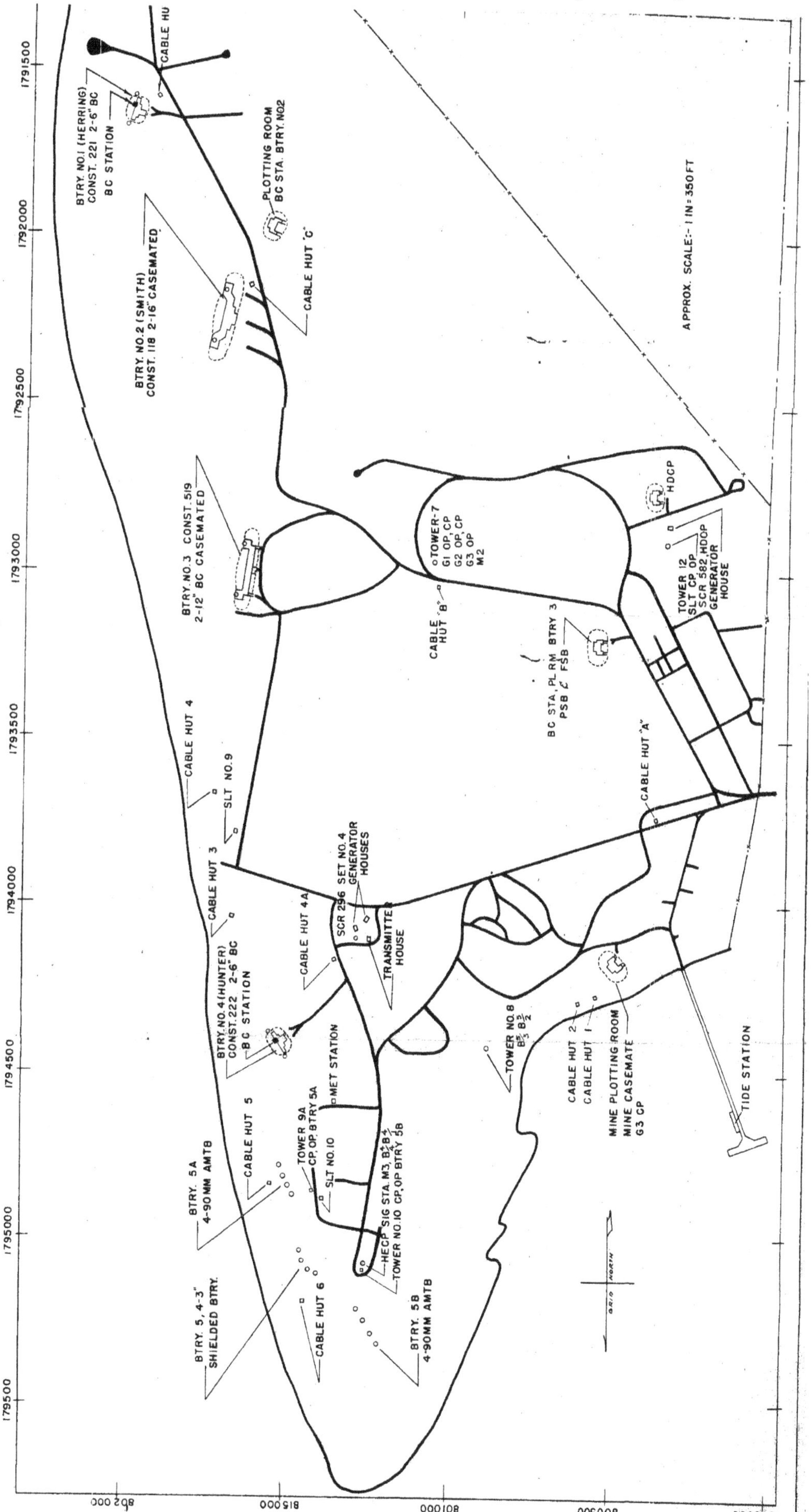

CABLE HU

BTRY. NO.1 (HERRING)
CONST. 221 2-6" BC
BC STATION

PLOTTING ROOM
BC STA BTRY. NO2

CABLE HUT 'G'

BTRY. NO.2 (SMITH)
CONST. 118 2-16" CASEMATED

BTRY. NO.3 CONST. 519
2-12" BC CASEMATED

CABLE HUT 4

SLT NO.9

CABLE HUT 3

TOWER-7
G1 OP, CP
G2 OP, CP
G3 OP
M2

CABLE HUT 'B'

HDCP

TOWER 12
SLT CP OP
SCR 582 HDOP
GENERATOR HOUSE

BC STA, PL RM BTRY 3
PSB & FSB

CABLE HUT "A"

APPROX. SCALE:- 1 IN = 350 FT

CABLE HUT 4A

SCR 296 SET NO.4
GENERATOR HOUSES

TRANSMITTER HOUSE

BTRY. NO.4 (HUNTER)
CONST. 222 2-6" BC
BC STATION

MET STATION

TOWER NO.8
B½ B½

CABLE HUT 1
CABLE HUT 2

MINE PLOTTING ROOM
MINE CASEMATE
G3 CP

TIDE STATION

BTRY. 5A
4-90MM AMTB

CABLE HUT 5

TOWER 9A
CP, OP, BTRY 5A

SLT NO.10

HECP SIG STA M3, B½ B4
TOWER NO.10 CP, OP BTRY 5B

BTRY. 5, 4-3"
SHIELDED BTRY.

CABLE HUT 6

BTRY. 5B
4-90MM AMTB

GRID NORTH

1791500 1792000 1792500 1793000 1793500 1794000 1794500 1795000 1795500

852 000
815000
801000
800500
800000

HARBOR DEFENSES OF THE (DELAWARE
FORT MILES
GENERAL MAP
DELAWARE

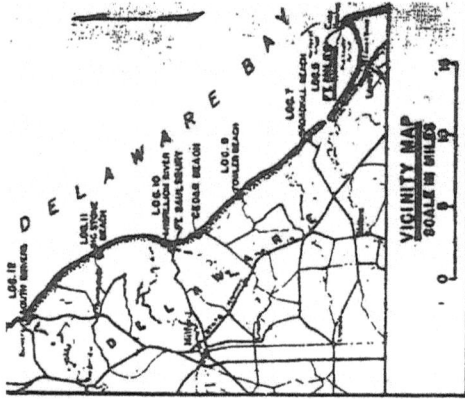

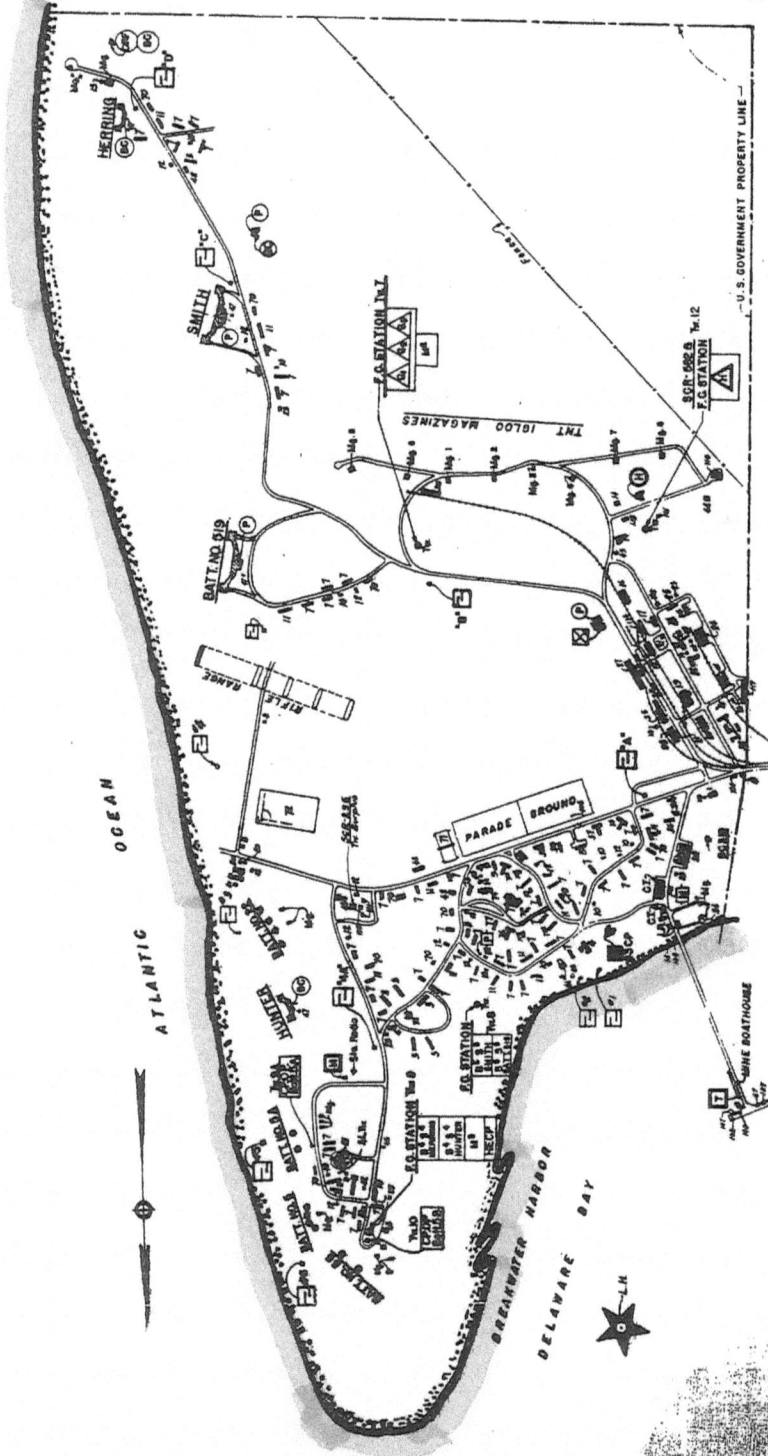

(ACTIVE STATUS)

REVISIONS   JAN. 1946
             JAN. 1944

LEWIS

SERIAL NUMBER

VICINITY MAP
SCALE IN MILES

DELAWARE BAY

SCALE IN FEET

U. S. GOVERNMENT PROPERTY LINE

ATLANTIC OCEAN

DELAWARE BAY

BREAKWATER HARBOR

PARADE GROUND

Battery Hall and Battery Haslett Fort Saulsbury (Glen Williford)

Battery Herring Cape Henlopen State Park (Glen Williford)

Battery Smith Cape Henlopen State Park (Glen Williford)

Battery #519 Cape Henlopen State Park (Glen Williford)

Fort Miles Gun Batteries

- A temporary battery for one 6-inch pedestal gun was emplaced at Cape Henlopen in 1917. The Model 1900 6-inch gun and carriage (#39/#38) were received here on May 23, 1917 from Battery Merrill at Fort St. Philip, New Orleans defenses. The emplacement was most likely just a simple gun block and adjacent magazine. It served only for a short while before being dismounted in 1918 or 1919. The emplacement no longer exists.

- **SMITH**: A new emplacement for heavy 14 or 16-inch guns had been contemplated for Cape Henlopen throughout the 1920s and 1930s. Finally, construction of a pair of modern emplacements for 16-inch barbette guns was approved on April 20, 1934. Appropriations, though, were not forthcoming until the 1940 Program and the FY-1941 and 1942 Budgets. This particular battery was started as the last of the five pre-1940 Program 16-inch batteries. It was incorporated in the 1940 series as Battery Construction No. 118. Like the other early batteries, it featured the recessed main gallery and a full burster cap. The work was signed national priority #6 on September 11, 1940, and increased to #3 on August 11, 1941. Basic concrete work was done from March 24, 1941 to October 31, 1942. It was transferred on December 21, 1943 for a cost of $1,326,000. By December 4, 1942 one gun was mounted, the second followed shortly after. The battery had two 16-inch guns MkIIM1 on Model M4 carriages with 4-inch shields (#44/#23 and #110/#24). A separate plotting (PSR) room and eight supporting base-end station assignments were also constructed. The battery was named on General Orders No. 11 of October 14, 1941 for Major General William Smith. It served throughout the war years, finally being deleted in 1948. The emplacement still exists at the Cape Henlopen State Park. It is closed to the general public, but special tours can be arranged.

- *Battery #119* (planned): A second projected 1940 Program dual 16-inch barbette battery was planned for Fort Miles. It was to be sited north of and generally in line with Battery Smith. It was assigned national priority #14 on September 11, 1940, but subsequently the Delaware River defenses were decreased in priority so that by August 11, 1941 it ranked only #31. On authority of November 14, 1942 the construction was cancelled altogether. Subsequently Battery #519 (for two relocated 12-inch guns) was substituted for #119.

- **Battery #519**: A wartime project for two 12-inch barbette guns designed to replace in function and coverage Battery #119 as the second major gun battery for the Delaware Capes. It was planned to move two excess 12-inch barbette guns from Fort Saulsbury to a new emplacement at Miles. At first one gun from each of the two batteries was to be moved, but eventually Battery Haslet provided both guns and carriages (12-inch Model 1895M1A4 on Model 1917 barbette carriages #51/#23 and #52/#24). Battery construction was done from November 15, 1942 to August 31, 1943 for transfer on February 18, 1944 at a cost of $857,000. The design was similar to a scaled-down 16-inch casemated 100-series battery, but with reduced gun centers, smaller gun houses and magazines, and less elaborate support structures. It was practically identical to the only other such World War II modern 12-inch battery—Battery #520 at Charleston, South Carolina. Though Battery Reed of the Harbor Defenses of San Juan is very similar to Battery #519 as it also mounts two 12-inch BC guns in a casemated battery (Reed has some design differences and was approved before the 1940 Program). The battery was never named and just was referred to as Battery Construction No. 519. That number apparently came from a decision to use numbers in the five hundred range for new 12-inch batteries, and carrying over the "19" from cancelled Battery #119 for the rest. The armament was in place by March 1943 and the battery served until deleted in May of 1948. The emplacement still exists on display at the Cape Henlopen State Park. This battery is now home to the Fort Miles Museum and open to the general public. The battery has been restored and mounts a navy 12-inch gun in its southern gun house.

- **HERRING**: A 1940 Program battery for two 6-inch barbette guns emplaced at the Cape Henlopen reservation of Fort Miles. It was planned as Battery Construction No. 221. The project was assigned national priority #7 initially and funded under the FY-1942 Budget. Herring was emplaced about 3000 yards to the southeast of Battery Smith. Work was done from January 25, 1942 to April 30, 1943, completed by the end of August 1943. Transfer was made on March 4, 1944 for $181,300. The battery consisted of two 6-inch Model 1903A2 guns on Model M1 barbette carriages (#4/#99 and #42/#98). It was of standard 200-series design, except it had the battery commander's station on the top of the traverse covering the magazines, though it was still detached with no connection to the battery interior itself. It was named on General Orders No. 46 of September 17, 1942 for Lieutenant Colonel Ralph E. Herring. The battery was disarmed in 1948, but the emplacement still exists at the Cape Henlopen State Park. The U.S. Navy later used this battery as part of SO-SUS System during the Cold War. The U.S. Navy removed the protective earth and destroyed the battery commander station. A large metal building was added to the battery. The SOSUS station has been removed, and the battery is abandoned and locked-up. The battery site is open, but the interior is closed to the public.

- **HUNTER**: A second 1940 Program battery for two 6-inch barbette guns. It was placed on the northern flank of the heavy guns, to the north of Battery #519. Within the new program it was assigned national priority #15 and built with funds from FY-1942 Budget. While being built it was known as Battery Construction No. 222. Construction work was done from April 15, 1942 to October 29, 1943. Transfer was made on December 13, 1943 for a cost of $180,200. The battery was armed with two 6-inch Model 1903A2 guns on Model M1 barbette carriages (#17/#102 and #32/#104). It was of typical 200-series design, though the battery commander station was on the earth cover over the traverse magazine. It was named on General Orders No. 46 of September 17, 1942 for Colonel Charles H. Hunter, U.S. Army. For some reason the battery is shown as carrying just one gun in May of 1945. In any event it was disarmed and deleted in 1948. The emplacement still exists at the Cape Henlopen State Park. The U.S. Navy used the emplacement during the Cold War for harbor defense and a radio station. Today, its locked-up while the battery commander station is used as a hawk viewing position. The battery site is open, but the interior is closed to the public.

- **Examination Battery**: Four 3-inch pedestal guns were emplaced on new simple gun blocks near the northeastern tip of Cape Henlopen to act as the examination battery in early World War II. It was authorized on April 9, 1942. Work was done on these blocks from May 15 to August 31, 1942 for transfer on December 11, 1943 for a cost of $11,400. Two guns came from Battery Hentig at Fort Delaware (Model 1903 #54/#52 and #87/#95) and two from Battery Turnbull in New York (Model 1903 #82/#42 and #81/#41). They were removed after the war under authority of February 15, 1946. The blocks still exist at the Cape Henlopen State Park, but are buried. The battery site is open to the public.

- **AMTB #5A**: A 1943 Program battery for two fixed and two mobile 90mm dual-purpose guns emplaced to the north of Battery Hunter under authority of November 20, 1942. Work was done on the blocks for the fixed guns starting in June 1943. The emplacement served until removed after the war. The blocks still exist at the Cape Henlopen State Park, but are buried. The battery site is open to the public.

- **AMTB #5B**: A second 1943 Program AMTB battery authorized on November 20, 1943. It was armed with two 90mm fixed and two 90mm mobile guns. It was emplaced at the northern extremity of the reservation on the tip of the cape itself. Work was done from June 15, 1942 to the beginning of 1944. Transfer was made on December 21, 1943 at a cost of $11,000. It was disarmed shortly postwar. The blocks still exist at the Cape Henlopen State Park, but are buried. The battery site is open to the public.

Cape May Military Reservation (1917-1946) is located at the mouth of the Delaware River, near the Cape May lighthouse in New Jersey, Cape May MR supported Fort Miles by provided coast artillery location for a 155mm GPF battery, a 6-inch shielded battery and a AMTB battery. It also provided searchlight and anti-aircraft installations. Now part of Cape May Point State Park. The exposed concrete structure of Battery 223 and several repurposed buildings that serve as the Visitor Center can still be seen. Fire control tower # 23 has been restored and visitors can climb to the top.

Cape May MR gun batteries

- A temporary battery for one 6-inch pedestal gun emplaced at Cape May in 1917. The Model 1900 6-inch gun and carriage were received from Battery Merrill at Fort St. Philip. The emplacement was a simple gun block and adjacent magazine. It served only for a short while before being dismounted in 1918 or 1919. The emplacement no longer exists.

- **Battery #223**: A 1940 Program dual 6-inch barbette battery approved for Cape May as part of the modern defenses of the Delaware River. Work was done under national priority #24 for new 6-inch batteries with the FY-1942 Budget. Construction was done from September 12, 1942 to June 23, 1943 for transfer on April 8, 1944 at a cost of $342,000. It was never named and just known as Battery Construction No. 223 during building and service. It was armed with two 6-inch Model M1(T2) guns on carriages Model M3 (#19/#13 and #18/#12). It served as Tactical Battery No. 6 in the defenses during the war. It was authorized for abandonment and the armament removed in about 1947. The emplacement has subsequently been heavily eroded by the surf with much undermining. As part of the periodic beach sand replenishment the battery has been re-supported by new sand. The battery is abandoned, but open and the concrete gun emplacements are now buried. As such it still exists at the Cape May Point State Park.

- **AMTB #7**: A 1943 Program AMTB battery authorized on November 20, 1942. It had two 90mm fixed and two 90mm mobile guns. Work on the concrete gun blocks was done from January 1 to June 15, 1943. It was transferred on December 21, 1943 at a cost of $15,300. The blocks were located just to the east (about 1000-feet) of Battery #223. The battery guns were removed after the war, and the blocks subsequently destroyed by the surf. Northing remains today.

Battery #223 Cape May (Glen Williford)

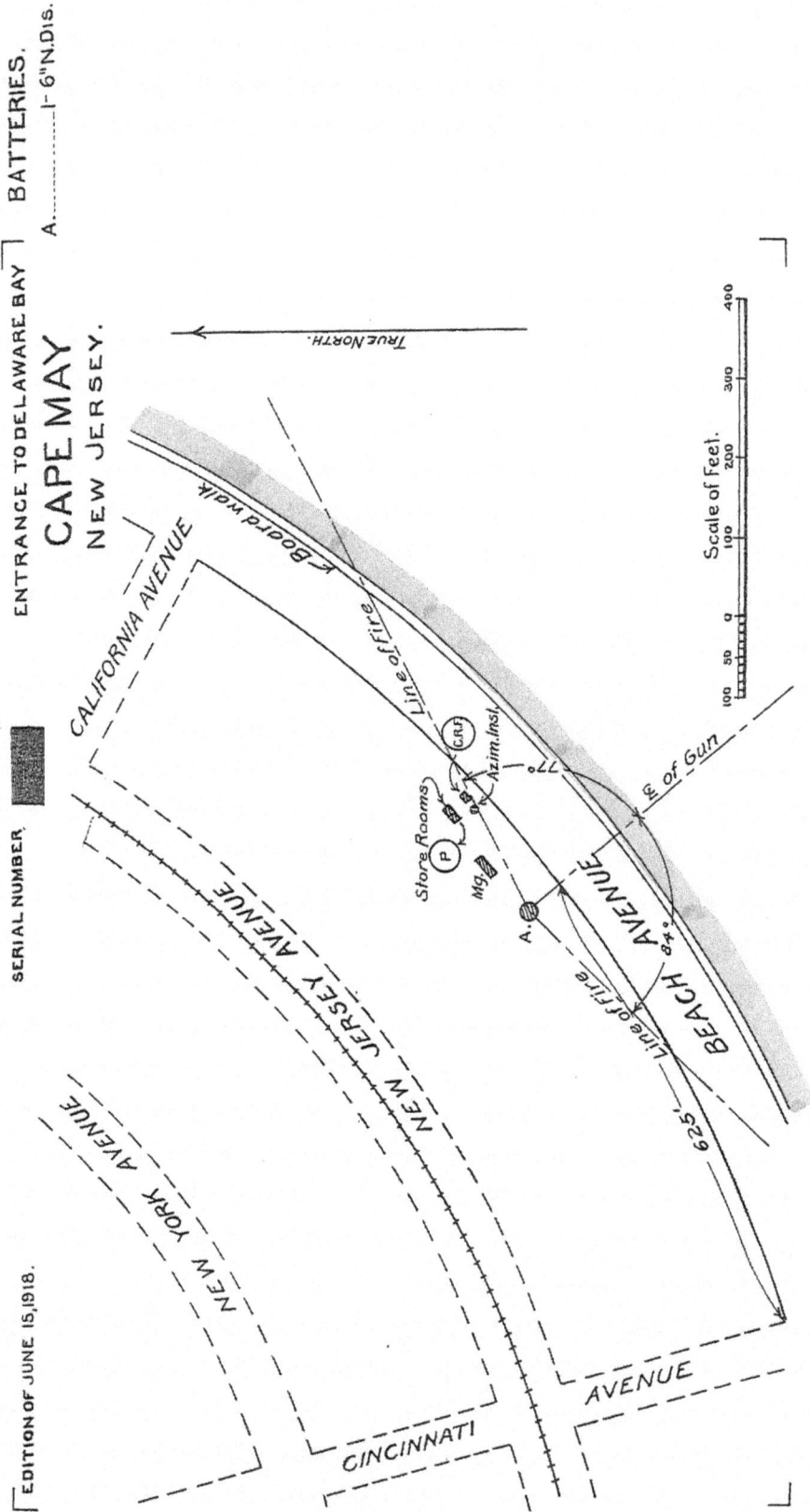

BATTERIES.

A............ 1- 6"N.Dis.

ENTRANCE TO DELAWARE BAY

CAPE MAY
NEW JERSEY.

SERIAL NUMBER.

EDITION OF JUNE 15,1918.

TRUE NORTH.

CALIFORNIA AVENUE

Board walk

Line of Fire

C.R.F.

Azim.Inst.

77°

Store Rooms

P

Mg.

BEACH AVENUE

of Gun

84°

A.

Line of Fire

625'

NEW JERSEY AVENUE

NEW YORK AVENUE

CINCINNATI   AVENUE

Scale of Feet.

100 50 0 100 200 300 400

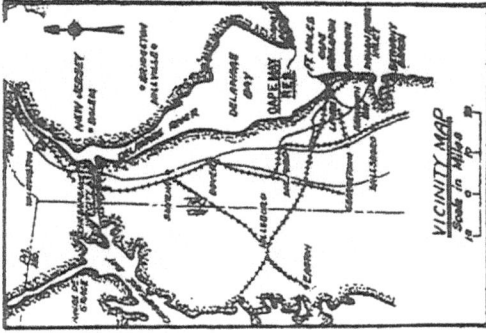

VICINITY MAP
Scale in Miles

BATTERIES

BATTERY NO. 223 --- 2-6" BARBETTE
*BATTERY NO. 25 --- 4-155MM,MOBILE
*BATTERY NO. 7 --- 2-90MM

* - DISMANTLED

(ACTIVE STATUS)

REVISIONS   JAN. 1945
                    JAN. 1946

HARBOR DEFENSES OF
THE DELAWARE

CAPE MAY RESERVATION
GENERAL MAP

CAPE MAY   NEW JERSEY

SERIAL NUMBER

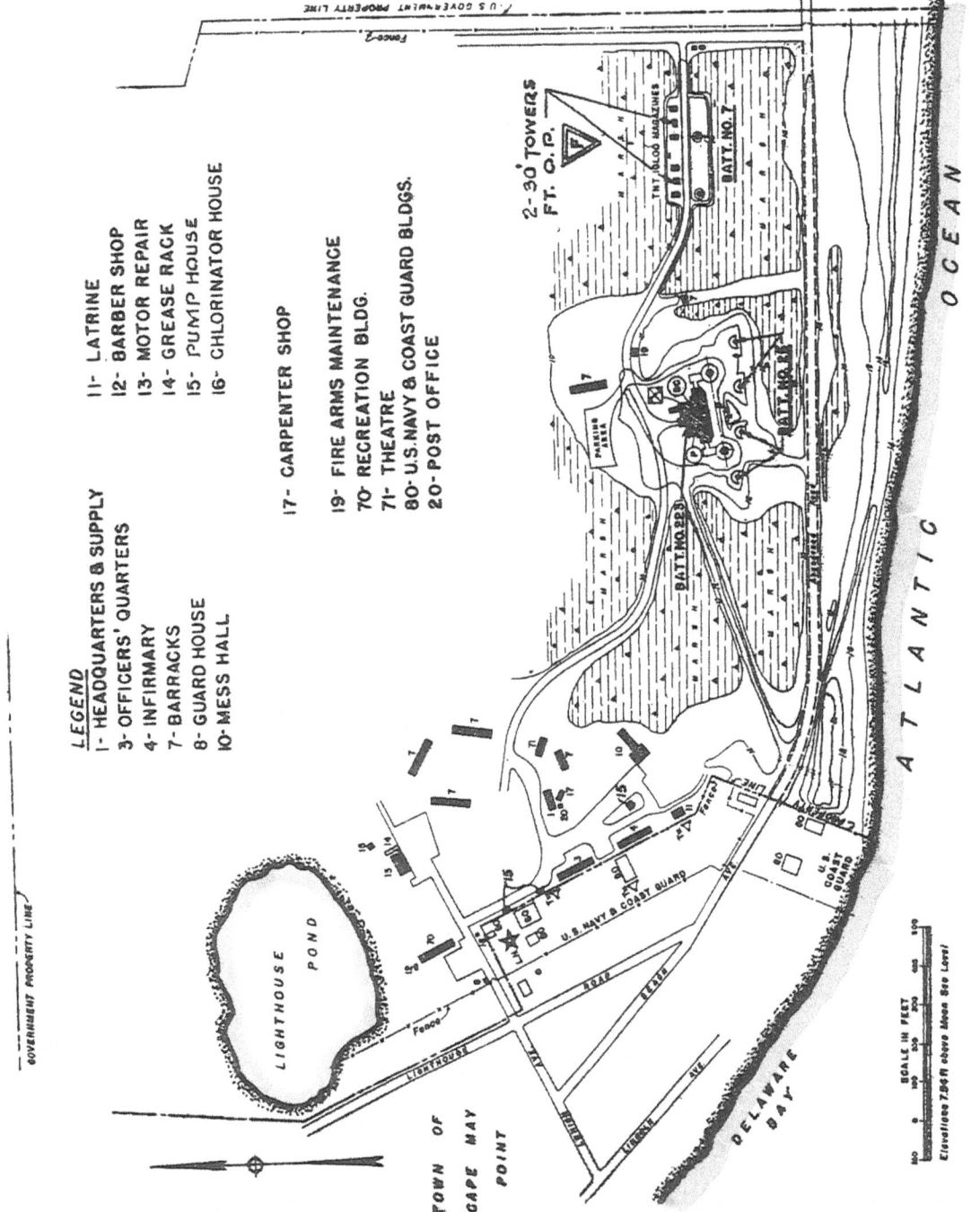

LEGEND

1- HEADQUARTERS & SUPPLY
3- OFFICERS' QUARTERS
4- INFIRMARY
7- BARRACKS
8- GUARD HOUSE
10- MESS HALL

11- LATRINE
12- BARBER SHOP
13- MOTOR REPAIR
14- GREASE RACK
15- PUMP HOUSE
16- CHLORINATOR HOUSE

17- CARPENTER SHOP

19- FIRE ARMS MAINTENANCE
70- RECREATION BLDG.
71- THEATRE
80- U.S. NAVY & COAST GUARD BLDGS.
20- POST OFFICE

GOVERNMENT PROPERTY LINE

TOWN OF
CAPE MAY
POINT

LIGHTHOUSE
POND

SCALE IN FEET

Elevation 7.94 ft above Mean Sea Level

Outside and inside views of a 12 inch navy gun barrel remounted in Battery #519, Cape Henlopen State Park
(Mark Berhow)

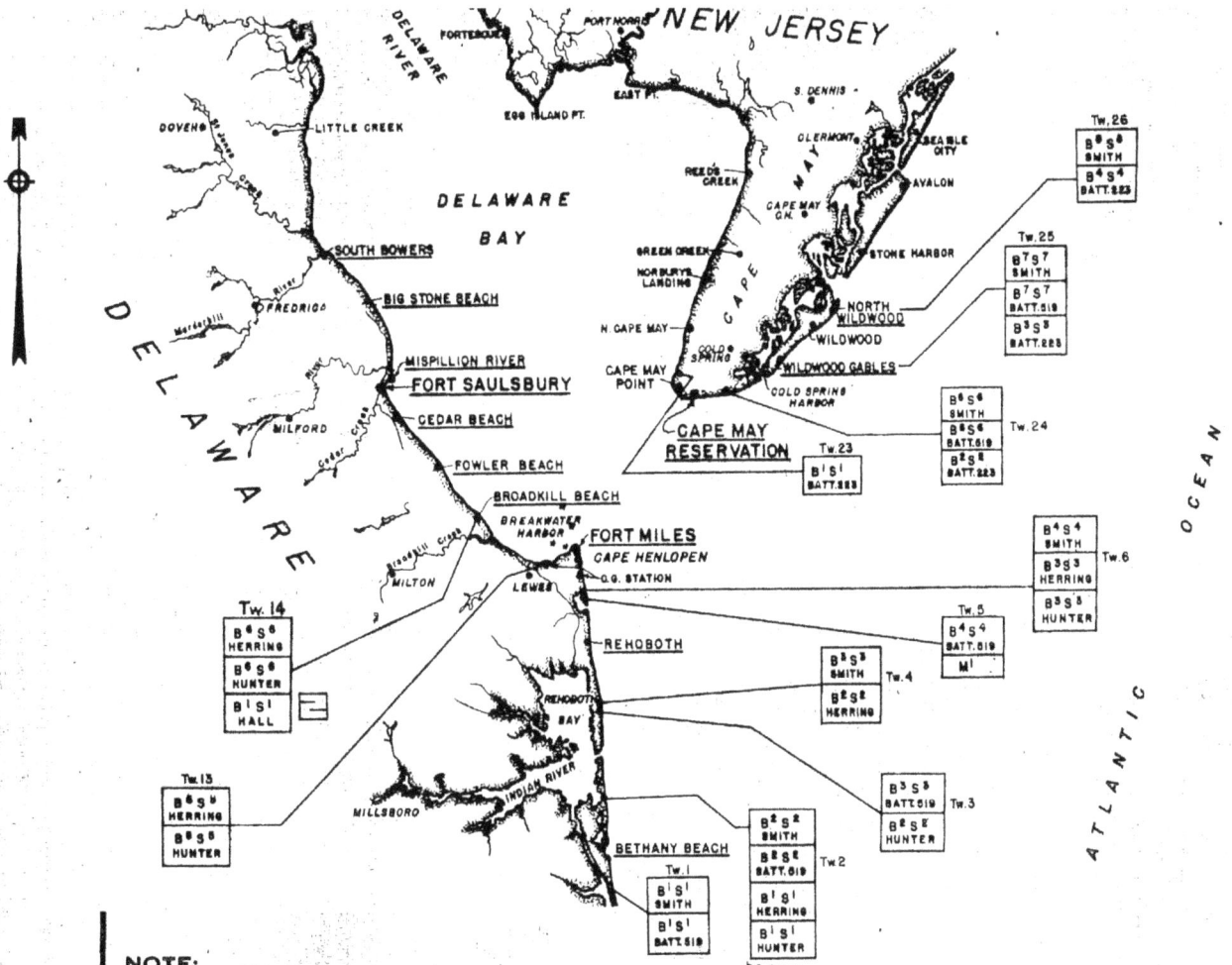

NOTE:
THE HARBOR DEFENSE ELEMENTS
INDICATED HEREWITH ARE   SHOWN
ON SEPERATE INDEX PLATS.

REVISIONS   JAN. 1948
              JAN. 1946

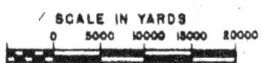

HARBOR DEFENSES OF
THE DELAWARE
**HARBOR DEFENSE
ELEMENTS**
LOCATION MAP
NEW JERSEY AND DELAWARE

SERIAL NUMBER

McGovern, Terrance. *The Chesapeake Bay at War! The Coastal Defenses of Chesapeake Bay during World War Two*. Three Sister's Press. McLean, VA, 2010.

Gaines, William C. "The Maritime Defenses of the Delaware, 1771-1950, *CDSG Journal* Vol. 9, No. 4, Nov. 1995 p. 4; Vol. 10, No. 1, Feb. 1996, p. 35. Vol. 10, No. 2, May 1996, p. 11. Vol. 10, No. 3, Aug. 1996, p. 4.

Delaware Bay World War II-era Site Locations. Stations housed in a single structure are connected by dashes (-)

location	Loc#	Purpose
Bethany Beach	1	BS1/118, BS1/519
Indian River Rehoboth	2	BS2/118, BS2/519, BS1/222, BS1/221, SCR296-1
Dewey Beach	3	BS3/519, BS2/222, BS3/118, BS2/221, SCR296-2
Fort Miles	4	BS4/519, M1, BS4/118, BS3/222, BS3/221
Fort Miles	5	Batt. Tact. #1 Herring, Batt. Tact. #2 Smith, Batt. Tact. #3 BCN 519, Batt. Tact. #4 Hunter, Batt. Tact. #5 3-inch, Batt. Tact. #5A AMTB, Batt. Tact. #5B AMTB, Batt. Tact. #8 mines, CPOP, BS5/519-BS5/118, HECP-HDCP-M3-BS4/222-BS4/221, CPOP-B C5A, MC, G3, HDOP-SCR 582, PSR 519, PSR 118, G1, G2, G3, M2, BC/519, BC/118, BC221 Herring, BC222 Hunter
Fort Miles	6	BS5/222-BS5/221
Broadkill Beach	7	BS6/222-BS6/221
Fowler Beach	8	Ex-Fire Control
Mispillion River	10	Ex-Fire Control
Big Stone Beach	11	Ex-Fire Control
South Bowers	12	Ex-Fire Control
Cape May Point	17	BS1/223
Cape May	18	Batt. Tact. #6 BCN 223, Batt. Tact. #7 AMTB, BC/223
	19	BS6/118-BS6/519-BS2/227, SCR296-4
Wildwood Gables	20	BS7/118-BS7/519-BS3/227, SL 16,17
North Wildwood	21	BS8/118-BS4/222, SL 18,19

Fire control tower at Fort Mott State Park (Mark Berhow)

THE HARBOR DEFENSES OF CHESAPEAKE BAY — MARYLAND AND VIRGINIA

Chesapeake Bay is a large estuary on the mid-east coast of North America that is fed by several major rivers. The initial British settlement was here on the James River in 1607 followed by several other settlements along the James, York, Rappahannock, Potomac, and Patapsco Rivers up the sheltered waters of the bay through the early 1700s. Several ports were defended with colonial fortifications as much from the neighboring natives as from pirates or Dutch, French or Spanish incursions. The wide entrance to the Chesapeake Bay could not be effectively sealed by 18th century seacoast artillery. As a result, the new fortifications to protect the Patapsco River (Baltimore), the Severn River (Annapolis), the Potomac River (Washington DC), and the Hampton Roads (Norfolk) were constructed during the First and Second Systems. The British seaborne invasion up the Potomac River and to Baltimore in 1814 (as well as the attack on New Orleans) demonstrated the need for new effective seacoast fortifications, resulting in the creation of the new Third System defenses built 1816-1860. Major new masonry fortifications were built at Baltimore Harbor, Potomac River, and the Hampton Roads. The defenses at Baltimore, Washington and the Hampton Roads were modernized with new concrete batteries mounting guns with 10 to 15 miles in range during the Endicott Program. In 1922, a battery of four 16-inch howitzers built at Cape Henry, Virginia (Fort Story) effectively covered the entrance to the bay, which made the defenses at Baltimore Harbor and the Potomac River redundant. Additional defenses were added at Cape Henry and Cape Charles during the 1940 Program to protect the Bay and the important U.S. Navy naval bases and shipyards in Norfolk, Newport News, and Washington as well as the important commercial ports up the bay.

THE HARBOR DEFENSES OF BALTIMORE — MARYLAND

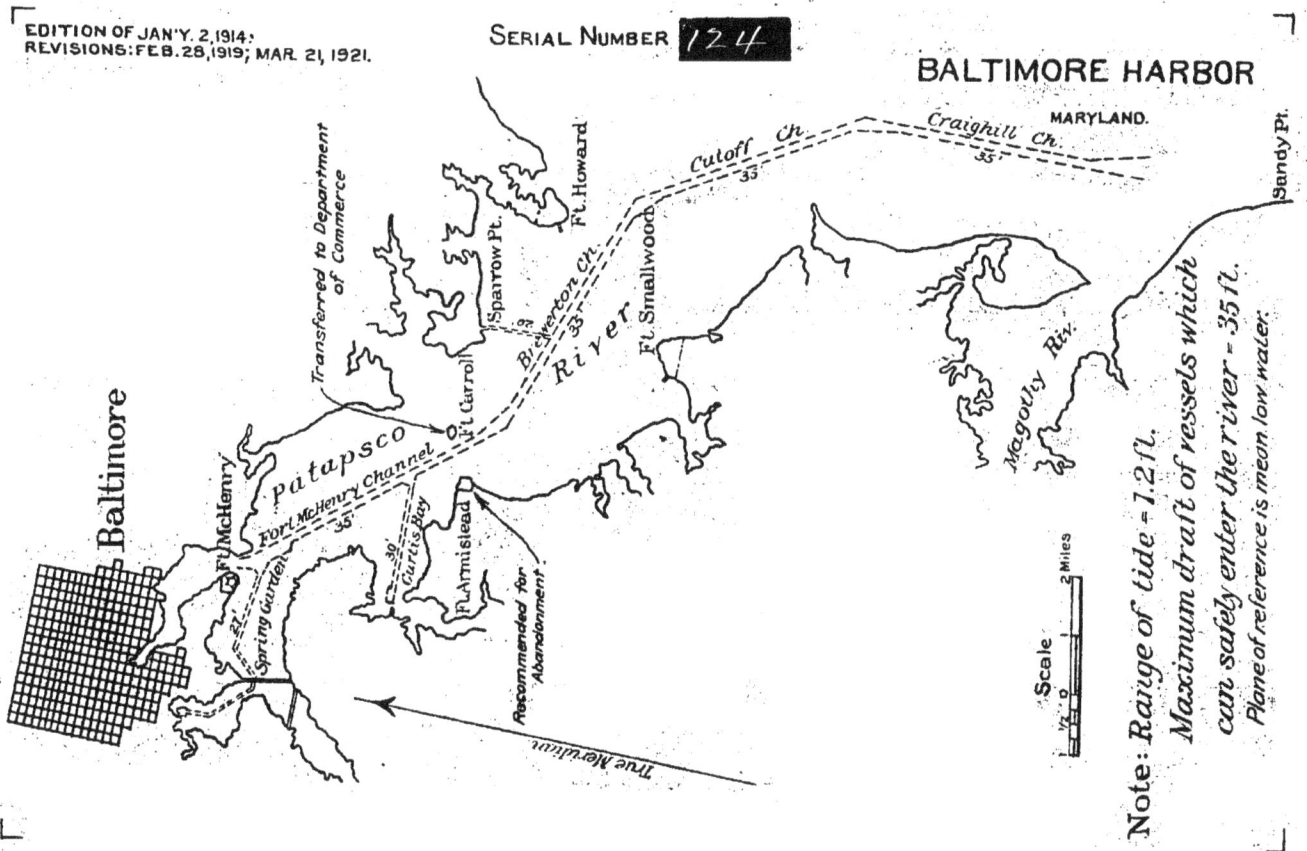

BALTIMORE HARBOR, MD.

FORT McHENRY

WHETSTONE POINT.

Occupied as U.S. Public Health
Service Hospital No.56.

SERIAL NUMBER

EDITION OF: MAR.14,1916.
REVISIONS: FEB. 28,1919; MAR. 21, 1921.

Plane of reference is mean low water.

100 50 0   1   2   3   4   5   6   7   800'

True Meridian

Sea Wall

Star Fort

Original

Channel Range

LEGEND

1 POST HEADQUARTERS.
3 OFFICER'S QRS.
4 HOSPITAL.
4a MORTUARY HOUSE.
6 N.C.OFFICER'S QRS.
7 BARRACKS.
8 GUARD HOUSE.
9 POST EXCHANGE.
10 PUMPHOUSE.
11 OFFICER'S STABLE.
12 STOREHOUSE.
13 FIRE STATION.
14 BLACKSMITH SHOP.
15 STORE-N.C. QRS.
16 ANIMAL HOUSE.
17 SCALE HOUSE.
18 WHARF-GUARD BLDG.
19 GARAGE.
20
21 Q.M.& COMSY. ST. HO.
22
23 MESS BLDG.
24 KITCHEN.
71 OFFICERS' CLUB.
72 RED CROSS.
73 REFRESHMENT STAND.
100 LAVATORY.
101 DERRICK.
102 ENGINE HOUSE.
103 TRANSFORMER PLAT.
104 SEWAGE PUMP HO.
105 BANDSTAND.
106 BAKERY.
107 NURSES QUARTERS.
108 SCHOOL.
109 ORTHOPEDIC SHOP.
110 SHOOTING GALLERY.
111 CHEMICAL LABOR-
     ATORY.
41 ENGINEER BUILDING.
70 CHAPEL.
74 NURSES' CLUB.
75 LIBRARY.
76 RECREATION HALL.
112 INCINERATOR.
113 POST OFFICE.
114 FEDERAL SCHOOL
     BOARD.

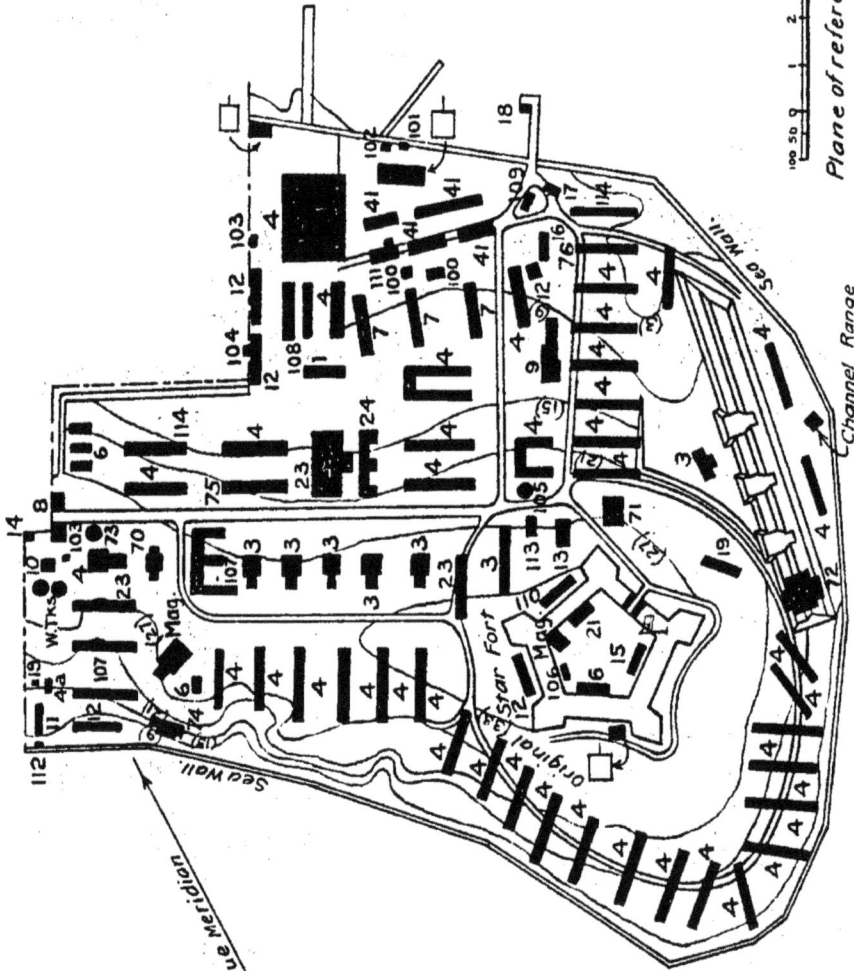

Fort McHenry 1932 (NARA)

Fort McHenry 1927 (NARA)

The grounds of **Fort McHenry,** the famous War of 1812 fort, were used as a military hospital facility during the first half of the 20th century.

Fort Howard (1896-1928) is located on North Point on the Patapsco River, about seventeen miles below Baltimore, Maryland. Constructed during the Endicott Program, Fort Howard's six batteries defended the approaches to Baltimore Harbor from warships that might have made their way up the Chesapeake Bay. It was named in General Orders 43 of 1900 for Col. John Eagar Howard of the Continental Army. In 1920 Fort Howard became the headquarters for the U.S. Army 3rd Corps and the Harbor Defense of Baltimore. As the coast defenses at the mouth of Chesapeake Bay improved in the Interwar Period, it allowed of the Endicott forts guarding Baltimore and the Potomac River to be withdrawn from service, including Fort Howard in 1928. The fort was then used by the U.S. Army as a regional headquarters and for an infantry regiment. The 150-acre reservation was transferred to the Veterans Administration in 1940 for use as a veteran's hospital, as well as prisoner-of-war camp during World War II. During the Cold War, part of Fort Howard was used a training area by Army Intelligence School. In 1975, about 61 acres were transferred to Baltimore County for use as a county park (which included most of the fort's coast artillery batteries). The Veterans Medical Center was closed in 2002 with plans to convert these 90 acres into a residential community, but due to local opposition this area remains abandoned and closed to the public. The park is open daily to the public.

Fort Howard Gun Batteries

- **STRICKER:** A battery for two 12-inch disappearing guns emplaced at the southern tip of the North Point reservation of Fort Howard. It was built using the funding made available by the National Defense Act of March 3, 1898. The plan followed the standard type plans. The two platforms separated by 150-feet had lower-level magazines on the left side. Ammunition service was by hoist. The two emplacements were both flank type. Trunnion elevation was just 36-feet. Work started in early April of 1898, and due to the Spanish American War was construction was pressed by the contractor even working at night. Most concrete work was done by the end of June; all work within another year. Transfer was made on July 5, 1899 for a cost of $113,499.04. It was named on General Orders No. 16 of February 24, 1902 for Brigadier General John Stricker of War of 1812 service. The armament was received and mounted by early 1902. It consisted of two 12-inch Model M1888MII Watervliet gun on Model LF 1896 disappearing carriages (#34/#24 and #36/#9). Modifications were made to the emplacement in 1905-1915, and the disappearing carriages were converted for higher elevation in 1916-1917. The guns were removed in March and April 1918 and sent to rearm Battery Brown at Fort Hamilton (which had lost its 12-inch M1895 tubes for railway mounts). The carriages were soon scrapped in place. The unarmed emplacement was used for several tactical structures—a CRF station for Battery Nicholson, a fire control switchboard room, and plotting rooms for Batteries Nicholson and Key. The emplacement still exists at the Fort Howard Park. The battery site is open, but the interior is closed to the public.

- **KEY:** The mortar battery emplaced at Fort Howard, 600-feet north of the shoreline in the eastern center part of the reservation. It would have been authorized in 1895 except for difficulties encountered in acquiring the additional land required at the site. Plan submission was eventually made on September 19, 1896. The plans were consisted with the design engineer mimeograph of 1896. It had the older, cramped pits, with magazines forward, under the front parapet. The two pits were side by side. A peculiarity of the design was the effort made to round off all the sharp corners of concrete in the construction. Work was begun on December 10, 1896 and reported completed by

SERIAL NUMBER

EDITION OF: MAR. 14, 1916.
REVISIONS: FEB. 28, 1919; MAR. 21, 1921.

BALTIMORE HARBOR, MD.

FORT HOWARD.

NORTH POINT.

Plane of reference is mean low water.

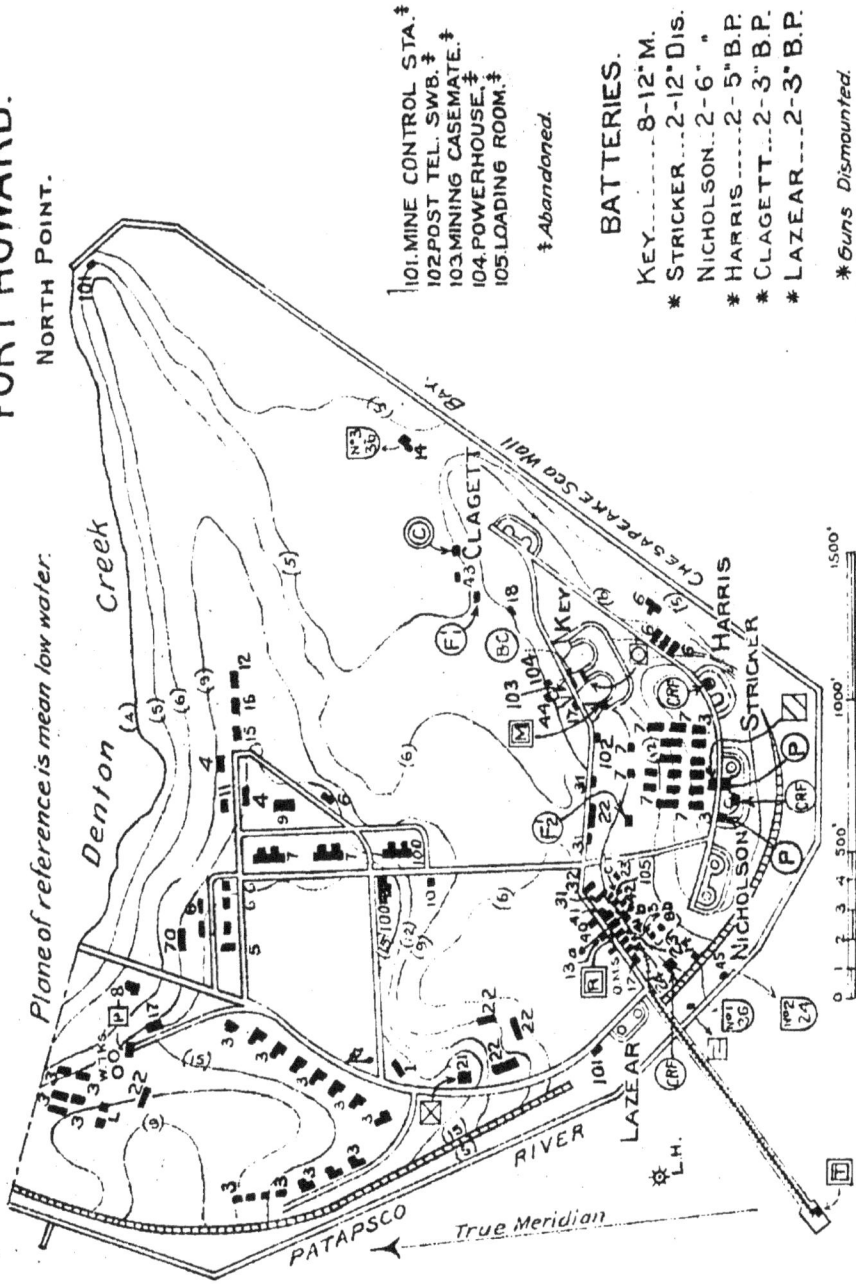

True Meridian

LEGEND.
1. ADMINISTR. BULD.
2.
3. OFFICER'S QRS.
4. HOSPITAL.
5. HOSPITAL STWD'S. QRS.
6. N.C. OFFICER'S QRS.
7. BARRACKS.
8. GUARD HOUSE.
9. POST EXCHANGE.
10. FIRE ENGINE HOUSE.
11. BAKERY.
12. BLACKSMITH SHOP.
13. WORK SHOP.
14. TELEPHONE BOOTHS.
15. STABLE.
16. WAGON SHED.
17. COAL SHED.
18. OIL HOUSE.
19.
21. Q.M. STOREHOUSE.
23. Q.M. & SUB. ST. HO.
23. COMMISSARY ST. HO.
24. Q.M. PAINT SHOP.
31. ORDNANCE ST. HO.
32. MINE ST. HO. & ARTILLERY.
40. ENGINEER OFFICE.
    AND STORE HOUSE.
41. ENGINEER ST. HO.
42. ENGR. BL. SMITH SHOP.
43. ENGINEER QRS.
44. OPERATING ROOM.
45. ENGINEER OIL HOUSE.
46.
70. READING ROOM AND
    CAFETERIA.
80. LIGHTHOUSE KEEPER.
100. HDQRS. THIRD ARMY
     CORPS AREA.

101. MINE CONTROL STA. ‡
102. POST TEL. SWB. ‡
103. MINING CASEMATE. ‡
104. POWERHOUSE. ‡
105. LOADING ROOM. ‡

‡ *Abandoned.*

BATTERIES.
KEY........8-12" M.
* STRICKER...2-12" DIS.
NICHOLSON..2-6 "
* HARRIS......2-5" B.P.
* CLAGETT...2-3" B.P.
* LAZEAR...2-3" B.P.

* *Guns Dismounted.*

Fort Howard 1920s (NARA)

Fort Howard 1920 (NARA)

June 30, 1898. Armament was mounted in 1898 and formal transfer made on January 8, 1900 for a cost of $112,999.43. It was named on General Orders No. 16 of February 14, 1902 for Francis Scott Key of Star-Spangled Banner fame. It was armed with eight 12-inch Model 1890M1 mortars on Model 1896 carriages (Watervliet tubes #15/#46, #20/#20, #25/#41, #29/#37, Bethlehem tubes #28/#34, Builders tubes #12/#44, #15/#39, and #16/#45). This armament continued to serve until the defenses of Baltimore were discontinued, the battery was not reduced to two guns per pit like many other such emplacements. All eight mortars were removed on September 2, 1927 and shipped to Aberdeen Proving Grounds. The carriages were scrapped in place. The emplacement was not subsequently used but still exists at the Fort Howard Park. The battery site is open, but the interior is closed to the public.

- **HARRIS**: A battery for two 5-inch balanced pillar guns emplaced on the right flank, to the south of the mortar emplacement. Plans were provisionally submitted on February 13, 1897, final plans were approved on May 1, 1897. It followed recommendations of type plans, with two platforms and gun centers of 56.5-feet. The low site had a trunnion height of just 22-feet. It fired to the southeast. The magazines were in the flank traverses on the left side, ammunition service by an endless-chain hoist that was hand operated. It was funded by the Act of March 3, 1897. Work was done from August 10 1897 to June 1, 1898. Transfer was made on October 27, 1900 for a cost of $17,400. It was named on General Orders No. 16 of February 14, 1902 for Colonel David Harris of War of 1812 service. On June 30, 1900 the carriages were received and mounted, but the gun tubes were delayed until August 1902. Eventually it was armed with two 5-inch Model M1897 Bethlehem guns on Model 1896 balanced pillar mounts (#10/#23 and #6/#22). These served until removed in September 1917 and shipped out for use on wheeled carriages. The old pillars were scrapped in place in December 1920. The battery's left (No. 2) platform was used as a site to build a new co-incidence rangefinder station for Battery Clagett in 1920. The emplacement still exists at the Fort Howard Park. The battery site is open, but the interior is closed to the public.

- **NICHOLSON**: A battery for two 6-inch disappearing guns emplaced on the same gun line on the right flank of 12-inch Battery Stricker. It faced and fired to the southwest. Plans were submitted on April 21, 1899 for the battery using funds from the Act of March 3, 1899. The plan was fairly standard in design. The two flank-type emplacements were spaced with gun centers of 125-feet. Trunnion height was 32-feet. The magazines were under the central traverse, ammunition service by hand-powered hoists. Concrete work was done from mid-1899 to early 1900. It was completed for transfer on May 8, 1900 for a cost of $37,028.73. It was named on General Orders No. 16 of February 24, 1902 for Judge Joseph Nicholson, Captain of Volunteers in 1814. It was armed with two 6-inch Model M1897M1 Watervliet guns on Model 1898 disappearing carriages (#13/#23 and #26/#22). The guns were mounted in August 1902 and then dismounted on August 4, 1927 at the end of the Baltimore defenses. Tube #28 was later utilized at Battery Dutton at Fort H. G. Wright to replace a damaged gun. The emplacement still exists at the Fort Howard Park. The battery site is open, but the interior is closed to the public.

- **CLAGETT**: A battery for two 15-pounder, 3-inch rapid-fire guns emplaced on the eastern side of the main gun line, pointing and firing to the southeast. Plans were submitted on June 14, 1900 using funds from the Fortification Act of May 25, 1900. Plans followed type designs; with guns at 30-foot centers and a trunnion height of 21.7-feet. The magazines were on a lower level on the right side of each platform. Both platforms were built as flank emplacements. Work was done from late 1900 to early 1901. Transfer was made on January 15, 1901 at a cost of $10,460. The armament was mounted in October 1904 and consisted of two 3-inch Model 1898 Driggs-Seabury guns and

Fort Howard VA Hospital and Park (Terry McGovern)

Fort Carroll (Terry McGovern)

masking parapet mounts (#68/#63 and #7/#50). The battery was named on General Orders No. 16 of February 14, 1902 for Lieutenant Levi Clagett, who was killed at Fort McHenry in 1814. The pillars were modified to the M1898M1 pedestal type about 1916. The armament served until removed on August 11, 1920 and shipped away to Watervliet Arsenal, the carriages were scrapped in place on December 2, 1920. The emplacement still exists at the Fort Howard Park. The battery site is open, but the interior is closed to the public.

- **LAZEAR**: A battery for two 15-pounder, 3-inch rapid-fire guns emplaced near the western shore of North Point, not far from the base of the wharf. It fired to the southwest. Plans were submitted on February 9, 1898 using funds from the Fortification Act of July 7, 1898. Plans generally followed the arrangement of type plans. Two platforms were separated by 30-feet. As the ground here only was at 4-feet elevation above mean low water, trunnion height was just 19.7-feet. Both platforms were shaped as flank emplacements, magazines were on the lower left of each one. Work was done from 1899-1900. By June 30, 1900 it was done excepting for parapet filling and sodding. Transfer was made on May 8, 1900 for a cost of $10,445. It was named on General Orders No. 16 of February 14, 1902 for Doctor Jesse W. Lazear, an early Yellow Fever researcher. The armament was mounted on December 1, 1903 and consisted of two 3-inch, 15-pounder Driggs-Seabury Model 1898 guns and masking parapet carriages (#78/#78 and #67/#67). The pillars were modified to the M1898M1 pedestal type about 1916. These in turn were removed on August 11, 1920 and turned over to the salvage office as scrap on December 2nd of that year. The emplacement itself was removed by the Veteran's Administration in later years.

Fort Carroll (1847-1920) is located on Soller's Point Flats, a shoal converted into an island in the middle of the Patapsco River, about five miles below Baltimore, Maryland. Fort Carroll was to be a three-level, nine-sided, masonry fort of the Third System, mounting 225 cannons. Due to length of time to construct the required foundation system, only the first tier was completed by the time the Civil War begun. It was named in General Orders 38 of 1850 for Charles Carroll a Maryland Statesman and a signer of the Declaration of Independence. The incomplete fort was armed to defend Baltimore Harbor during Civil War, but the advance of military technology cause further construction efforts on the Third System fort to be halted. The Endicott Program saw the construction of three concrete batteries into the existing walls of Fort Carroll. The small island fort was only 3.4-acres in size, so permanent quarters for garrison was at Fort McHenry in Baltimore. Fort Carroll was abandoned by the U.S. Army in 1920 and transferred a portion to the U.S. Lighthouse Service. During World War II, the U.S. Army used the fort as a firing range. In 1958, the island fort was purchased by the Eisenberg Family for use as a casino and restaurant, which never occurred. Today, the fort is still abandoned but is owned by the Eisenberg Family and closed to the public.

Fort Carroll Gun Batteries

- **TOWSON**: A battery for two 12-inch guns on barbette mounts emplaced within the structure of old Fort Carroll in Baltimore Harbor. In 1895 this location was slated to receive a battery of three 12-inch guns on gun lift carriages. Tests of the foundation proved that the site had been carefully prepared with sand fill for the Third System fort begun here and could well take the weight of a large lift battery. However, the cost estimated in August 1895 of $700,000 seemed out of proportion to the benefit of a battery of this size. Work was not authorized to proceed. Subsequently in 1898 Fort Carroll was selected to receive a pair of the limited number of new Model 1892 barbette carriages quickly ordered at the start of the Spanish American War. An emplacement for two guns was designed to fit into front No. 6 of the old fort. To conserve space the standard plan for disappearing

BALTIMORE HARBOR, MD.

FORT CARROLL

SOLLERS POINT FLATS.

SERIAL NUMBER. *124*

This fort has been transferred to the Department of Commerce for use by the Lighthouse Service.
See 680.4(Balto-Carroll)Fl.

Plane of reference is mean low water.

EDITION OF JAN'Y. 2, 1914.
REVISIONS MARCH. 4, 1914; DEC. 7, 1915;
JAN'Y. 24, 1916; MAR, 14, 1916; FEB. 28, 1919;
MAR. 21, 1921.

LEGEND
7a SQUAD ROOM.
10 KITCHEN & MESS ROOM.
11 CISTERNS.
80 LIGHT-KEEPER'S QRS.

BATTERIES.
Towson .... 2-12"Dis
Heart .... 2-5"B.P.
Augustin .... 2-3"B.P.

All guns removed.

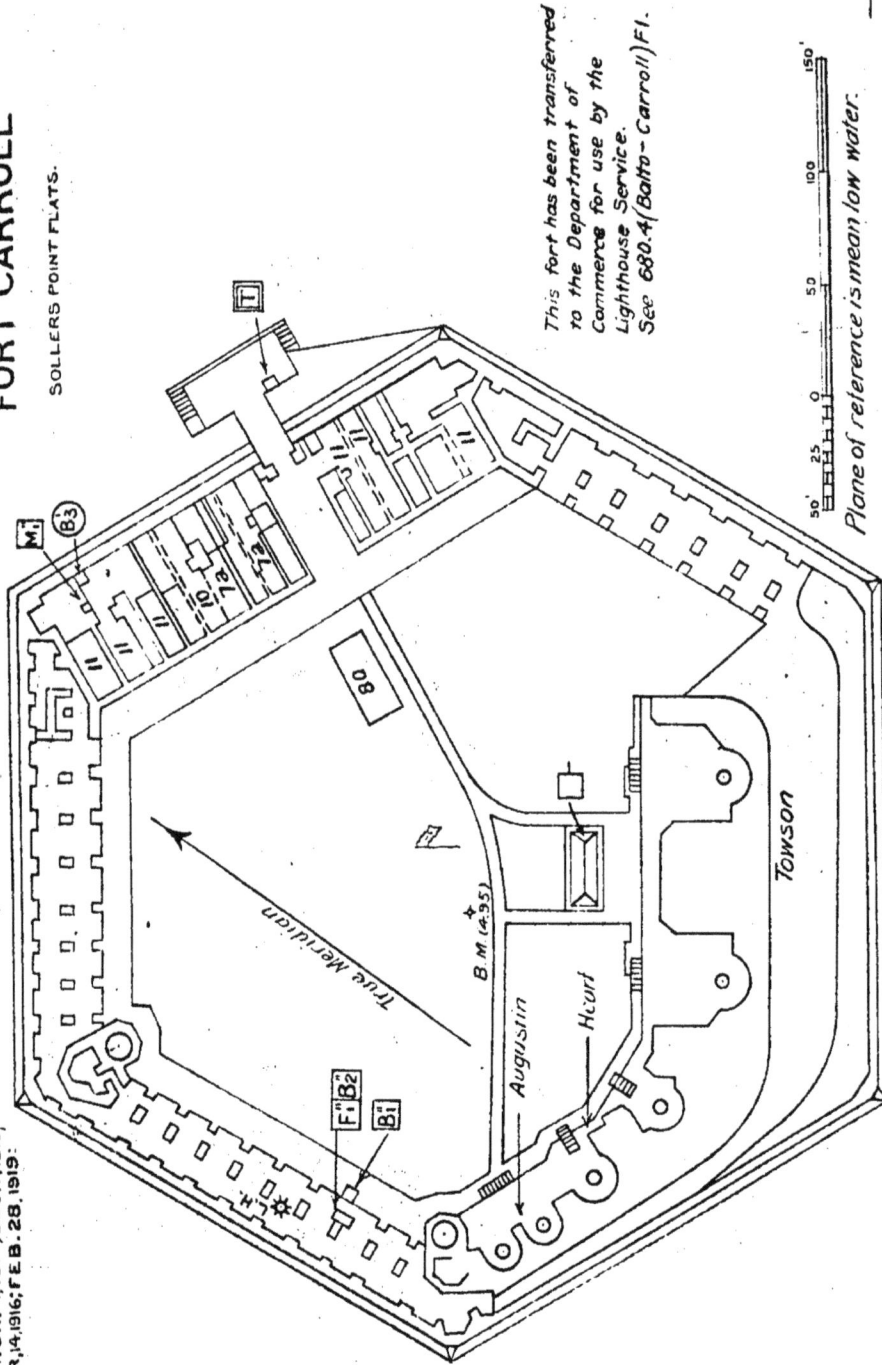

True Meridian

Towson

Augustin

Heart

B.M.(4.95)

Fort Carroll 1920s (NARA)

Fort Armistead 1920s (NARA)

carriages was modified to allow just 125-feet between gun centers and to have a single magazine in the traverse between the guns. Ammunition service was by hoist. Trunnion elevation was just 25.5-feet. Additional overhead protection was given by a steel plate instead of the additional concrete in most other batteries. The battery fired towards the main channel to the southwest. Work was done in 1898 and was reported completed on June 26, 1900. The carriages were mounted by June 30, 1899 and the gun tubes received shortly after. Transfer was made on July 5, 1900 at a cost of just $79,937. It was named on General Orders No. 78 of May 25, 1903 for General Nathan Towson of War of 1812 and Mexican War service. It carried two 12-inch Model M1888MII guns on Model 1892 barbette carriages (#45/#18 and #26/#17). The guns served until dismounted on May 28, 1918 and were removed and shipped to Fort Wadsworth to be used in Battery Ayres. The barbette carriages were scrapped in 1919 and 1920. The fort was turned over to the city in later years and eventually became private property. It still exists but is difficult to access due to the condition of the island fort and private ownership. The battery is closed to the public.

- **HEART**: A battery for two 5-inch balanced pillar guns emplaced adjacent to and immediately to the west (right flank) of Battery Towson. Plans were submitted on August 2, 1898. The emplacement had a field of fire to the south towards the main ship channel. The emplacement was somewhat withdrawn from the parapet so as to be out of the field of fire of the 12-inch battery. It was a modified type plan. The support in the rear of the loading platform was with a retaining wall of granite. Building the battery necessitated the removal of masonry of the upper tier of the old fort and shifting some Rodman guns that were mounted there. The gun centers were just 47-feet, and the trunnion height was the same 25.5-feet of the adjacent Battery Towson. Work was done in 1898-1899, the battery being transferred on August 1, 1900 for a cost of $12,300. It was named on General Orders No. 78 of May 25, 1903 for Major Jonathan Heart killed in the Indian Wars in Ohio in 1791. The battery was armed with two 5-inch Model 1897 Watervliet guns on Model 1896 balanced pillar mounts (#10/#26 and #9/#21). The battery served until being decomissioned during World War I. The guns were removed on September 27, 1917 and the carriages were subsequently scrapped. The emplacement still exists on the island fort which is private property. The battery is closed to the public.

- **AUGUSTIN**: A battery for two 3-inch, 15-pounder masking parapet guns emplaced just on the right flank of Battery Heart inside old Fort Carroll. The plans were submitted in September 1898 with funds from the Fortification Act of July 7, 1898. Due to the constructed space available within the walls of old Fort Carroll it had a modified, compact design. Gun centers were reduced to just 21-feet. A single magazine under the front cover below the loading platforms was adopted, with separate stairs to each platform. The two emplacements were rounded-front flank emplacements. The battery fired towards the mine field to the south in the main shipping channel. Work was done from 1898-1899. By June 30, 1899 it was done except for railings and doors. The guns were mounted in January 1901 having been received on August 24, 1900. The battery was transferred on August 1, 1900 for a total cost of $7,600. It was named on General Orders No. 78 of May 25, 1903 for 2nd Lieutenant Joseph N. Augustin who was killed in 1898 at the Battle of San Juan Hill in Cuba. It was armed with two 3-inch, 15-pounder Model 1898 Driggs-Seabury guns and masking parapet mounts (#17/#17 and #16/#16). These served until dismounted on August 11, 1920. The emplacement was not subsequently used for armament but still exists on the island fort which is private property. The battery is closed to the public.

Fort Armistead (1898-1923) is located on Hawkins Point on the Patapsco River, about five miles below Baltimore, Maryland. Constructed during the Endicott Program, Fort Armistead's four concrete batteries and controlled mine casemate covered the south side of the shipping channel into Baltimore Harbor. It was named in General Orders 134 of 1899 for Maj. George Armistead, commander of Fort McHenry during the bombardment of 1814. All of Fort Armistead's armament was removed by 1920. After the property was declared surplus, it was transferred to the City of Baltimore. During World War II, the U.S. Navy used the property as am ammunition depot and then the U.S. Army used it for an anti-aircraft battery from 1952-1954. Today, Fort Armistead is a day use city park with a boat launch and shoreline fishing pier. The batteries are overgrown and neglected, filled with trash and graffiti and is subject to a high rate of crime. Recently, it was impacted by the collapse of the Franics Scott Key Bridge that carried the interstate highway over the Patapsco River.

Fort Armistead Gun Batteries

- **WINCHESTER:** An emplacement for a single 12-inch disappearing gun emplaced as the No. 4, left-most unit of the four-gun main battery at the Hawkins Point reservation of Fort Armistead. The original three-gun Battery McFarland had been submitted on May 25, 1896. After some discussion, a contiguous fourth emplacement for a heavier 12-inch gun was submitted on October 28, 1896. It was a single platform, emplaced at the bend of the linear battery, firing slightly to the northeast of east. The magazine was on the lower right flank with ammunition service by hoist. Interior crest was at the 41.5-foot height. Work was done by contract from January 1897 to April 1898. It was complete by April 20, 1898 and transferred to artillery service on January 15, 1900. Engineering cost was included in the $140,383.10 for both batteries Winchester and McFarland. It was named on General Orders No. 78 of May 25, 1903 for Brigadier General James Winchester of War of 1812 service. Armament was one 12-inch Model 1888MII Watervliet gun on disappearing carriage LF Model 1896 (#21/#3). Changes were made in 1903 to the hoists; the carriage was modified for increased elevation in about 1915-1917. The battery served until being disarmed on June 26, 1918 under orders of April 23rd. The gun tube itself was shipped to Fort Wadsworth for use in Battery Hudson on July 13, 1918, and the carriage scrapped in place in 1921 (though major parts remained at the fort for years). The emplacement still exists at the Fort Armistead Park but is covered with graffiti and trash. The battery is open to the public.

- **McFARLAND:** A battery for three 8-inch guns on disappearing carriages emplaced as the first armament at this reservation at Hawkins Point. Plans were submitted on May 25, 1896. Originally envisioned for four emplacements, authorization was initially provided to build the first two. However, the contractor bid was low enough to allow the building three, and with the change of emplacement No. 4 to a 12-inch battery (Battery Winchester), authority was given to proceed with the entire battery. The constricted space available at Hawkins Point forced the local engineers to design tight, compact emplacements varying somewhat from type plans. These three emplacements were very close together, gun centers separately by just 84-feet. Magazines were placed in front, under the parapet rather than entirely in flank traverses. With the earth protection in front, just 13.5-feet of vertical protection was needed over the magazines. Original ammunition handling was by cranes only, though hoists were added later and transferred in 1908. Loading platforms were also enlarged about 1910. Work was done from 1896-1897, transfer being made with Battery Winchester on January 15, 1900 at a cost (all four units) of $140,383.10. It was named on General Orders No. 78 of May 25, 1903 for Major Daniel McFarland who was killed in action at Lundy's Lane in 1814. Armament consisted of three 8-inch Model 1888M1 Watervliet guns on Model LF M1894 disap-

BALTIMORE HARBOR, MD.

FORT ARMISTEAD

HAWKINS POINT.

SERIAL NUMBER 124

EDITION OF: MAR. 14, 1916.
REVISIONS: FEB. 28, 1919; MAR. 21, 1921.

This fort is recommended for abandonment.
See C. of E. 662 L (Balto) F4

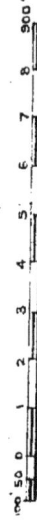

LEGEND.

7. BARRACKS & MESS HALL.
10. STABLE.
31 ORDNANCE ST. HO.
41 ENGINEER'S QRS.
42 COAL CHUTE.

BATTERIES.

WINCHESTER....1-12"Dis
McFARLAND....3-8" "
IRONS........2-4.7"P.
MUDGE........2-3"B.P.

All guns removed.

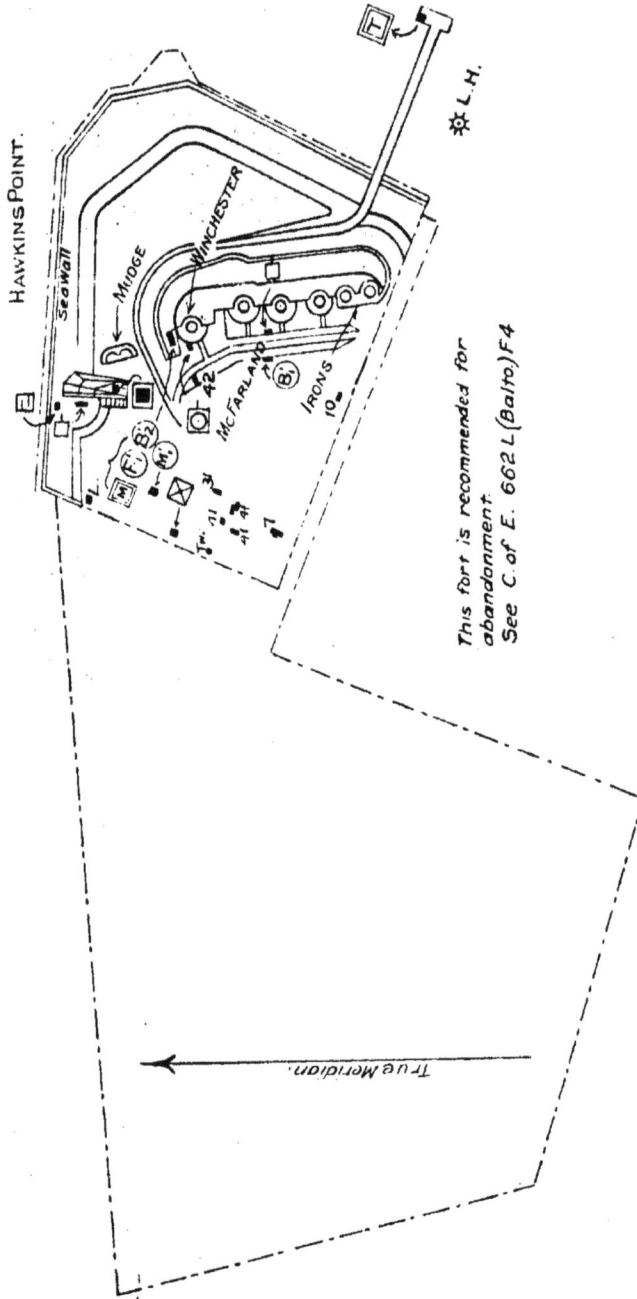

Plane of reference is mean low water.

True Meridian.

L.H.

pearing carriages (#18/#18, #16/#16, and #22/#14). Armament was mounted in March and April 1898 with the urgency of the start of the Spanish American War. The armament served until being dismounted under orders of August 25, 1917, the tubes being shipped to Watervliet on October 19, 1917. The disappearing carriages were scrapped in 1920. The emplacement still exists at the Fort Armistead Park, but covered with graffiti and trash. The battery is open to the public.

- **IRONS:** A battery for two recently acquired 4.7-inch rapid-fire guns emplaced on the southern flank contiguous to Battery McFarland. It pointed and fired to the east. Battery plans were submitted on March 30, 1898 after local engineers were notified that two of the guns recently purchased from Armstrong of England would be made available. Contract builder Jones, Pollard & Co., who was finishing work on the main gun line, was available and had the plant to build this unit immediately. It differed somewhat from standard types due to the constricted space available. The two platforms were placed just 36-feet apart (type recommendation was for 45-feet), and a single shared magazine was placed in the parapet in front of emplacement No. 2. Hand hoists were incorporated to facilitate ammunition service. By April 3, 1898 work was well underway, and it was reported complete as soon as May 10, 1898. The guns were received and quickly mounted in the summer of 1898. Transfer was made on January 5, 1900 for a cost of just $10,616.08. It was named on General Orders No. 78 on May 25, 1903 for 1st Lieutenant Joseph F. Irons, an artillerist who was mortally wounded at Churubusco Mexico in 1847. The battery carried two 4.7-inch guns and pedestals of Armstrong manufacture (#11001/#11015 and #11005/#11016). These were dismounted relatively soon; being shipped away on July 24, 1913 to Hawaii for use there. The emplacement was not subsequently used. The emplacement still exists at the Fort Armistead Park but is covered with graffiti and trash. The battery is open to the public.

Fort Armistead City Park (Terry McGovern)

- **MUDGE**: A battery for two 3-inch, 15-pdr masking parapet guns emplaced on the west side of the reservation, near the new mine casemate and mine wharf. This was the site of the old 1870s Rodman battery and building required destruction of most of the older work. Plans were submitted on December 8, 1898. It was a half-sunken plan and had just 21-feet between gun centers on the constricted site. The work had just a single, shared magazine under the platform between both guns, with a separate doorway entry for each platform. Work was done in early 1899, finishing by June 30, 1899. Transfer was made on May 8, 1900 for $6860. It was named on General Orders No. 78 of May 25, 1903 for 2nd Lieutenant Robert R. Mudge, 3rd U.S. Artillery, who was killed in action with the Seminole Indians in 1835. The armament was mounted in late 1901 and consisted of two 3-inch Model 1898 Driggs-Seabury guns and masking parapet mounts (#4/#4 and #3/#3). These were modified to the M1898M1 pedestal standard in about 1916. The guns were dismounted at the same time as most other such guns, on August 11, 1920. The carriages were soon scrapped in place. The emplacement still remains, though partially destroyed and mostly buried at the Fort Armistead Park. The battery site is open to the public.

Fort Smallwood (1896-1926) is located on Rock Point on the Patapsco River, about twenty miles below Baltimore, Maryland. Constructed during the Endicott Program, Fort Smallwood's two small concrete batteries covered the south side of the shipping channel up the Patapsco River into Baltimore Harbor, while Fort Howard covered the north side of the channel. It was named in General Orders 43 of 1900 for Maj. Gen. William Smallwood, Continental Army, and a Governor of Maryland. The 100-acre reservation was sold to the City of Baltimore as a city park in 1926 and was city park until 2006, when Anne Arundel County took over management of the park. Today the park is a large day use picnicking and boating area. Only one building, the post hospital, and one 6-inch battery remain at the park today. The hospital serves as staff offices and a small visitors center.

Fort Smallwood Gun Batteries

- **HARTSHORNE**: A battery for two 6-inch disappearing guns emplaced at the Rock Point reservation of Fort Smallwood. Plans were submitted on January 24, 1899. The site was particularly low, the entire battery had to be built above ground. It fired to the north, 40-degrees east to cover the main channel. It was of fairly conventional design, following the mimeograph type plans. Both emplacements were flank types, but an enlarged magazine serving both guns was centrally located underneath the center traverse between the guns. Building was done with funds in the May 7, 1898 Fortification Act. Work was done in 1899. Transfer was made on May 8, 1900 at a cost of $48,173.14. It was named on General Orders No. 120 of November 22, 1902 for Captain Benjamin M. Hartshorne Jr., 7th U.S. Infantry who was killed during the Philippine Insurrection in 1902. It carried two 6-in Model 1897M1 guns on Model 1898 disappearing carriages (#20/#2 and #21/#3). These were mounted in July of 1902. They served until dismounted on June 18, 1927; being shipped to Aberdeen on August 4, 1928. The emplacement was not utilized again but still exists at the Fort Smallwood Anne Arundel County Park. The battery is open, but the interior is closed to the public.

- **SYKES**: A battery for two 3-inch rapid-fire pedestal guns emplaced at Fort Smallwood on the right flank of the 6-inch battery recently built there. Plans were submitted on August 14, 1902 using funds allocated from the Act of June 6, 1902. It was of mimeograph type plan for 1902, with gun centers of 42-feet and having its two magazines in the central traverse. Work was done in 1902-1903, for transfer on December 23, 1905 at a cost of $11,856.90. It was named on General Orders No. 194

FORT SMALLWOOD
BALTIMORE HARBOR, MD.

ROCK POINT.

True Meridian

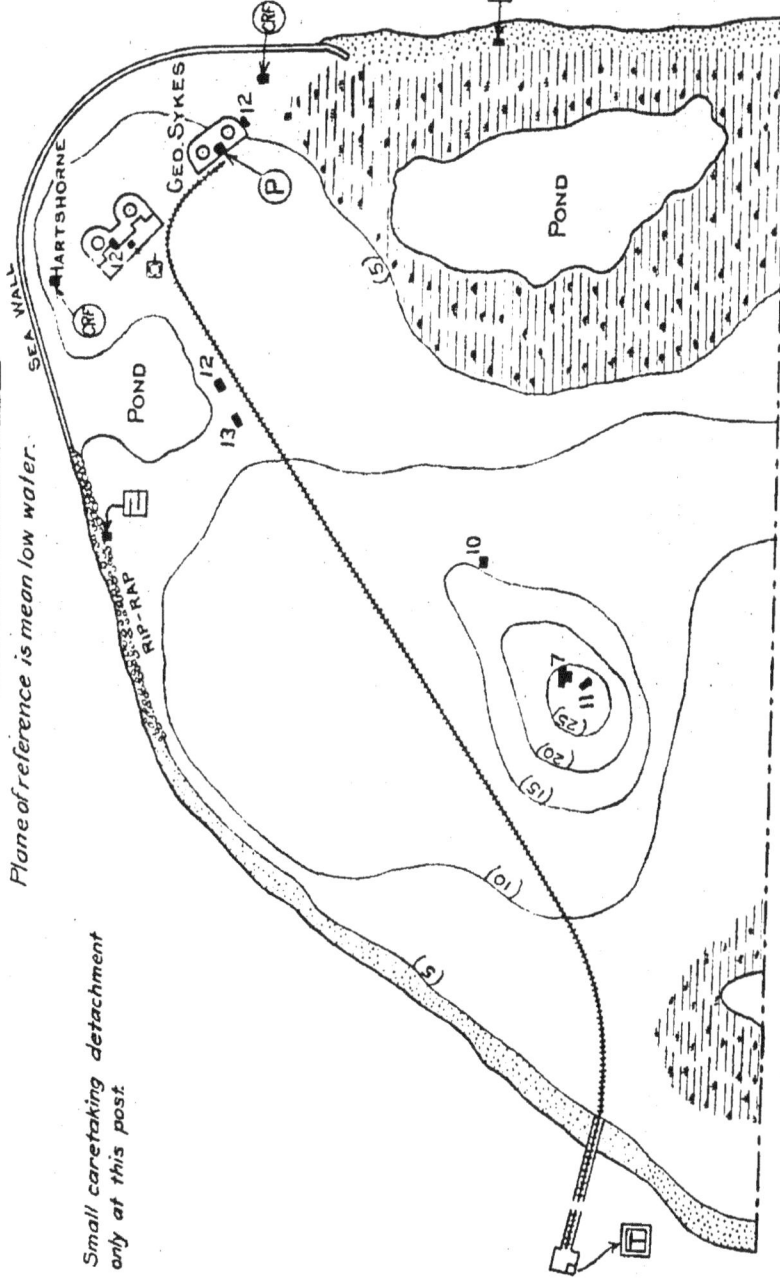

B"(key)

Plane of reference is mean low water.

Small caretaking detachment only at this post.

SERIAL NUMBER **124**

EDITION OF: MAR.14.1916.
REVISIONS: FEB.28,1919; MAR.21,1921.

LEGEND.
7. BARRACKS.
10. WINDMILL & WELL.
11. COAL SHED.
12. FIRE CONTROL STA.‡
13. POWERHOUSE.‡

‡ *Abandoned.*

BATTERIES
HARTSHORNE 2-6"DIS.
SYKES .... 2-3"P.

SEA WALL

RIP-RAP

POND

GEO. SYKES

HARTSHORNE

CRF

POND

1000'

500'

0

500'

500'

Fort Smallwood 1920 (NARA)

Fort Smallwood 1920 (NARA)

of December 27, 1904 for Major General George Sykes, US Volunteers, who served in the Mexican and Civil Wars. It was armed with two 3-inch Model 1902 Bethlehem guns and M1902 pedestal mounts (#12/#12 and #13/#13). These were dismounted on June 18, 1927 and subsequently held by the Ordnance Department as spares. The emplacement was not utilized again, and at some time the concrete emplacement was destroyed and removed. No traces remain of it today.

16-inch Navy MkIII gun on a Army proof mount at Aberdeen Proving Ground. While not actually emplaced for coast artillery use, this is the last remaining example of the Army 16-inch mounts (Mark Berhow)

Battery Nicholson, Fort Howard County Park (Mark Berhow)

THE HARBOR DEFENSES OF THE POTOMAC RIVER –
MARYLAND AND VIRGINIA

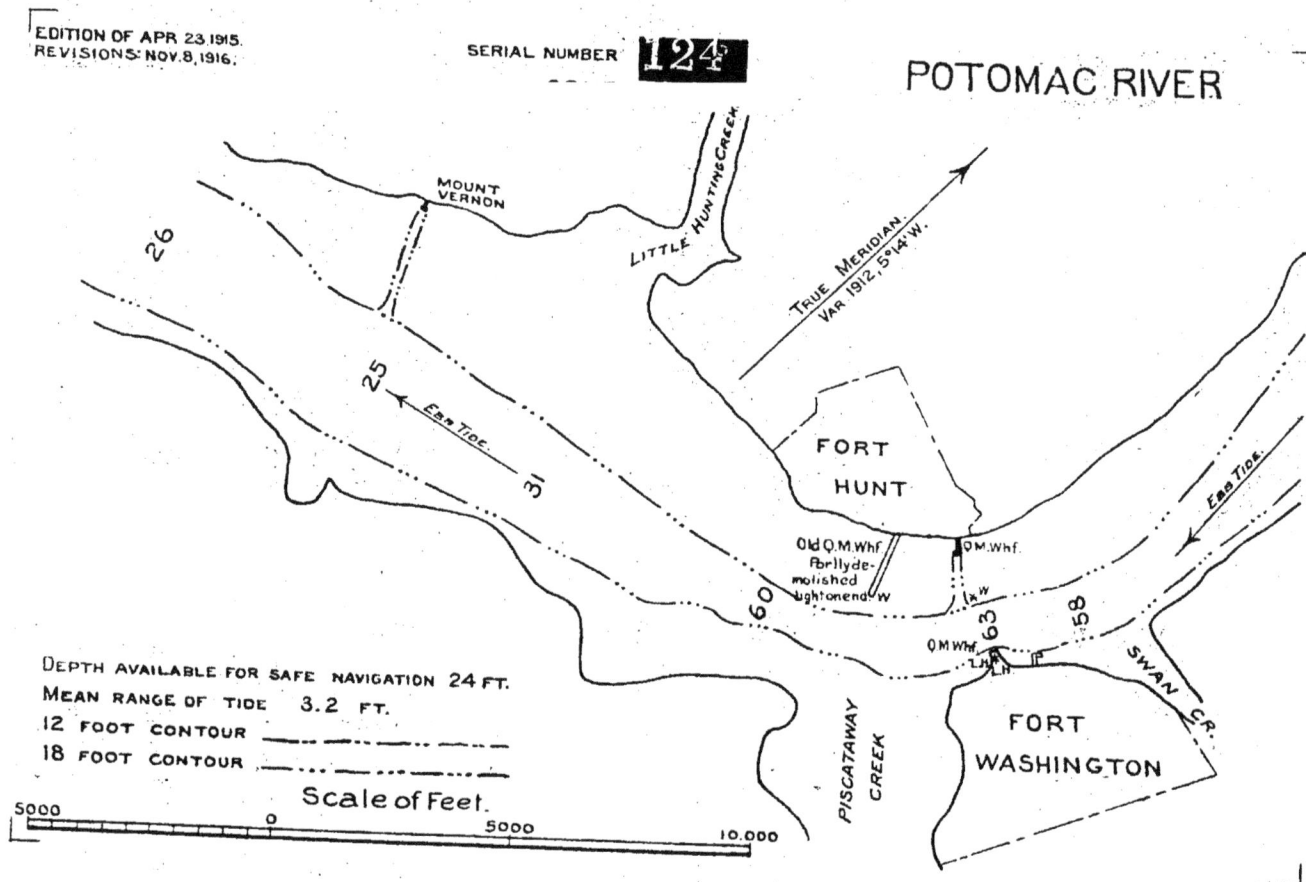

Fort Washington (1808-1928) is located on Digges' Point, Maryland where the Piscataway Creek joins the Potomac River, about twelve miles below the Washington, D.C. The first fort on this site was Fort Warburton, a Second System fort constructed in 1809 and destroyed in 1814. The replacement fort, called Fort Washington, was built from 1816 to 1824. In the 1840s the fort underwent an extensive remodeling program to bring it up to the standards of the third generation of coastal fortifications. Work crews constructed 88 permanent gun platforms (though the first guns were not emplaced until 1846), increased the height of the east wall, rebuilt the drawbridge, strengthened the powder magazines, and added a caponier to protect the approaches from Piscataway Creek. The casemate fort of stone and brick served as the primary defense of the nation's capital via the Potomac River up to and through the Civil War. Replaced by earthen batteries mounting 15-inch Rodman guns in front of the masonry fort and by guns at the Civil War era Fort Foote in the 1870s Period, Fort Washington's 341-acre reservation was used during the Endicott Program for eight concrete batteries starting in 1891. The reservation also contains controlled mine facilities and a temporary Spanish-American emplacement of a 10-inch gun on a barbette carriage. It was officially named in General Orders 1 of 1937 for General George Washington, the First President of the United States. In 1925, the coast artillery activities at Fort Washington ceased, but the fort used for various U.S. Army functions until 1946 when it was transferred to the National Park Service. Most of the fort's buildings have been removed. Since that time, it has been a public park commemorating the long history of coastal fortifications and serving as a recreational area for park visitors. Most of the Endicott batteries are closed to the public, while the 3rd System fort is open on certain days. Today, it is known at the Fort Washington National Park and open to the public on a day use basis.

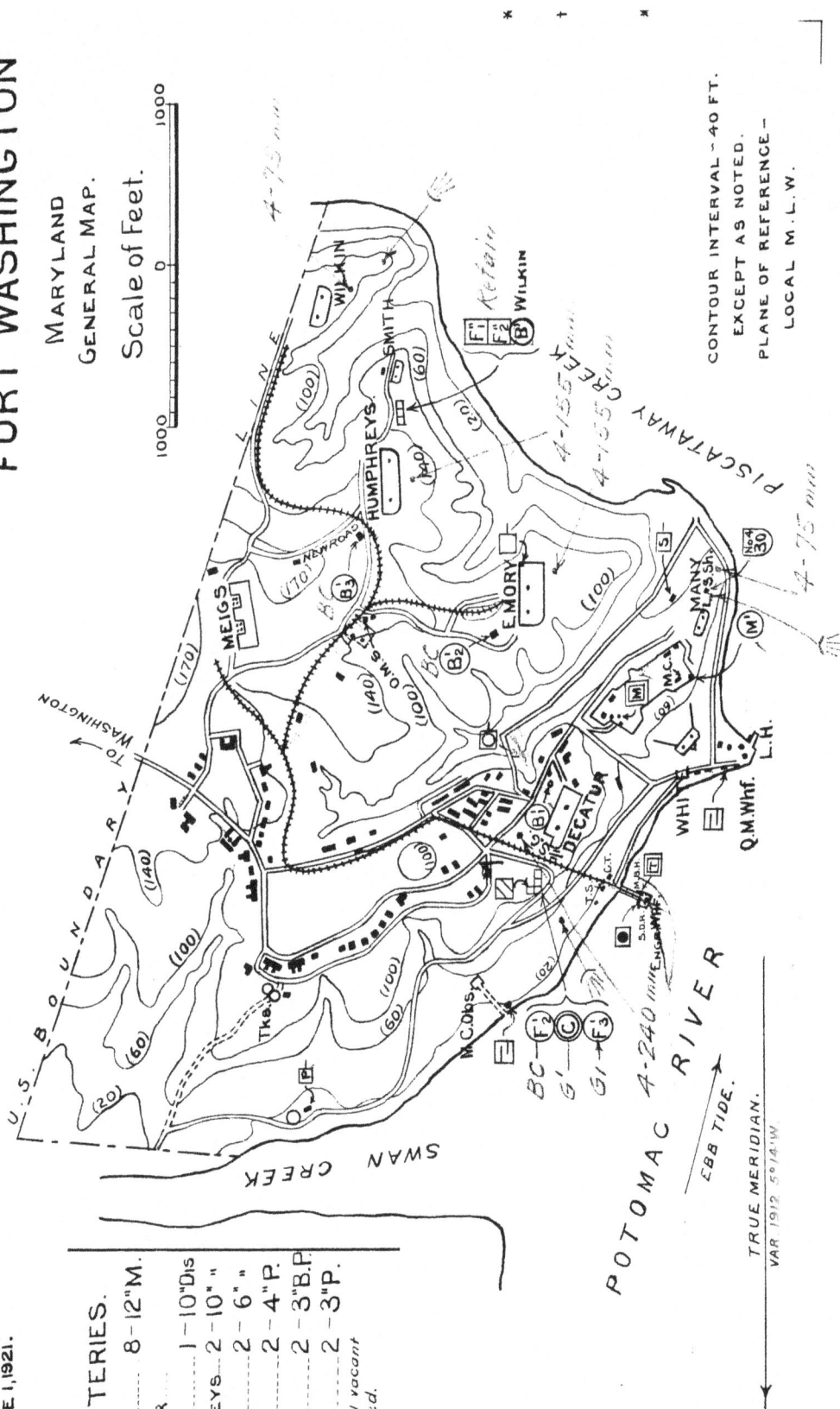

POTOMAC RIVER

FORT WASHINGTON

MARYLAND
GENERAL MAP.

Scale of Feet.

1000   1000

SERIAL NUMBER

EDITION OF MAY 5, 1919.
REVISIONS: FEB. 28, 1920;
JUNE 1, 1921.

BATTERIES.

*MEIGS .... 8 - 12" M.
DECATUR
†EMORY .... 1 - 10" Dis.
HUMPHREYS .. 2 - 10" "
WILKIN .... 2 - 6" "
*WHITE .... 2 - 4" P.
SMITH .... 2 - 3" B.P.
MANY .... 2 - 3" P.

† Empl. No 1 vacant
* Abandoned.

CONTOUR INTERVAL ~ 40 FT.
EXCEPT AS NOTED.
PLANE OF REFERENCE ~
LOCAL M.L.W.

TRUE MERIDIAN.
VAR 1912 5° 14' W.

EBB TIDE.

POTOMAC 4-240 RIVER

SWAN CREEK

U. S. BOUNDARY LINE

TO WASHINGTON

PISCATAWAY CREEK

POTOMAC RIVER
FORT WASHINGTON D-1.
MARYLAND.

Scale of Feet.

SERIAL NUMBER

EDITION OF MAY 5, 1919.
REVISIONS: JUN. 25, 1920;
JUNE 1, 1921.

TRUE MERIDIAN.
Var. 1912, 5°14'W.

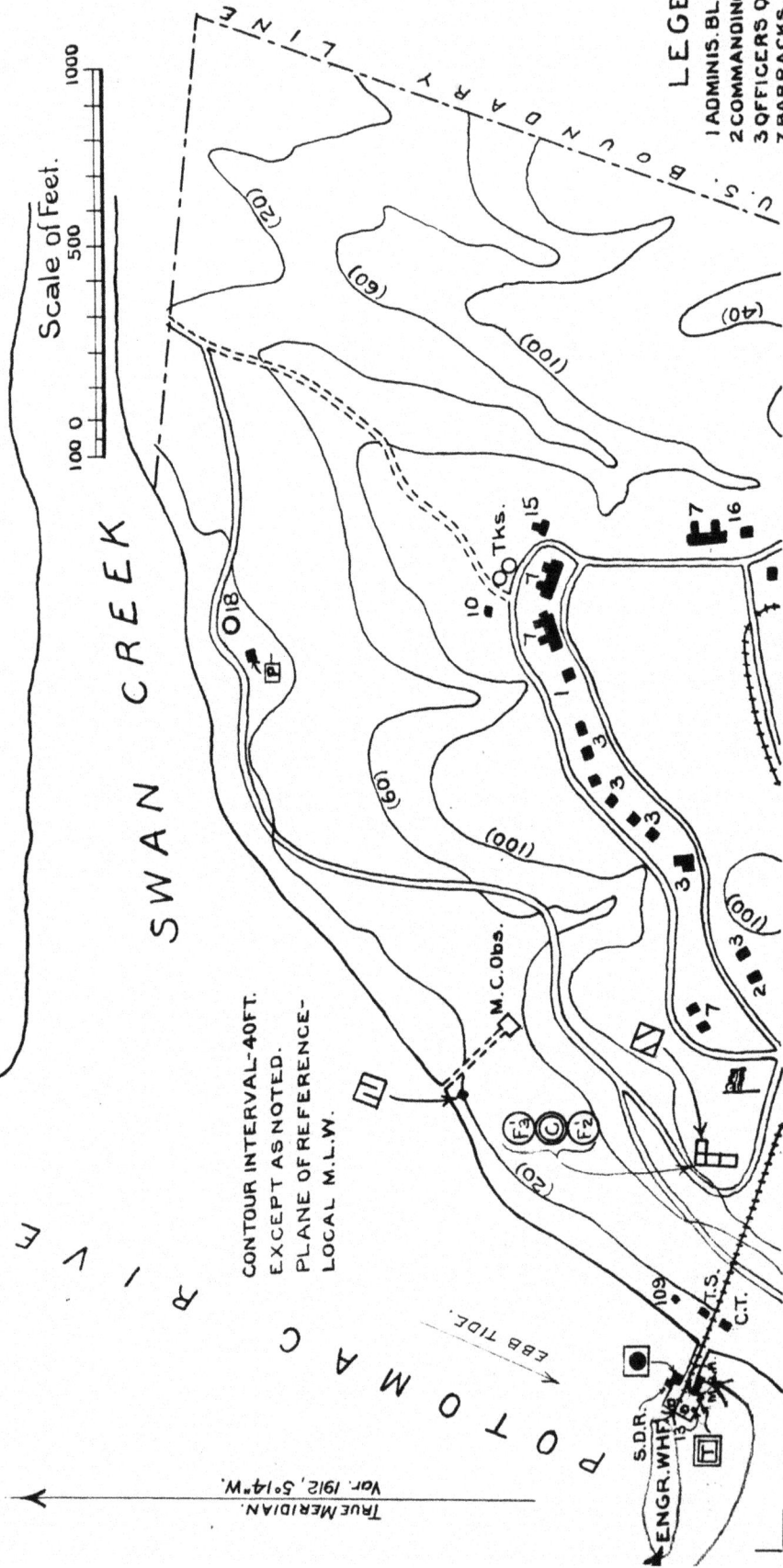

CONTOUR INTERVAL-40FT.
EXCEPT AS NOTED.
PLANE OF REFERENCE-
LOCAL M.L.W.

POTOMAC RIVER

SWAN CREEK

EBB TIDE.

M.C. Obs.

ENGR. WHF.
S.D.R.
T.S.
C.T.
109

LEGEND
1 ADMINIS. BLDG.
2 COMMANDING OFF. QRS.
3 OFFICERS QRS.
7 BARRACKS.
10 C.D.A.E. OFFICE.
13 ENGINE HOUSE.
15 FIRE APPARATUS.
16 BAKERY.
18 RESERVOIR.
109 OIL HOUSE.

U.S. BOUNDARY LINE

FORT WASHINGTON D-2.

POTOMAC RIVER

MARYLAND.

POTOMAC RIVER

EBB TIDE.

TRUE MERIDIAN
Var. 1912, 5°14'W.
TO WASHINGTON.

U.S. BOUNDARY

MEIGS

EMORY

JAMES MANY

WHITE

DECATUR

Salvaged

Engineer Reservation

SERIAL NUMBER

Scale of Feet.
1000   500   0   100   1000

CONTOUR INTERVAL-40FT
EXCEPT AS NOTED.
PLANE OF REFERENCE-
LOCAL M.L.W.

BATTERIES.

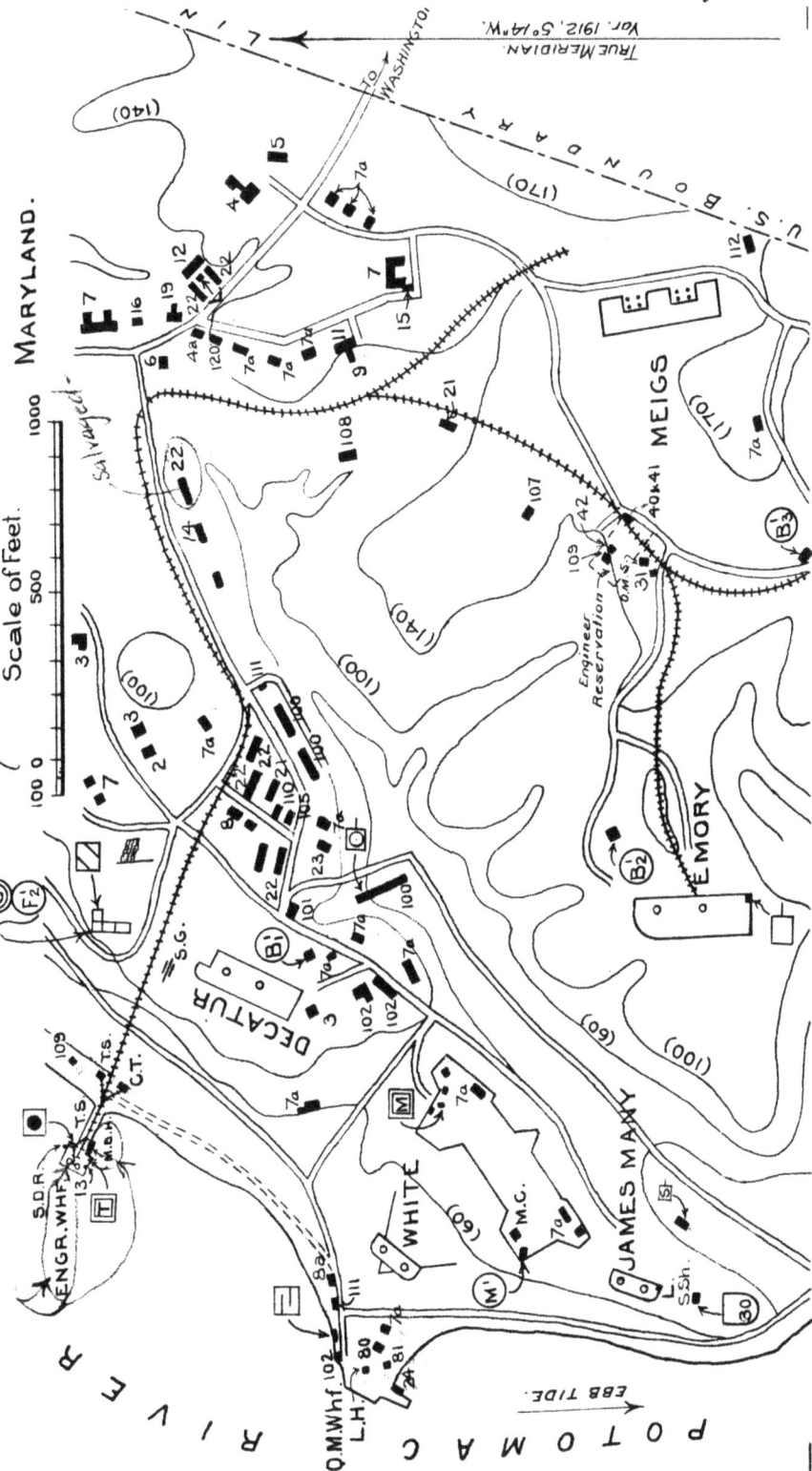

* MEIGS ..... 8-12"M.
  DECATUR ..... 1-10"Dis
† EMORY ..... 2-4"P.
* WHITE ..... 2-4"P.
  MANY ..... 2-3"P.
† Empl. No.I vacant.
* Abandoned.

LEGEND

EDITION OF MAY 5, 1919.
REVISIONS: JUN. 25, 1920.
JUNE 1, 1921.

2 COMMANDING OFF. QRS.
3 OFFICERS QRS.
4 HOSPITAL.
4a DENTAL OFFICE.
5 HOSPITAL STWD'S QRS.
6 N.C.O. QUARTERS.
7 BARRACKS.
7a ENLISTED MENS QRS.
8 GUARD HOUSE.
8a SENTRY BOX.
9 POST EXCHANGE.
11 GYMNASIUM.
12 MESS HALL.
13 ENGINE HOUSE.
14 GARAGE.
15 FIRE APPARATUS.
16 BAKERY.
19 BOWLING ALLEY.
100 COAL SHED.
101 POST OFFICE.
102 STORE HOUSE.
105 BLACKSMITH, CARPEN-
    TER & PLUMBER SHOP.

107 LOCOMOTIVE HOUSE.
108 SAW MILL.
109 OIL HOUSE.
110 TOOL HOUSE.
111 SCALES.
120 FIELD HEADQUARTERS.
21 Q.M. SHOPS.
22 Q.M.&C. STOREHOUSE.
23 Q.M. STABLES.
24 Q.M. BOAT HOUSE.
31 ORDNANCE ST. HO.
40 ENGINEER OFFICE.
41 ENGINEER ST. HO.

42 ENGINEER SHOP.
80 LIGHT KEEPER'S HO.
81 L.H. PAINT HOUSE.

POTOMAC RIVER

FORT WASHINGTON D-3.

SERIAL NUMBER

MARYLAND.

112

7a.

B3

New Road

HUMPHREYS

(140)

(100)
(60)
(20)

SMITH

109

WILKIN

(100)

(100)
(60)
(20)

(100)
(60)

(101)

CREEK

PISCATAWAY

EBB TIDE.

True Meridian.
Var. 1912, 5°40'W.

BATTERIES.
HUMPHREYS 2-10"D
WILKIN   2-6"
SMITH   2-3"E

Scale of Feet.

100 0   500   1000

EDITION OF MAY 5,1919.
REVISIONS: FEB. 28,1920.
JUN. 25, 1920.

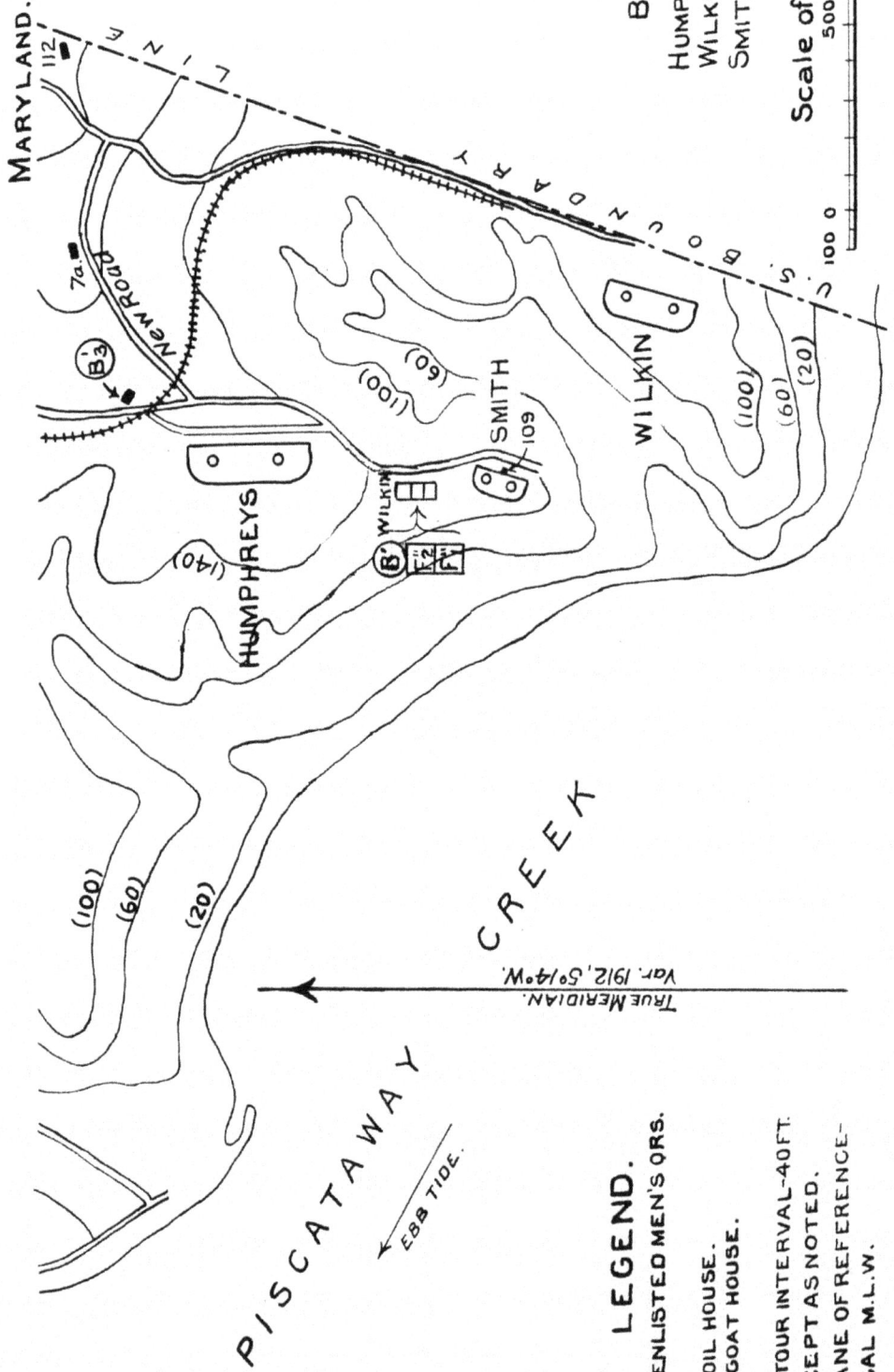

LEGEND.
7a. ENLISTED MEN'S QRS.
108 OIL HOUSE.
112 GOAT HOUSE.

CONTOUR INTERVAL-40FT.
EXCEPT AS NOTED.
PLANE OF REFERENCE-
LOCAL M.L.W.

Fort Washington 1920s (NARA)

Fort Washington 1937 (NARA)

Fort Washington Gun Batteries

- **DECATUR:** The first new Endicott battery for the old Fort Washington reservation on the west bank of the Potomac River. It was located on the bluff just north of the old fort, firing to the north-west. The project was first submitted on June 26, 1891. It was one of the first of the new, large-bore batteries for disappearing guns—in fact neither the bore size nor carriage type was yet decided upon, the battery initially described as being for 8-inch guns. Its final plan reveals earlier elements dating from an older design concept. Two separate emplacements had circular, restricted loading platforms. Magazines were on the lower left of each platform, ammunition service by crane. The battery was not equipped for electricity, and on the southern rear flank was a small rifleman gallery with loopholes for rearward land defense. Interior crest was 134-feet. Work was started in 1891 and suspended in 1893 pending decisions on the armament. The change to 10-inch gun size was made in 1895, and the carriage plans for the LF Model 1894 disappearing type received on March 2, 1896. Work resumed for completion in 1897. On June 30, 1897 it was reported wired for electricity. Transfer was made on July 6, 1899 at a cost of $128,492. It was named on General Orders No. 43 of April 7, 1900 for Commodore Stephen Decatur, U.S. Navy. The armament consisted of two Model 1888 10-inch Watervliet guns on Model 1894 disappearing carriages (#20/#1 and #9/#28). Extensive modernization took place in 1905-1912, including widened platforms, modern hoists, battery commander's station and communications galleries. The guns were removed after World War I, in March 1919. The carriages were scrapped in place. The emplacement still exists at the Fort Washington National Park. The battery is closed to the public.

- **EMORY:** A second 10-inch disappearing battery for Fort Washington. It was located to the east of the old fort and fired to the west. It was planned and authorized in parts. The first platform (designed for a M1894 disappearing carriage) was funded with the Act of June 6, 1896. Work on that section was completed within a year, in 1897. The adjacent second platform (this one for an M1896 disappearing carriage) had plans submitted on May 8, 1897 from funds of the Act or March 3, 1897. The emplacement design generally followed type plans, with separate platforms and lower-level magazines on the left of each served by hoists for ammunition delivery. Both platforms were laid out as internal emplacements. Transfer was made on July 6, 1899 for a cost of $91,432. Changes to platforms and hoists were made during the immediately following years. It was named on General Orders No. 43 on April 7, 1900 for Major General William Emory of Civil War service. The guns were mounted by the end of 1897. It carried two 10-inch Model 1888 guns on one Model 1894 LF and one Model 1896 LF disappearing carriage (Watervliet tube M1888 #6/carriage #17 and Watervliet tube M1888MII #56/carriage #27). The latter gun and M1896 carriage were removed in July 1916 to arm a "target" battery being tested at Fort Morgan, Alabama. The last gun and carriage served alone until being dismounted in 1929 at the termination of the Potomac defense project. The emplacement still exists at the Fort Washington National Park. The battery is closed to the public.

- **HUMPHREYS:** The third 10-inch battery erected on the southern end of the reservation, 2000-feet to the east of the old masonry fort. It was located to fire to the west towards the river. Plans were submitted on April 4, 1898 for this emplacement using Model 1896 10-inch disappearing carriages. It was immediately begun with initial funding from the National Defense Act of March 1898—in fact orders were issued to build the platforms, if possible, for immediate arming, even before completion of the rest of the battery. The battery had a crest of 170-feet and closely complied with designs of Mimeograph No. 22. Platforms, which were both of internal configuration, had gun centers spaced at 104-feet. Magazines were on the lower left with ammunition service by hoists. Work was done from March 21, 1898; as promised platforms were ready by April 13 and

18. The battery was reported completed on June 30, 1899. Transfer was made on July 6, 1899 at a cost of $93,207. It was named on General Orders No. 43 of April 7, 1900 for Major General Andrew Humphreys, former Chief of Engineers 1866-1879. The armament consisted of two 10-inch Model 1888MII Bethlehem guns on disappearing carriages Model 1896 (#17/#34 and #19/#57). Certain elements of the emplacement were modified in the early 1900s. The armament served until removed on October 4, 1929 at the end of the defenses of the Potomac. The emplacement still exists at the Fort Washington National Park. The battery is open, but the interior is closed to the public.

- **Water Battery**: A temporary Spanish-American War 10-inch barbette battery used at Fort Washington. Previous to the war a barbette mounting had been emplaced at Fort Washington in a ravine close behind the old work to test fire at new types of concrete wall construction. Apparently, this was done in early 1897. The firing weapon was a 10-inch Model 1888MII gun (Bethlehem #6) on barbette carriage Model 1893 (Watertown #2). During the scare that occurred early in the war, the gun and carriage were relocated onto a new platform near Battery Humphrey to augment the coverage for the Potomac River. It consisted of just a concrete gun base with hold-down bolts for the carriage and a simple an earthen parapet in front of it. It served only a short while in this capacity, before being shifted back to its original location to conclude the tests on authority of September 16, 1898. The tests were completed on June 29, 1899 and the gun and carriage dismounted. Eventually the gun was emplaced at Battery Randol at Fort Worden, and the carriage went to the PanAm Exposition in Buffalo in 1901. The wartime, emergency gun block does still exist, in a parking lot between Battery Humphrey and Smith at the Fort Washington National Park. The battery site is open to the public.

- **WHITE**: An emplacement for two 4-inch rapid-fire guns emplaced during the emergency at the start of the Spanish American War. Originally slated to receive a pair of the recently purchased British 4.7-inch guns, the local engineers were informed on May 3, 1898 that there was an insufficient number of these available, and that two navy-type 4-inch guns would be substituted. An emplacement of two platforms had already been hastily erected but construction was suspended before completion on April 11th. The platforms were located in the salient of the Demilune located between the old fort and the river. Older 1870s brick and concrete magazines were to be used to serve the new wartime, temporary battery. The plan was relatively simple. Two interior platforms, separated by 45-feet had lower-level old magazines on their left flanks. Ammunition service was by hand. Work was done by June 1898. Transfer was made on July 6, 1899 for $13,125. It was named on General Orders No. 78 of May 25, 1903 for Major William White, an army surgeon who was killed in action at Antietam in 1862. It carried two 4-inch Driggs-Schroeder guns and navy-type pedestals (#3/#4 and #4/#3). Only Battery White and Battery Plunkett at Fort Warren were armed with this expedient type of armament. The armament was declared obsolete in 1919 and were soon removed and sent away as memorial guns. The emplacement still exists at the Fort Washington National Park. The battery is open to the public.

- **MEIGS**: The mortar battery for Fort Washington was emplaced east of old Fort Washington just about at the edge of the reservation boundary. The site was on a gentle reverse slope of the plateau that made up most of the fort; set back some 2400-feet from the edge of the river bluff. Plans were submitted on July 27, 1898. It was of typical mimeograph mortar design of the time, with two wide pits side by side. Magazines were not under the front parapet, they were located on the flanks and in the shared central traverse. Work was commenced on August 25, 1898 and done by May 1900 except for some earth covering, Transfer was made on October 27, 1902 for a cost of $117,000. The battery was named on General Orders No. 43 of April 7, 1900 for Major General

Montgomery Meigs, the Quartermaster General from 1861-1882. It was armed with eight 12-inch Watervliet mortars Model 1890M1 on Model 1896 carriages (#110/#208, #81/#175, #125/#174, #117/#172, #111/#173, #122/#169, #109/#171, and #112/#170). This armament served only a little over a decade. The gun tubes, carriage, and even base rings were removed and shipped out for the modernization of Battery Kellogg at Fort Banks in 1913. The emplacement was abandoned except for occasional storage by May 1914. The emplacement still exists at the Fort Washington National Park. The battery is closed to the public.

- **WILKIN:** A battery for two 6-inch disappearing guns emplaced at the southern extreme of the fort reservation along the shore of Piscataway Creek. Plans were submitted on February 28, 1899. It was sited on a plateau at the southeast corner of the reservation. Plans closely followed the recommended mimeograph No. 31. It had two platforms with wells for the disappearing carriages and lower-level magazines on the left flank. Ammunition service was by hoists. Work was done from June 1899 to early 1901. Transfer was made on October 27, 1902 for a cost of $64,630. It was named on General Orders No. 78 on May 25, 1903 for Captain Alexander Wilkin who was killed in action at Tupelo, MS in 1864. Armament consisted of two 6-inch Model 1897M1 guns on Model 1898 LF disappearing carriages (#25/#12 and #12/#11). These served until removed in 1928 or 1929 with the closure of the defenses of the Potomac River. One tube (#25) was later used as a replacement in the Long Island Sound. The emplacement still exists at the Fort Washington National Park. The battery is open to the public.

- **SMITH:** A battery for two 3-inch rapid-fire guns emplaced on the southern part of the reservation, between batteries Wilkin and Humphreys. Plans were submitted on December 8, 1898. Plans closely followed those for masking parapet batteries in mimeograph No. 30. Crest height of the battery was 117.5-feet. It had internal platforms with barrel niches, gun centers separated by 29-feet. A magazine for each gun was on its lower left flank, ammunition service being by hand. Work was approved to commence on December 28, 1898 and conducted from January to October 1899. Transfer was made on September 15, 1903 for a cost of $9,500. It was named on General Orders No. 78 of May 25, 1903 for 2nd Lieutenant Joseph Smith who was killed in action at Chapultepec Mexico in 1847. The battery was armed with two 3-inch, 15-pounder Model 1898 Driggs-Seabury guns on Model 1898 masking parapet mounts (#99/#99 and #100/#100). In 1916 the carriages were modified to the Model 1898M1 barbette pedestal standard. The guns were declared obsolete in June 1920 and removed in common with all other emplacements of this type. The emplacement still exists at the Fort Washington National Park. The battery is open to the public.

- **MANY:** A battery for two 3-inch rapid-fire guns emplaced just to the immediate south of old Fort Washington, firing across the river to the west, northwest. Plans were submitted on May 9, 1903. The site was considered excellent for coverage of the mine field. It was recommended that the scarp wall of the old fort behind the battery be dismantled, but Army management thought this would be unsightly, and such demolition could always be left to the start of any later potential conflict. The plan was for the standard mimeograph type 1903 for 3-inch pedestals, with 62-foot gun centers, two platforms and two magazines and a storeroom under the central traverse. Work was done from 1903-1904 for transfer being made on May 19, 1905 at a cost of $25,904. It was named on General Orders No. 194 of December 27, 1904 for Colonel James Many of the War of 1812 service. It carried two 3-inch Model 1902 Bethlehem guns and pedestal mounts (#3/#3 and #5/#5). These were removed in 1928 or 1929 with the abandonment of the defenses of the Potomac River. The emplacement still exists at the Fort Washington National Park. The battery is open to the public.

Fort Hunt (1898-1928) is located on Sheridan Point, Virginia on the Potomac River, about twelve miles below the Washington, D.C. Established during the Endicott Program to support the batteries at Fort Washington across the river in Maryland, Fort Hunt had four concrete batteries. It was named in General Orders 71 of 1899 for Bvt. Maj. Gen. Henry J. Hunt, US Army, an artillery commander during the Civil War. Fort Hunt was deactivated as a coast artillery post in 1925, but continued to serve other military functions, such as a CCC camp and a "high-value" prison-of-war camp in World War II. Today, Fort Hunt Park is a day-use unit of the National Park Service off of the George Washington Parkway near Mount Vernon. The four batteries are open to the public on a day use basis, but the only remaining garrison structure is an NCO quarters which is closed to public.

Fort Hunt Gun Batteries

- **MOUNT VERNON:** An Endicott battery for three 8-inch disappearing guns. It was located on the central plateau or bluff north of the Potomac River at Sheridan Point, firing to the south. Plans were submitted on October 2, 1896 for two guns; however, the conservative bid of the winning contractor allowed the construction of all three guns of the intended battery. On February 16, 1897 the revised plans for a three-gun battery were submitted, still using funds from the Act of June 6, 1896. It was built to adhere to standard mimeograph plans. The three separate platforms were all configured as internal types. Magazines were on a lower level, to the left flank of each platform. Ammunition service was by hoist. Changes were made to emplacement loading platforms and commander's station in the early 1900s. Excavation was started on June 30, 1897 and concrete work was done by August 15, 1898. Transfer was made on January 13, 1900 for a cost of $91,948. The battery was named for George Washington's estate, located just to the south of this fort on General Orders No. 43 of April 4, 1900. The armament was delivered and mounted by April 1, 1898 and consisted of three 8-inch Model 1888MII Watervliet guns on Model LF 1894 disappearing carriages (#53/#15, #35/#17, and #31/#19). These served until removed during World War I in 1917, the guns being shipped away in December of that year. The emplacement was not again used for coast defense armament. It still exists at the Fort Hunt Park. The battery is open, but the interior is closed to the public.

- **PORTER:** An emplacement for a single 5-inch balanced pillar gun emplaced on the left flank of Battery Mount Vernon, firing with it to the south. Plans for two single 5-inch emplacements were submitted together on September 29, 1898. It closely followed type plans for rapid-fire guns, with a single rounded platform and magazine on the lower left flank. Ammunition service was by hand. Work was started in November 1898 and finished by September 6, 1899. Transfer came on August 26, 1902 for a cost of $8,000. It was named on General Orders No. 78 of May 25, 1903 for Lt. James E. Porter, 7th U.S. Cavalry, killed in action at the Little Big Horn battle of 1876. The gun was mounted in June 1902 and was 5-inch gun Model 1897 Bethlehem gun on Model 1896 balanced pillar mount (#1/#16). This was dismounted on September 28, 1917, but apparently prematurely as it was remounted on October 26, 1917. Final authority to remove came on November 12, 1917 and the gun was shipped out to Morgan Engineering for use on a wheeled field mount. The carriage was removed and turned over to the salvage officer on August 27, 1920. The emplacement was not again used for coast defense armament. It still exists at the Fort Hunt Park. The battery is open to the public.

POTOMAC RIVER
FORT HUNT
SHERIDAN POINT, VA.
GENERAL MAP.

BATTERIES
MT VERNON
PORTER ....
ROBINSON ....
SATER ....

CONTOUR INTERVAL - 20 FT.
PLANE OF REFERENCE = SAME
AS HARBOR CHART - LOCAL M.L.W.

Removed Light maintained on end of wharf.

SERIAL NUMBER 124

POTOMAC RIVER

EBB TIDE

O.M.WRF.
B.H.

SATER
O.R.ROBINSON
SETER
Sh.600 Mobile Cadillac
PORTER
MT.VERNON
No.3
No.1
S.

OLD WHARF

SHERIDAN POINT

TRUE MERIDIAN
VER. 1912, 5°14′W.

U.S. RESERVATION

BOUNDARY OF

2000 Ft.   1000   0   500   1000

EDITION OF MAY 5, 1919.
REVISIONS: JUN. 25, 1920;
JUNE 1, 1921.

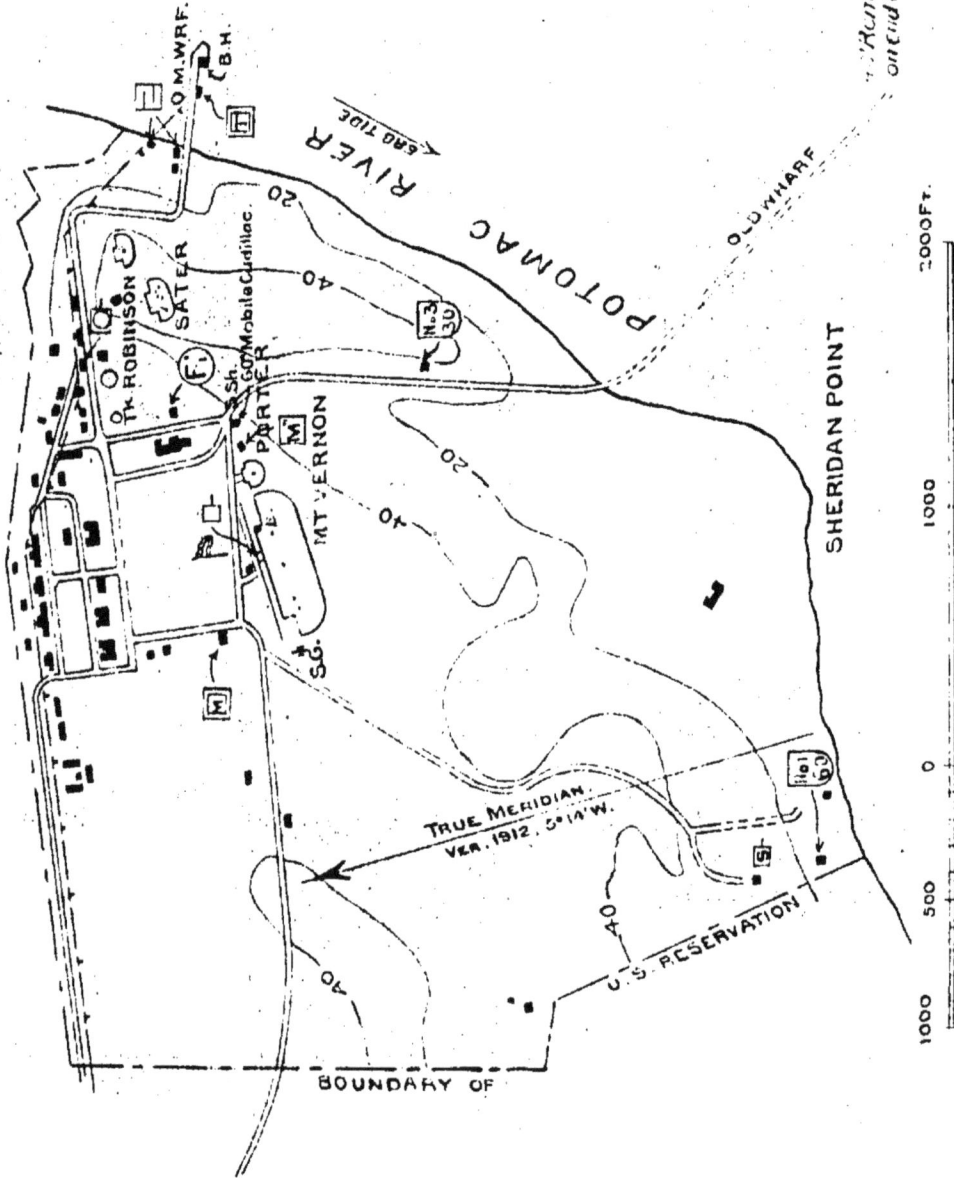

POTOMAC RIVER
FORT HUNT D-1.
SHERIDAN POINT, VA.

CONTOUR INTERVAL = 20 FT.
PLANE OF REFERENCE = SAME
AS HARBOR CHART = LOCAL M.L.W.

BATTERIES
MT.VERNON.
PORTER.
ROBINSON.
SATER.

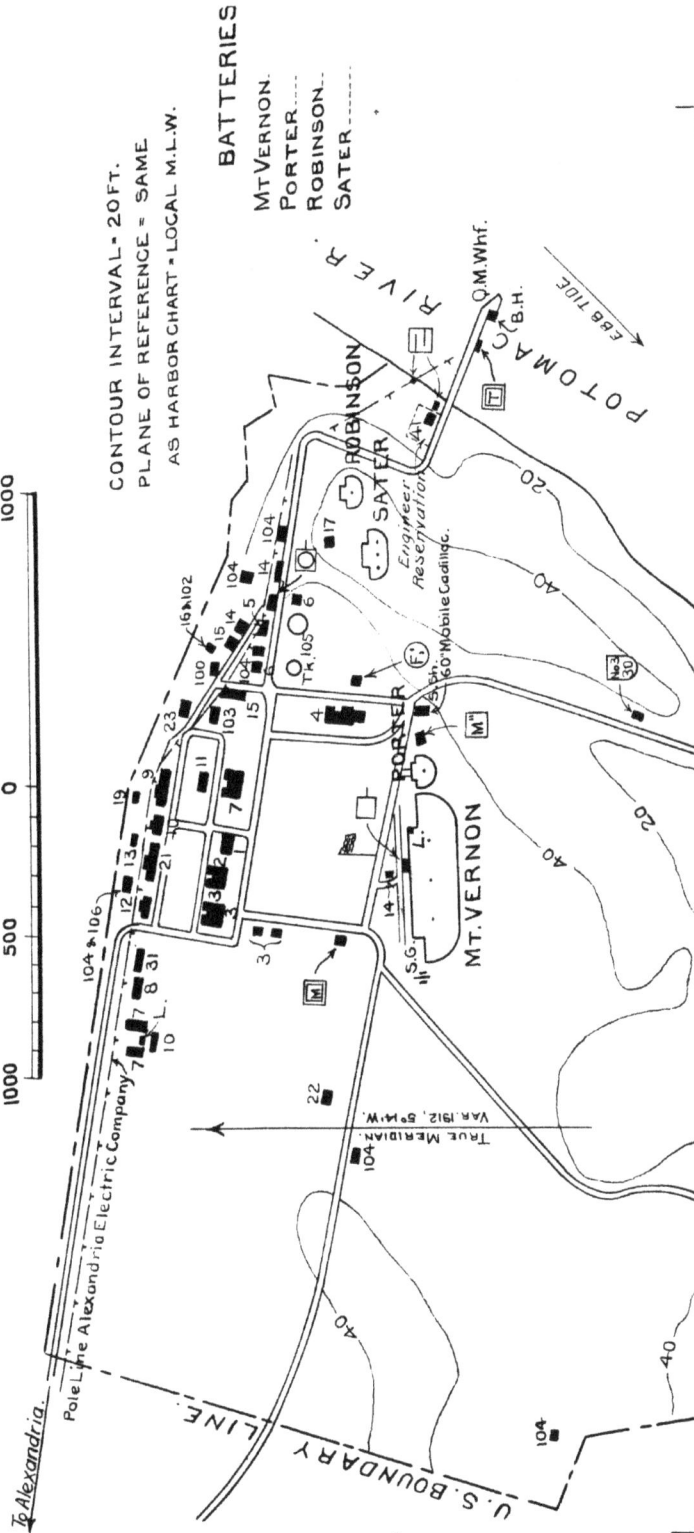

EDITION OF MAY 5, 1919.
REVISIONS: JUNE 1, 1921.

SERIAL NUMBER

Scale of Feet.
1000   500   0   1000

POTOMAC RIVER

ROBINSON
SATER
PORTER
MT. VERNON

Engineer Reservation.

Q.M.Whf.
B.H.
EBB TIDE

To Alexandria.
Pole Line Alexandria Electric Company.

TRUE MERIDIAN
VAR. 1912, 5° 14' W.

U.S. BOUNDARY LINE

LEGEND

1 ADMINISTRATION BLDG.
2 COMMANDING OFF.QRS.
3 OFFICER'S QRS.
4 HOSPITAL.
5 HOSPITAL STWD'S QRS.
6 N.C.OFFICER'S QRS.
7 BARRACKS.
8 GUARD HOUSE.
9 POST EXCHANGE.
10 MESS HALL.
11 LAVATORY.
12 BAKERY.
13 OIL HOUSE.
14 COAL SHED.
15 WAGON SHED.
16 TOOL HOUSE.
17 SAW MILL.
19 STORE HOUSE.
100 SCALES.
101
102 PLUMBER'S SHOP.
103 CARPENTER'S SHOP & TEAMSTER'S QRS.
104 ENLISTED MENS QRS.
105 RESERVOIR.
106 FIRE APPARATUS.
21 Q.M.& C. STOREHOUSE.
22 Q.M. STOREHOUSE.
23 Q.M. STABLE.
31 ORDNANCE ST. HO.
41 ENGINEER ST. HO.

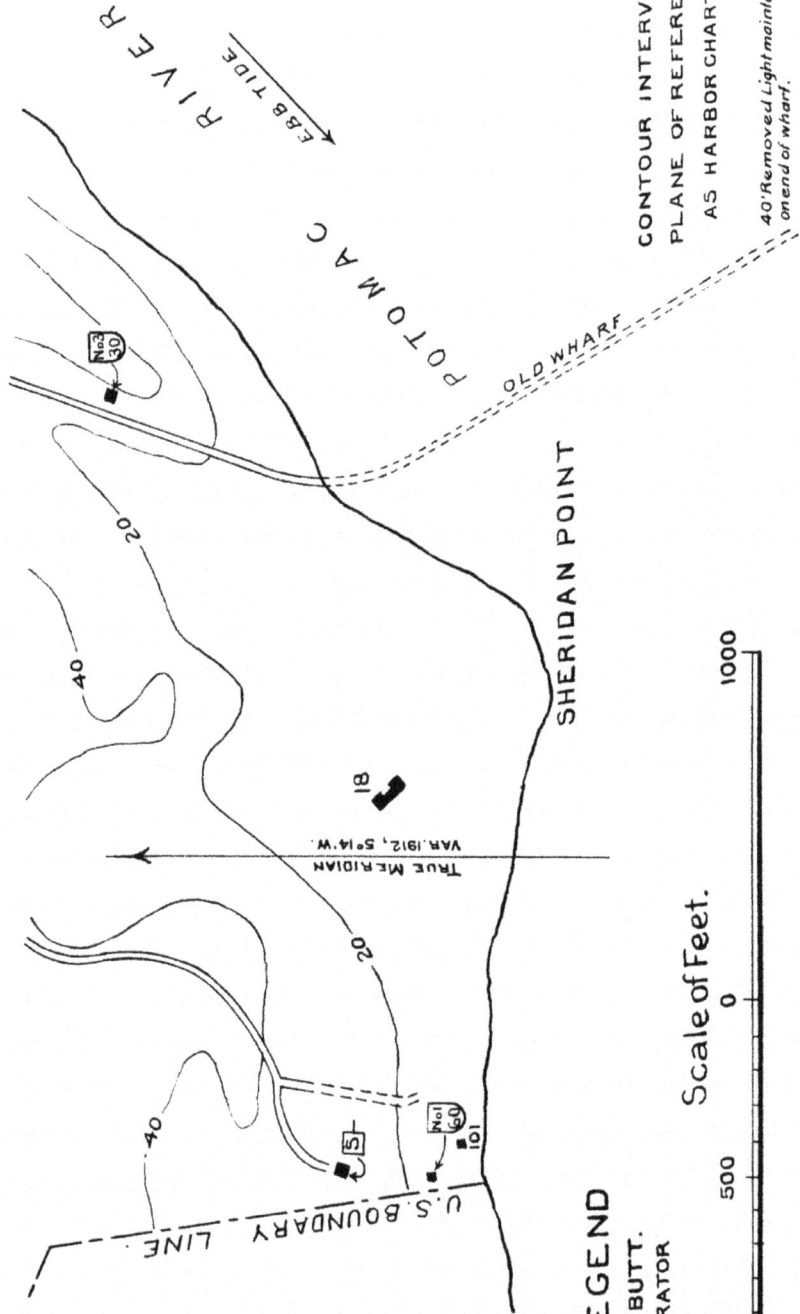

POTOMAC RIVER

FORT HUNT D-2.

SHERIDAN POINT, VA.

SERIAL NUMBER

EDITION OF MAY 5, 1919.
REVISIONS, JUNE 1, 1921.

CONTOUR INTERVAL = 20 FT.
PLANE OF REFERENCE = SAME
AS HARBOR CHART = LOCAL M.L.W.

40' Removed Light maintained
on end of wharf.

POTOMAC RIVER

EBB TIDE

OLD WHARF

No.3
30

No.1
60
101

5

U.S. BOUNDARY LINE

SHERIDAN POINT

18

TRUE MERIDIAN
VAR. 1912, 5° 14'. W.

20

40

20

40

Scale of Feet.

1000   500   0   500   1000

LEGEND

18 RIFLE BUTT.
101 INCINERATOR

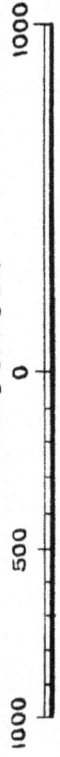

- **ROBINSON**: A second emplacement for a single 5-inch balanced pillar gun emplaced on a slight ridge close to the river and wharf on the east side of the reservation. It pointed and fired to the south. Plans for both single 5-inch emplacements were submitted together on September 29, 1898. Porter was built near the 8-inch battery, and Robinson was intended to be on the flank of a 12-inch disappearing battery that was subsequently never authorized or built. It was of type design, with lower level magazine on the right flank of the single gun platform. Work was done from late 1898 to 1899 and fully completed by June 20, 1901. It was transferred on August 26, 1902 for a cost of $9,200. It was named on General Orders No. 78 of May 25, 1903 for 1st Lieutenant Levi H. Robinson who was killed by hostile Indians in Wyoming Territory in 1874. The armament was mounted by June 30, 1902, and consisted of one 5-inch Model 1897 Bethlehem gun on a Model 1896 balanced pillar mount (#2/#17). This was removed under authority of August 24, 1917 and sent to Morgan Engineering for use on a field carriage along with Battery Porter's gun in November 1917. The emplacement was not again used for coast defense armament. It still exists at the Fort Hunt Park. The battery is open to the public.

- **SATER**: An emplacement for three 3-inch masking parapet guns emplaced on the right flank and a little forward of 5-inch Battery Robinson on the east side of the reservation. It also fired directly to the south towards the Potomac River. Plans were submitted on June 14, 1900 for the emplacement. It was of conventional mimeograph design, with two separate interior platforms and one rounded, flank platform on the left flank; gun centers being separated by 29-feet. One magazine for each gun was placed on the lower left flank of the platforms, ammunition service being by hand. Initial funds were allocated on June 25, 1900, but work was delayed until August. Concrete work was done within a year, but the guns were not received for mounting until October and November 1903. Transfer was made on January 24, 1904 at a cost of $15,100. It was named on General Orders No. 78 of May 25, 1903 for 1st Lieutenant William A. Sater killed in action at San Juan Hill Cuba in 1898. It was armed with three 3-inch, 15-pounder Model 1898 Driggs-Seabury guns on masking parapet mounts (#105/#105, #106/#106, and #107/#107). These were modified about 1916 into M1898M1 pedestal mounts. The guns were removed and shipped out of July 9, 1920 per authority of June 21, 1920. The emplacement was not again used for coast defense armament, though in the early 1940s its magazine was used for storage of National Archives photographic materials. It still exists at the Fort Hunt Park. The battery is open to the public.

Battery Decatur, Fort Washington Park (Mark Berhow)

Fort Washington Park (Terry McGovern)

Fort Hunt Park (Terry McGovern)

Bolling Smith, "Fort Hunt, Virginia, Coast Artillery on the Potomac River," *Coast Defense Journal* Vol. 32, Issue 4 (Fall 2018) pp. 4 - 51

Washington Barracks: A Headquarters post in Washington DC that also was used for housing soldiers posted at the Potomac River forts.

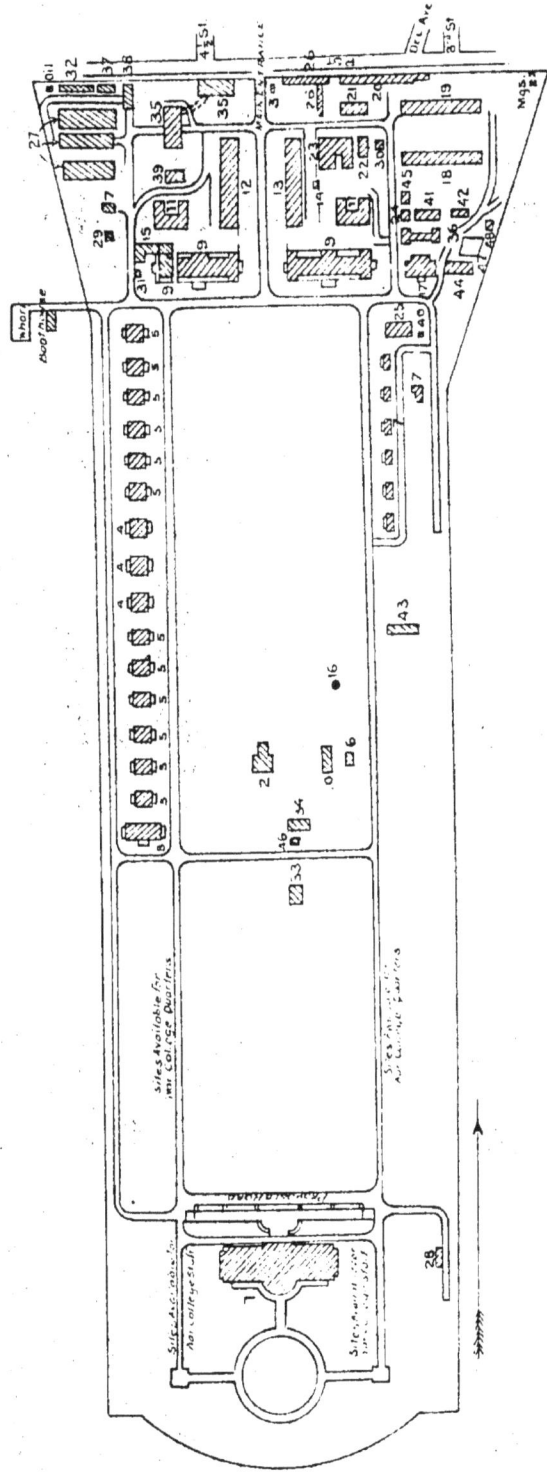

WASHINGTON BARRACKS
WASHINGTON, D.C.

SERIAL NUMBER 124

EDITION OF : MAR. 14 1916.

100 0 200 300 400 500 600 700 800 900 1000 ft.

LEGEND

1. WAR COLLEGE.
2. BACHELOR OFFICER'S QRS.
3. GATE HOUSE.
4. COM. & FIELD OFF. QRS.
5. OFFICERS QRS.
6. GARAGE FOR
   SEARCHLIGHT TRUCKS.
7. N.C. OFFICERS QRS.
8. OFFICERS MESS.
9. BARRACKS.
10. LITHOGRAPHIC SHOP.
11. MESS HALL
12. ENGR. STORE HOUSE.
13. Q.M. STORE HOUSE.
14. SCALE HOUSE.
15. HOSPITAL.
16. BAND STAND.
17. BAND QUARTERS.
18. ENGINEERS STABLES.
19. Q.M. & ENGR. STALES.
20. Q.M. WAGON SHED.
21. STABLE, GUARD HOUSE.
22. Q.M. SHOPS.
23. POST BAKERY.
24. HOOK & LADDER SHED.
25. POST EXCHANGE.
26. GREEN HOUSES.
27. PONTOON SHEDS.
28. WAR COLLEGE BOILER HO.
29. ENGR. PAINT SHOP.
30. TEAMSTERS QRS.
31. DEAD HOUSE.
32. ENGR. WAGON SHED.
33. OLD TRADE SCHOOLS,
    PAINT SHOP, ENGR.
    MUSEUM PROPERTY.
34. PHOTOGRAPHY.
35. HEADQUARTERS
    ENGR. SCHOOL.
36. SHOOTING GALLERY.
37. ENGINEER.
    BLACKSMITH SHOP.
38. ENGR. CARPENTER SHOP
39. GARAGE.
40. GASOLINE TANK.
    (UNDER GROUND)
41. COMPANY B.S. SHOP.
42. Q.M. PAINT SHOP.
43. CHAPEL.
44. POST LAUNDRY.
45. PACK TRAIN ST. ROOM.
46. MASONRY SHOP & "
47. COAL BINS.
48. INCINERATOR.

THE HARBOR DEFENSES OF THE ENTRANCE TO CHESAPEAKE BAY — VIRGINIA

Fort Monroe (1819-1946) is located on Old Point Comfort in Hampton, Virginia, where the Chesapeake Bay turns into Hampton Roads. The first fort of many to be built on this site was Fort Algernourne in 1609. The existing Third System fort, called Fort Monroe, was built from 1819 to 1836. This very large casemate fort of stone and brick covered 63-acres in an irregular hexagon shape that was surrounded by a moat was designed by BG Simon Bernard. The plans called for a total of 412 cannons to be mounted in casemates, including a large casemated water battery, or on the fort's terreplein and would require a wartime garrison of 2,625 men. It was officially named in General Orders 11 of 1832 for James Monroe, the fifth President of the United States. Fort Monroe served as the primary defense of Hampton Roads, Norfolk and entrance to the James River up to and through the Civil War. Fort Monroe's 565-acre reservation (about half of the reservation is reclaimed land) was used during the Endicott Program for fifteen concrete batteries and controlled submarine mine facilities. Besides being one of the largest Endicott Program defenses, Fort Monroe served as the Headquarters for the Coast Artillery Corps and the Coast Artillery School from 1907 to 1946 (before that date known as the Artillery School of the U.S. Army). The fort's coast artillery weapons were heavily used for training purposes. Fort Monroe also hosted the Old Point Comfort Proving Ground for testing artillery and ammunition from the 1830s to 1861. The 1940 Program brought several anti-aircraft batteries, including a AMTB (#23) battery, but the focus for new coast artillery defenses switched to the Virginia Capes. Between 1942 and 1944, most of Endicott Period coast artillery was scrapped. After the war, Fort Monroe became the headquarters of the U.S. Army Ground Forces from 1955 to 1973. Due to the construction of military housing to the north of old fort, the land defenses for 3rd System fort and several Endicott Program batteries were destroyed in the early 1950s. Through various U.S. Army re-organizations, Fort Monroe finally was home to the headquarters of the U.S. Army Training and Doctrine Command. Fort Monroe was decommissioned on September 15, 2011. On November 1, 2011, portions of Fort Monroe were designated as a national monument under the control of the National Park Service. The rest of the reservation is under the control of the Fort Monroe Authority, which is a state agency, that is implementing a mixed-use plan for the fort's 190 historic buildings. The old Coast Artillery School Library building has recently been renovated and is now the Fort Monroe's visitors center. The Casemate Museum, the ex-U.S. Army Museum in the casemates of old Fort Monroe has an extensive collection of artifacts and displays on the history of Fort Monroe and coast artillery is operated by the Fort Monroe Authority. Outside the moat of the large and impressive Third-System stone fort is Battery Irwin, which retains two rare 3-inch guns on pedestal mounts, and Battery Parrott, which features a rare, fixed mount 90 mm gun. Fort Monroe is open to visitors and residents; the Casemate Museum is open daily except certain holidays.

#7, Battery Bomford, Ft. Monroe, Va.

ENTRANCE TO CHESAPEAKE BAY

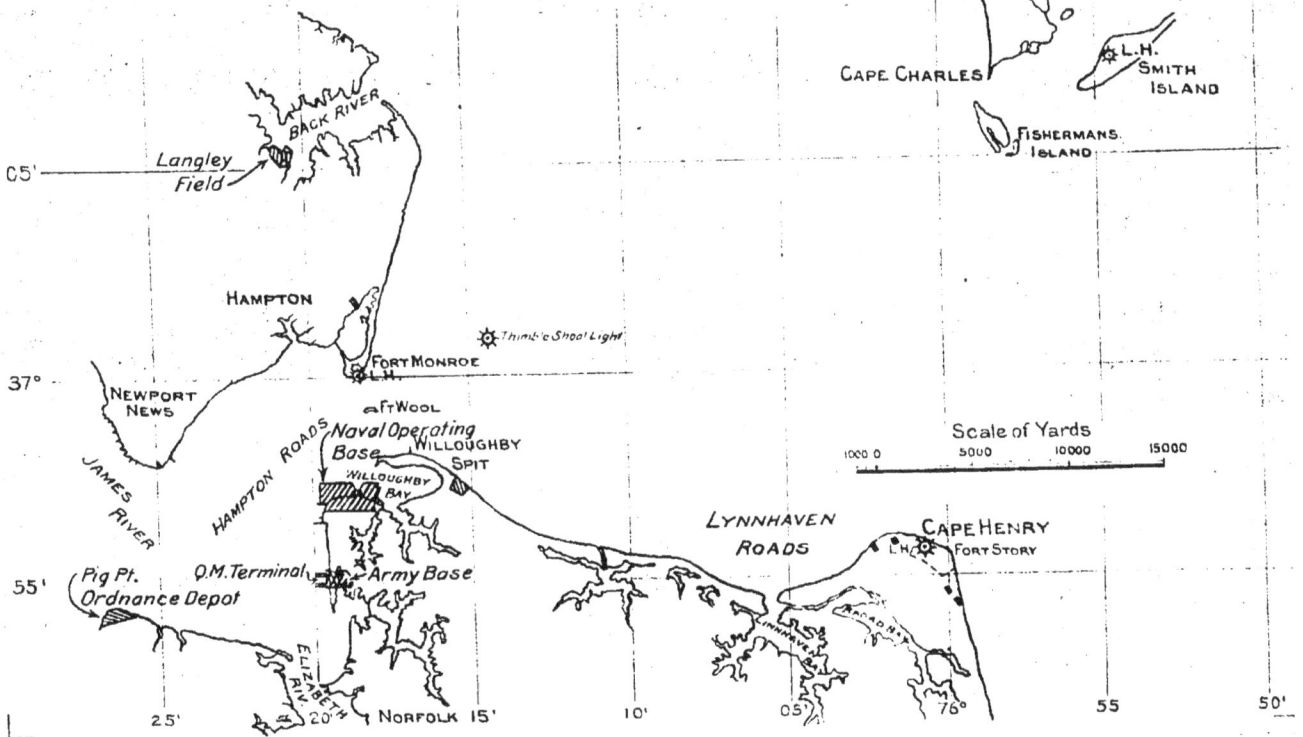

EDITION OF NOV. 8, 1916.
REVISIONS: MAR. 21, 1919.

SERIAL NUMBER 124

CAPE CHARLES

L.H. SMITH ISLAND

FISHERMANS ISLAND

BACK RIVER

Langley Field

05'

HAMPTON

Thimble Shoal Light

37°

FORT MONROE L.H.

NEWPORT NEWS

Ft WOOL

HAMPTON ROADS

Naval Operating Base

WILLOUGHBY SPIT

WILLOUGHBY BAY

Scale of Yards

1000 0   5000   10000   15000

LYNNHAVEN ROADS

CAPE HENRY
Fort Story
L.H.

JAMES RIVER

55'

Pig Pt.
Ordnance Depot

Q.M. Terminal   Army Base

ELIZABETH RIV.

25'   20'   NORFOLK 15'   10'   05'   76°   55   50'

HAMPTON ROADS, VA.

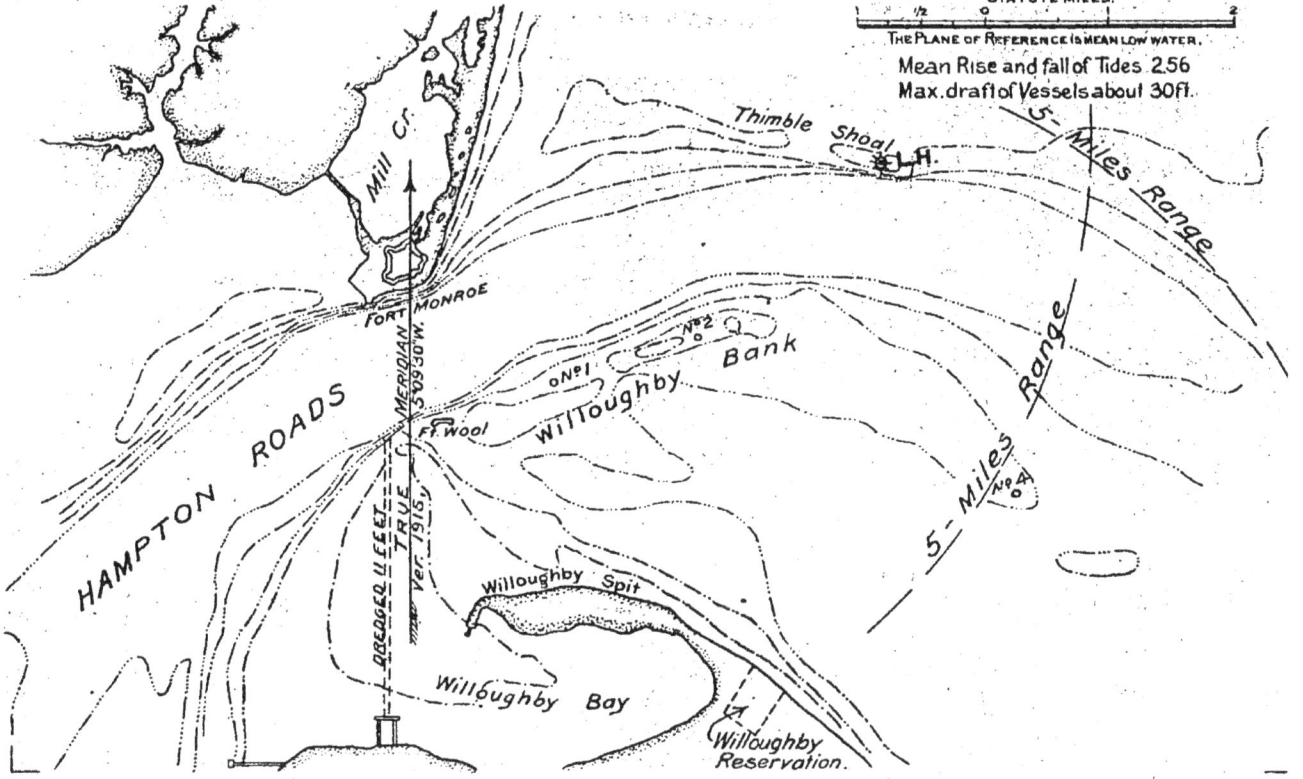

EDITION OF JAN. 14, 1915.
REVISIONS: JUNE 9, 1916.

SERIAL NUMBER 124

STATUTE MILES
1   1/2   0   1   2

THE PLANE OF REFERENCE IS MEAN LOW WATER.
Mean Rise and fall of Tides 2.56
Max. draft of Vessels about 30 ft.

Thimble Shoal L.H.

5-Miles Range

Mill Cr.

FORT MONROE

Willoughby Bank

No 2

No 1

HAMPTON ROADS

MERIDIAN 3° 09' 30" W.
DREDGED II FEET
TRUE Ver. 1916.

Ft. Wool

No 3
No 4

5-Miles Range

Willoughby Spit

Willoughby Bay

Willoughby Reservation.

Fort Monroe Gun Batteries

- **DeRUSSY:** The first 12-inch disappearing battery for Fort Monroe, DeRussy was located near the shore between Redoubts B and C. It was built with funds appropriated on July 2, 1898—which included a $150,000 allocation for building this emplacement. The plans were submitted on June 25, 1898. It complied with standard mimeograph type designs. Work for the emplacement of 12-inch guns was done in 1898-99, armament being mounted in June of 1901. Transfer was made on June 7, 1904 for a cost of $142,348.08. It carried three 12-inch Model 1895 gun tubes made at Watervliet Arsenal on Model 1897 disappearing carriages (#35/#19, #36/#20 and #37/#17). This ordnance was shipped to the site in April 1901. It was named for Colonel René E. DeRussy, engineer and former superintendent at Fort Monroe, in General Orders No. 105 of October 9, 1902. However, in General Orders No. 15 of January 28, 1909 the citation was changed to Brigadier General Gustavus A. DeRussy of Civil War service when the previous name was taken for use to name a new fort in the Hawaiian Islands. On July 21, 1910 a serious accident occurred during target practice, gun No. 1 discharged prematurely resulting in the deaths of eleven crew members and minor damage to the gun itself. Repairs were made and the battery continued in service. The battery received the usual modernization to platforms, stations, and hoists in 1912-1913. Elevation of the gun carriages was increased by 5-degrees around 1916. In subsequent years the armament was changed, gun serial #35 staying but the other two guns were replaced with new tubes with less wear (new Model 1895 guns #38 and #29) in the mid-1920s. The battery was an important element of the fort's defenses and remained active throughout the world wars. It was finally authorized for removal on April 18, 1944. The emplacement still exists under the control of National Park Service. The battery is closed to the public.

- **PARROTT:** A relatively modern Endicott emplacement for two 12-inch disappearing guns emplaced on the shore near Battery Irwin, just outside old Fort Monroe. Plans were submitted on January 31, 1902. It was constructed between May 1902 and early 1904. Being on the beach, gun trunnion elevations were just 35.5-feet when in battery. Concerns over sinkage demanded considerable soil testing and deep piles to reach stable subsurface layers. Guns were mounted in 1905, and it was temporarily transferred to troops in April 1905 so it could serve in upcoming maneuvers. Formal transfer to service troops made on June 29, 1906 for a cost of some $211,500. It was named on General Orders No. 78 of May 23, 1903 for Captain Robert Parrott, the famous American ordnance officer and inventor of the Parrott Rifle. Of fairly conventional design for its time, its location at Fort Monroe and modernity allowed it to be a showcase battery for the Coast Artillery School. It was armed with the technically advanced Model 1900 12-inch Watervliet Arsenal guns (#3 and #5) on disappearing carriages Model LF 1901 (#1 and #10). The disappearing carriages were modified to allow additional elevation around 1916. In 1920 these guns were removed and replaced with two 12-inch Model 1895 Watervliet guns (#31 and #67). This armament then served until deleted under authority of December 29, 1942. In 1943 the emplacement was modified by building platforms on the former gun blocks for the 90mm anti-motor torpedo boat battery at Fort Monroe. The emplacement still exists under the control of the Fort Monroe Authority. The battery interior is closed to the public.

- **ANDERSON – RUGGLES:** This was the mortar battery for Fort Monroe. It was one of the first "inline" design types that followed the earlier quadrangular or Abbot type. Initial submission of plans was made on October 5, 1895 by engineer Captain Thomas Casey, based on recent new Board of Engineer Typical Battery Mortar Emplacement plans of September 1894. At the time it was so far

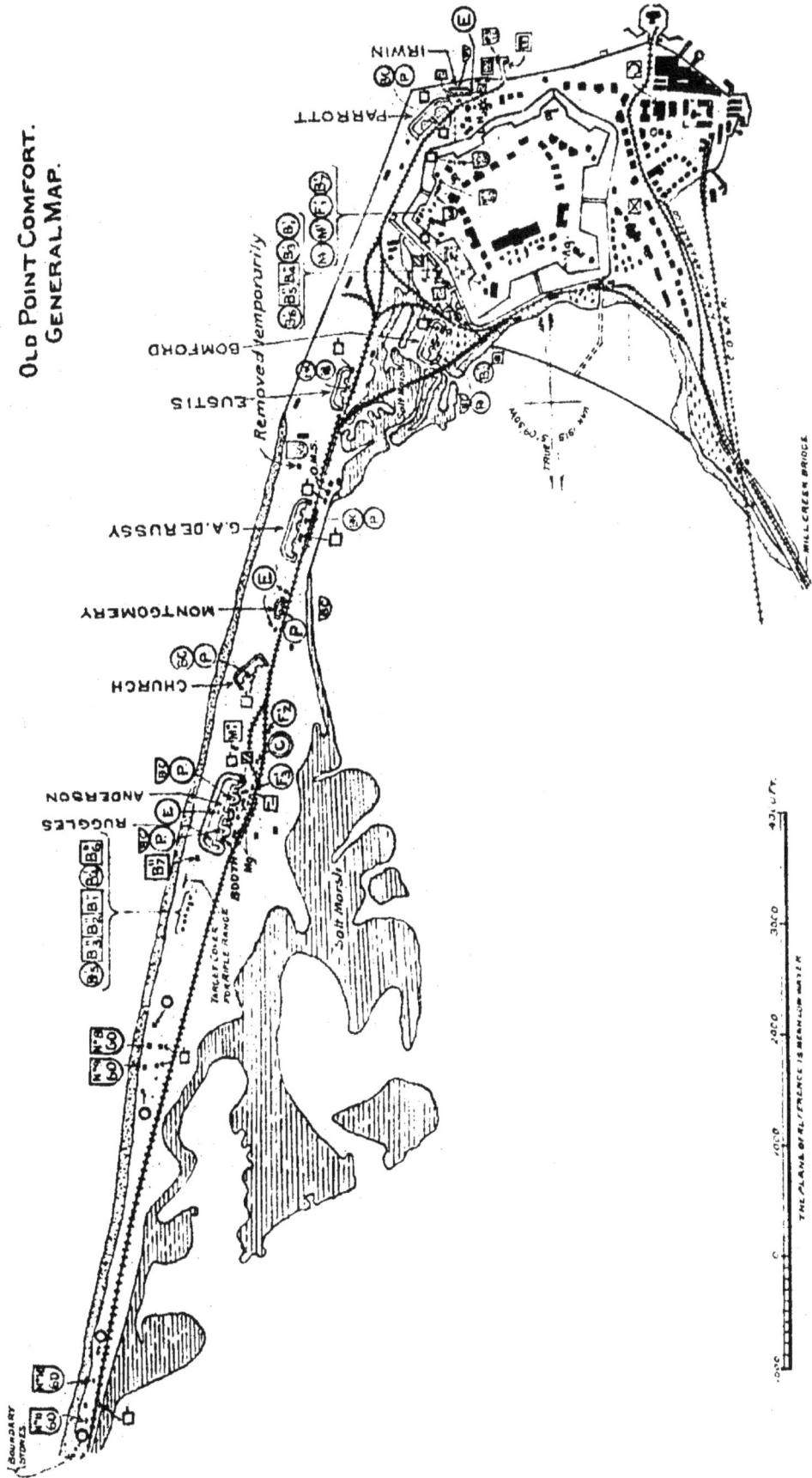

HAMPTON ROADS, VA.
FORT MONROE
OLD POINT COMFORT.
GENERAL MAP.

SERIAL NUMBER

124

EDITION OF JUNE 9,1916.
REVISIONS: APR.1,1921.

CHESAPEAKE BAY
FORT MONROE D-1.

SCALE OF FEET
400 200 0 200 400 600 800

SERIAL NUMBER

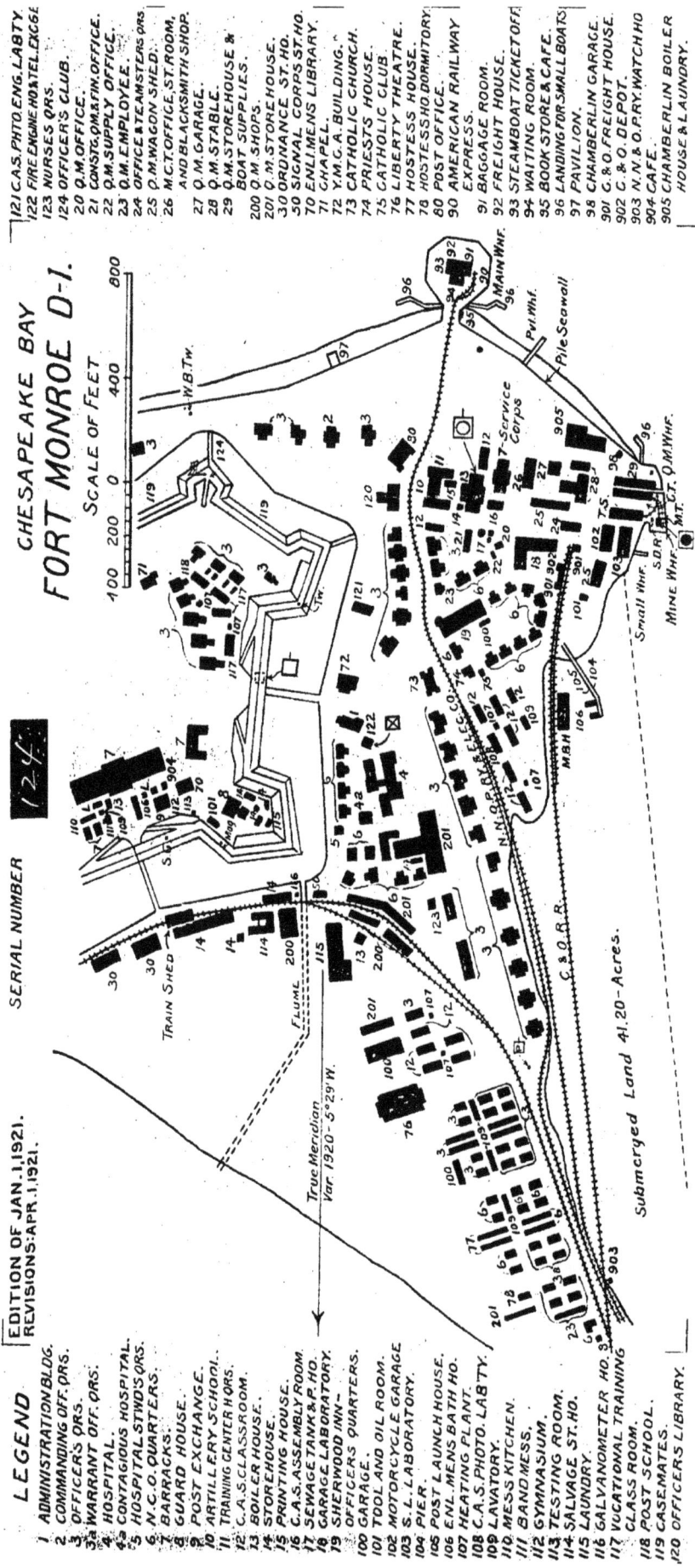

EDITION OF JAN.1,1921.
REVISIONS: APR.1,1921.

True Meridian
Var 1920-5°29'W.

Submerged Land 41.20-Acres.

LEGEND

1 ADMINISTRATION BLDG.
2 COMMANDING OFF. QRS.
3 OFFICER'S QRS.
3a WARRANT OFF. QRS.
4 HOSPITAL.
4a CONTAGIOUS HOSPITAL.
5 HOSPITAL STWD'S QRS.
6 N.C.O. QUARTERS.
7 BARRACKS.
8 GUARD HOUSE.
9 POST EXCHANGE.
10 ARTILLERY SCHOOL.
11 TRAINING CENTER HQRS.
12 C.A.S. CLASSROOM.
13 BOILER HOUSE.
14 STOREHOUSE.
15 PRINTING HOUSE.
16 C.A.S. ASSEMBLY ROOM.
17 SEWAGE TANK & P. HO.
18 GARAGE LABORATORY.
19 SHERWOOD INN -
100 GARAGE.
101 TOOL AND OIL ROOM.
102 MOTORCYCLE GARAGE.
103 S.L. LABORATORY.
104 PIER.
105 POST LAUNCH HOUSE.
106 ENL. MENS BATH HO.
107 HEATING PLANT.
108 C.A.S. PHOTO. LAB'TY.
109 LAVATORY.
110 MESS KITCHEN.
111 BAND MESS.
112 GYMNASIUM.
113 TESTING ROOM.
114 SALVAGE ST. HO.
115 LAUNDRY.
116 GALVANOMETER HO.
117 VOCATIONAL TRAINING CLASS ROOM.
118 POST SCHOOL.
119 CASEMATES.
120 OFFICERS LIBRARY.

121 C.A.S. PHTO. ENG. LAB'TY.
122 FIRE ENGINE HO. & TELE. EXCHGE.
123 NURSES QRS.
124 OFFICER'S CLUB.
20 Q.M. OFFICE.
21 CONSTG. Q.M. & FIN. OFFICE.
22 Q.M. SUPPLY OFFICE.
23 Q.M. EMPLOYEE.
24 OFFICE & TEAMSTERS QRS.
25 Q.M. WAGON SHED.
26 M.C.T. OFFICE, ST. ROOM, AND BLACKSMITH SHOP.
27 Q.M. GARAGE.
28 Q.M. STABLE.
29 Q.M. STOREHOUSE & BOAT SUPPLIES.
200 Q.M. SHOPS.
201 Q.M. STOREHOUSE.
30 ORDNANCE ST. HO.
50 SIGNAL CORPS ST. HO.
70 ENLIMENS LIBRARY.
71 CHAPEL.
72 Y.M.C.A. BUILDING.
73 CATHOLIC CHURCH.
74 PRIEST'S HOUSE.
75 CATHOLIC CLUB.
76 LIBERTY THEATRE.
77 HOSTESS HOUSE.
78 HOSTESS HQ. DORMITORY.
80 POST OFFICE.
90 AMERICAN RAILWAY EXPRESS.
91 BAGGAGE ROOM.
92 FREIGHT HOUSE.
93 STEAMBOAT TICKET OFF.
94 WAITING ROOM.
95 BOOK STORE & CAFE.
96 LANDING FOR SMALL BOATS.
97 PAVILION.
98 CHAMBERLIN GARAGE.
901 C. & O. FREIGHT HOUSE.
902 C. & O. DEPOT.
903 N.N. & O. P'RY. WATCH HO.
904 CAFE.
905 CHAMBERLIN BOILER HOUSE & LAUNDRY.

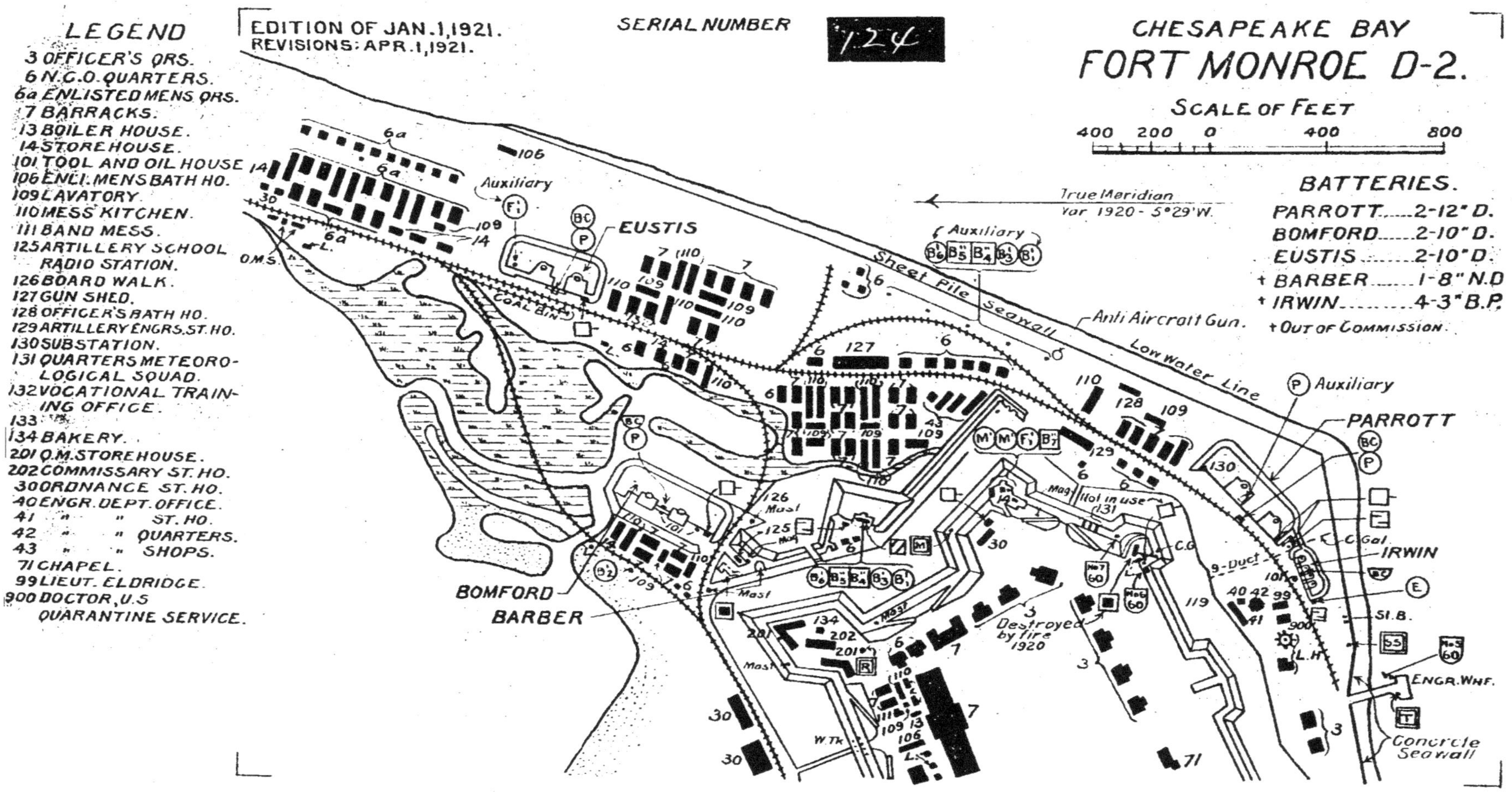
LEGEND
3 OFFICER'S QRS.
6 N.C.O. QUARTERS.
6a ENLISTED MENS QRS.
7 BARRACKS.
13 BOILER HOUSE.
14 STOREHOUSE.
101 TOOL AND OIL HOUSE
106 ENL'L. MENS BATH HO.
109 LAVATORY.
110 MESS KITCHEN.
111 BAND MESS.
125 ARTILLERY SCHOOL RADIO STATION.
126 BOARD WALK.
127 GUN SHED.
128 OFFICER'S BATH HO.
129 ARTILLERY ENGRS. ST. HO.
130 SUBSTATION.
131 QUARTERS METEOROLOGICAL SQUAD.
132 VOCATIONAL TRAINING OFFICE.
133
134 BAKERY.
201 Q.M. STOREHOUSE.
202 COMMISSARY ST. HO.
300 ORDNANCE ST. HO.
40 ENGR. DEPT. OFFICE.
41 " " ST. HO.
42 " " QUARTERS.
43 " " SHOPS.
71 CHAPEL.
99 LIEUT. ELDRIDGE.
900 DOCTOR, U.S QUARANTINE SERVICE.
EDITION OF JAN. 1, 1921.
REVISIONS: APR. 1, 1921.
SERIAL NUMBER
124
CHESAPEAKE BAY
FORT MONROE D-2.
SCALE OF FEET
400 200 0 400 800
BATTERIES.
PARROTT......2-12" D.
BOMFORD......2-10" D.
EUSTIS......2-10" D.
+ BARBER......1-8" N.D
+ IRWIN......4-3" B.P.
+ OUT OF COMMISSION.
True Meridian
Var. 1920 - 5°29'W.
Auxiliary
Sheet Pile Seawall
Anti Aircraft Gun.
Low Water Line
EUSTIS
PARROTT
IRWIN
BOMFORD
BARBER
COAL BIN
O.M.S.
Destroyed by fire 1920
9-Duct
C.Gal.
ENGR. WHF.
Concrete Seawall
Not in use
Mast
W. Tk.
L.H.
Sl.B.

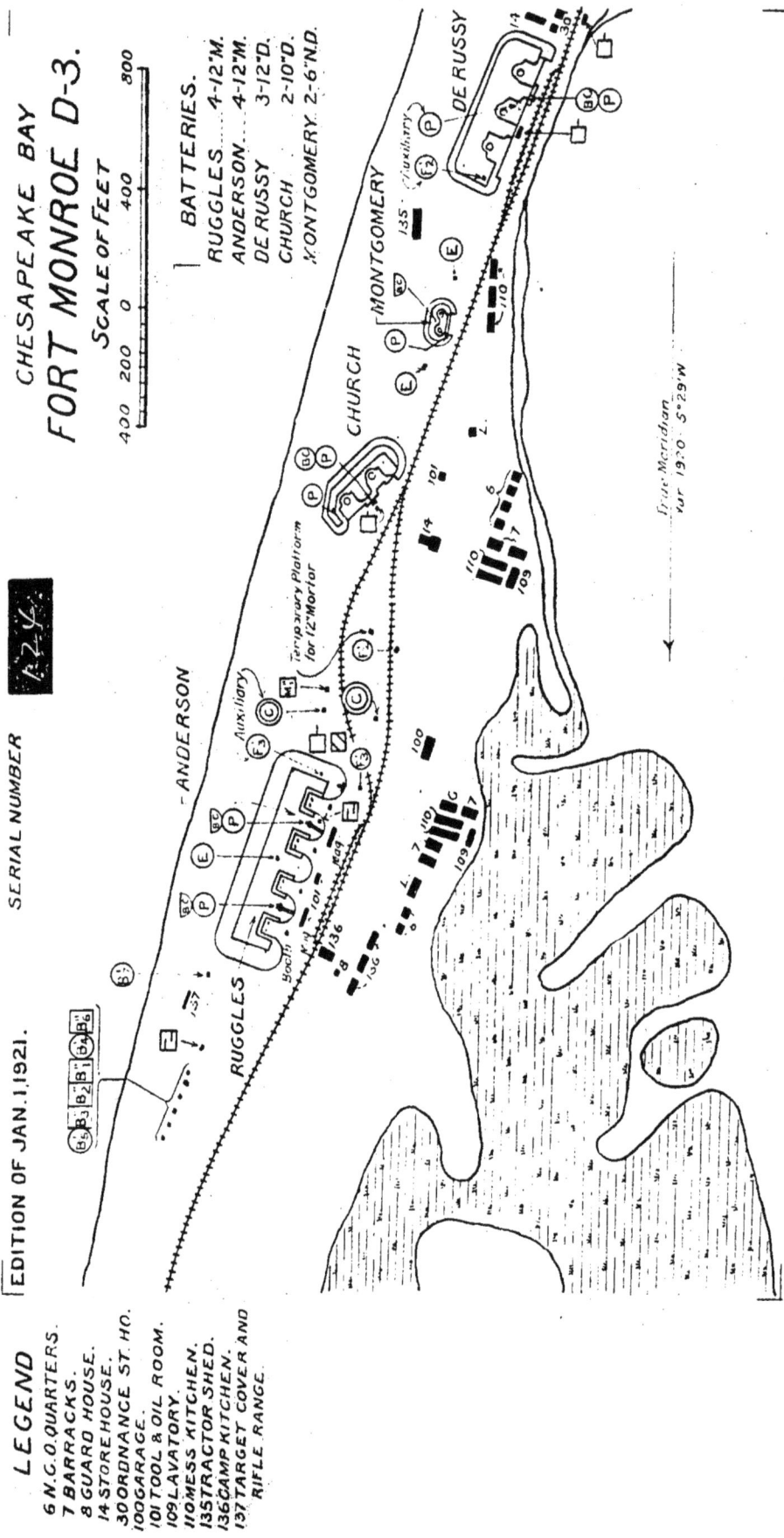

CHESAPEAKE BAY
FORT MONROE D-3.
SCALE OF FEET

BATTERIES.
RUGGLES........4-12"M.
ANDERSON.......4-12"M.
DE RUSSY......3-12"D.
CHURCH........2-10"D.
MONTGOMERY. 2-6"N.D.

SERIAL NUMBER

EDITION OF JAN. 1.1921.

LEGEND
6 N.C.O. QUARTERS.
7 BARRACKS.
8 GUARD HOUSE.
14 STORE HOUSE.
30 ORDNANCE ST. HO.
100 GARAGE.
101 TOOL & OIL ROOM.
109 LAVATORY.
110 MESS KITCHEN.
135 TRACTOR SHED.
136 CAMP KITCHEN.
137 TARGET COVER AND
    RIFLE RANGE.

CHESAPEAKE BAY
FORT MONROE D-4.
SCALE OF FEET

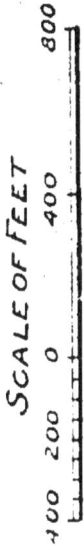

400 200 0 400 800

SERIAL NUMBER

124

EDITION OF JAN.1,1921.

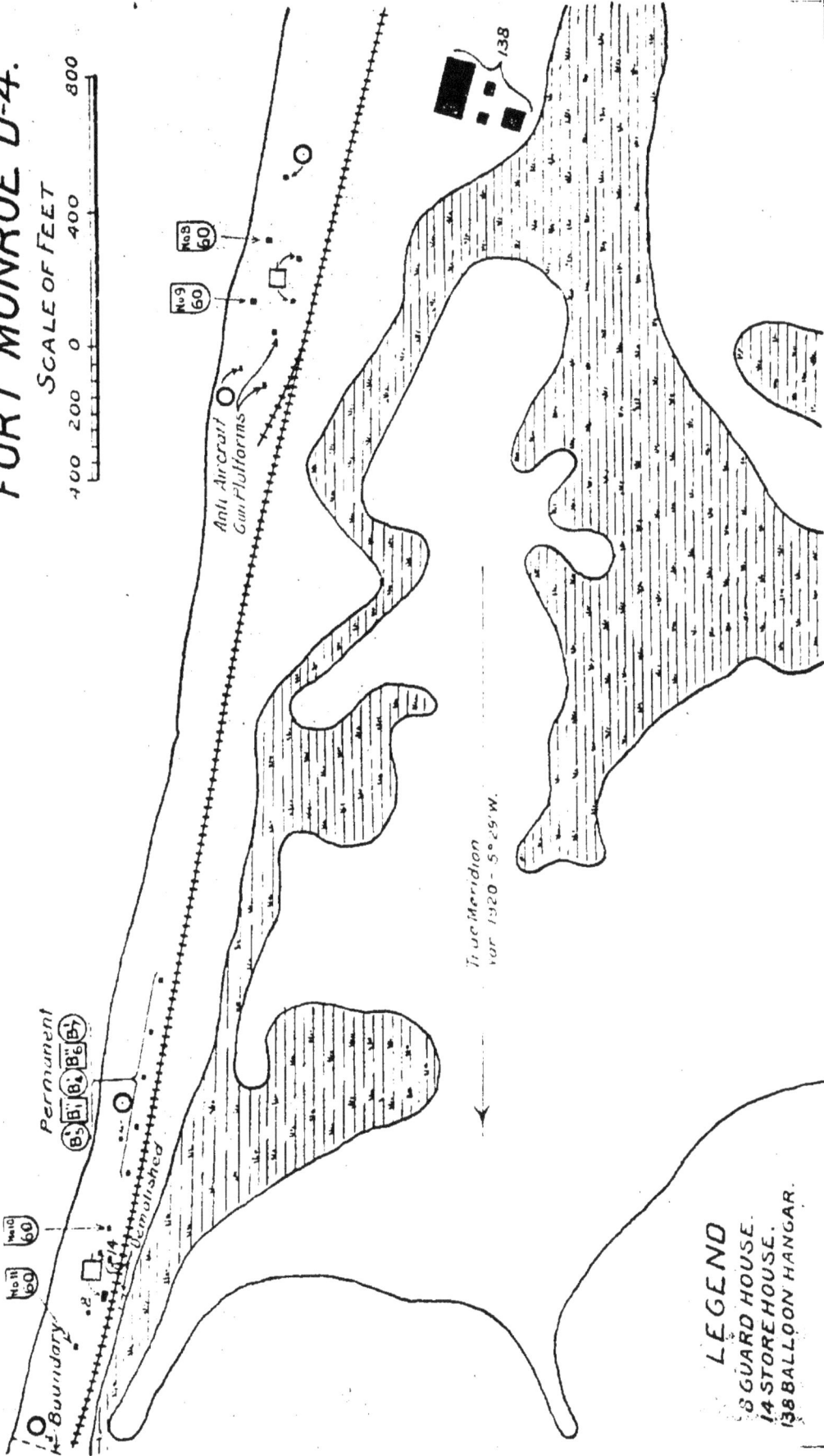

Anti Aircraft
Gun Platforms

No8
60

No9
60

Permanent

B3 B1 B4 B6 B7

Demolished

Boundary

No10
60

No11
60

138

True Meridian
var 1920 - 5°-25'W.

LEGEND
8 GUARD HOUSE.
14 STOREHOUSE.
138 BALLOON HANGAR.

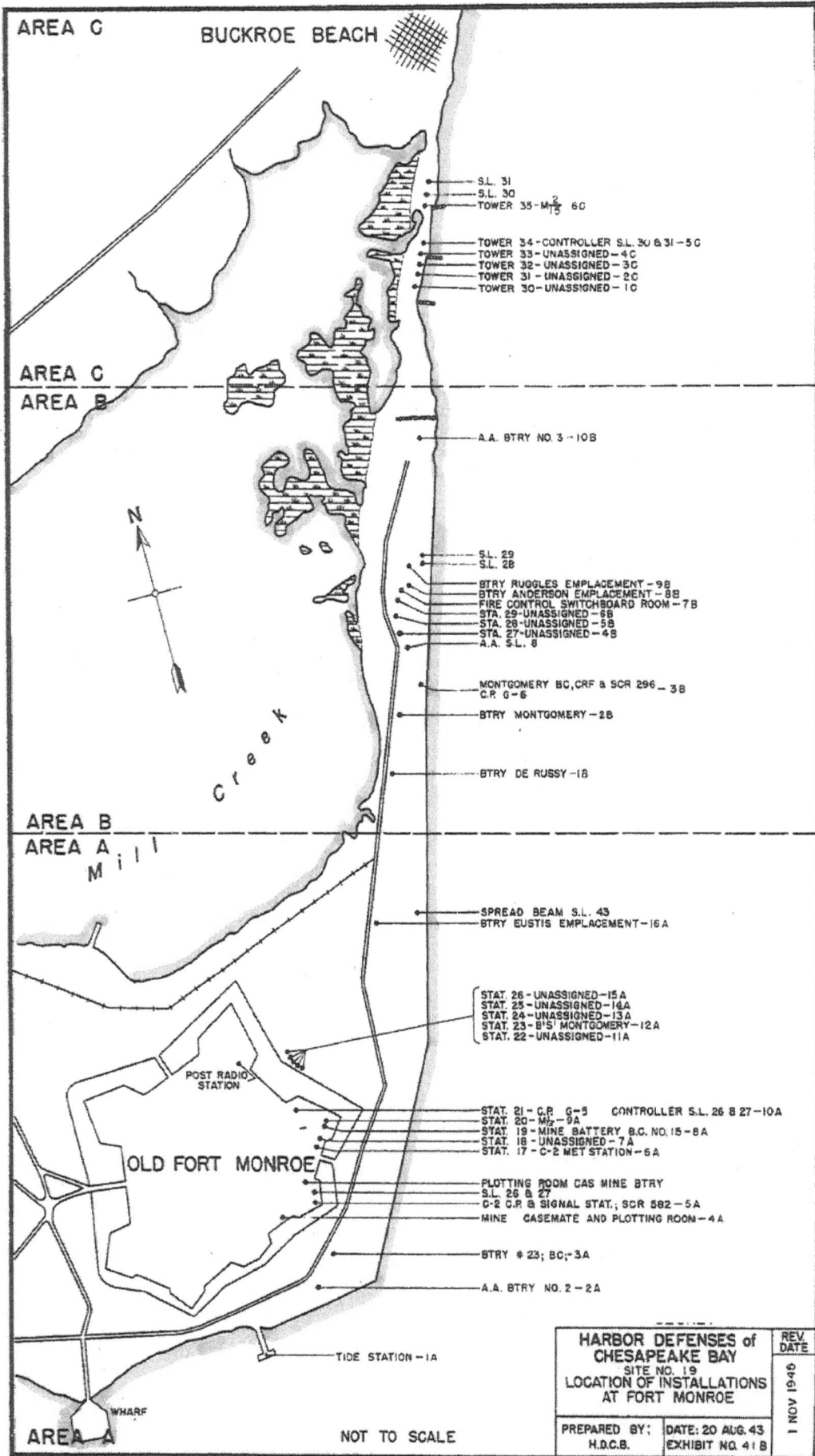

AREA C

BUCKROE BEACH

S.L. 31
S.L. 30
TOWER 35-M²/1.5 6C

TOWER 34-CONTROLLER S.L. 30 & 31-5C
TOWER 33-UNASSIGNED-4C
TOWER 32-UNASSIGNED-3C
TOWER 31-UNASSIGNED-2C
TOWER 30-UNASSIGNED-1C

AREA C
AREA B

A.A. BTRY NO. 3-10B

N

S.L. 29
S.L. 28
BTRY RUGGLES EMPLACEMENT-9B
BTRY ANDERSON EMPLACEMENT-8B
FIRE CONTROL SWITCHBOARD ROOM-7B
STA. 29-UNASSIGNED-6B
STA. 28-UNASSIGNED-5B
STA. 27-UNASSIGNED-4B
A.A. S.L. 8

MONTGOMERY BC,CRF & SCR 296_3B
C.P. G-6

BTRY MONTGOMERY-2B

C r e e k

BTRY DE RUSSY-1B

AREA B
AREA A

M i l l

SPREAD BEAM S.L. 43
BTRY EUSTIS EMPLACEMENT-16A

STAT. 26-UNASSIGNED-15A
STAT. 25-UNASSIGNED-14A
STAT. 24-UNASSIGNED-13A
STAT. 23-B¹S¹ MONTGOMERY-12A
STAT. 22-UNASSIGNED-11A

POST RADIO
STATION

STAT. 21-C.P. G-5   CONTROLLER S.L. 26 & 27-10A
STAT. 20-M¹/2-9A
STAT. 19-MINE BATTERY B.C. NO.15-8A
STAT. 18-UNASSIGNED-7A
STAT. 17-C-2 MET STATION-6A

OLD FORT MONROE

PLOTTING ROOM CAS MINE BTRY
S.L. 26 & 27
C-2 C.P. & SIGNAL STAT.; SCR 582-5A
MINE   CASEMATE AND PLOTTING ROOM-4A

BTRY # 23; BC;-3A

A.A. BTRY NO. 2-2A

TIDE STATION-1A

WHARF

AREA A

NOT TO SCALE

HARBOR DEFENSES of CHESAPEAKE BAY SITE NO. 19 LOCATION OF INSTALLATIONS AT FORT MONROE	REV. DATE
PREPARED BY: H.D.C.B.	DATE: 20 AUG. 43 EXHIBIT NO. 41 B

1 NOV 1946

Fort Monroe 1942 (NARA)

Fort Monroe 1935 (NARA)

Fort Monroe National Monument and Fort Monroe Authority (Terry McGovern)

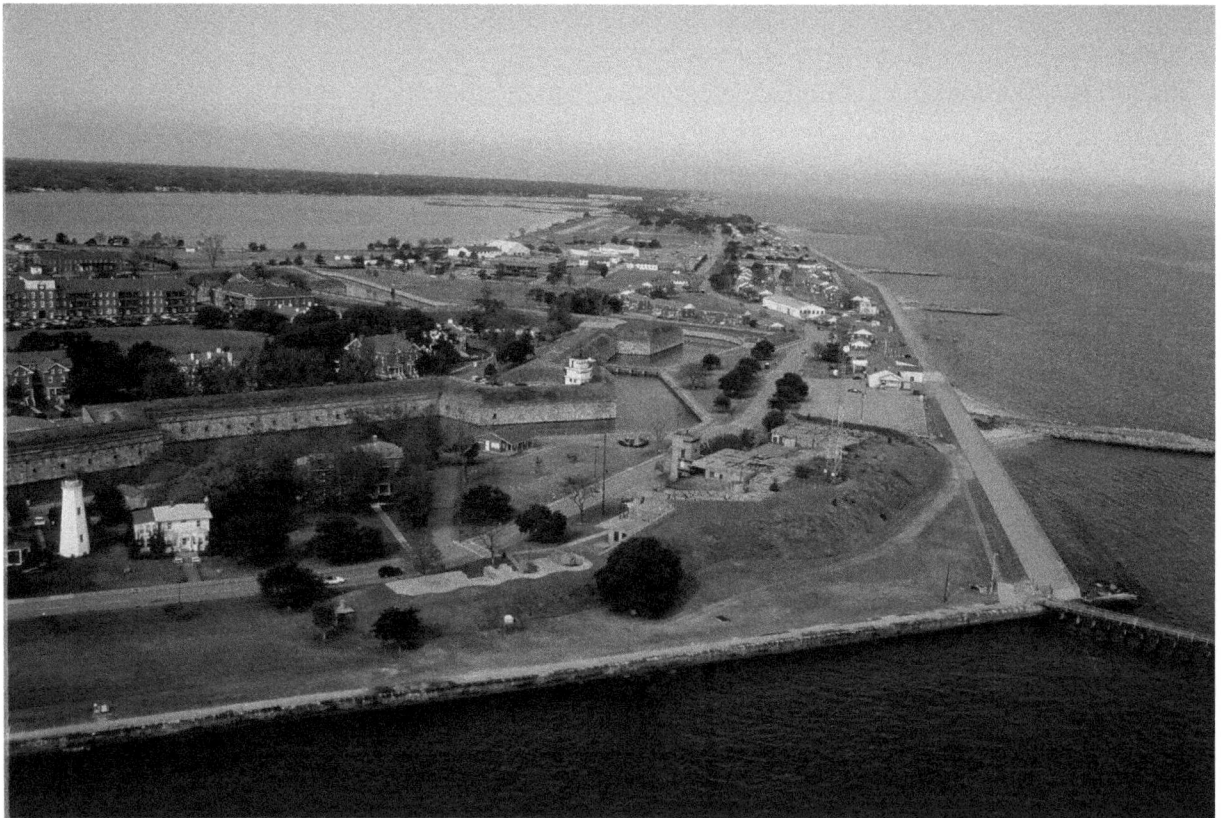

Battery Irwin and Battery Parrott, Fort Monroe (Terry McGovern)

north of the old fort (1.5-miles) that provision had to be made for a new road or even the railroad to carry materials to the construction site. After considerable debate on design, it was approved for an in-line, four-pit, 16-mortar battery with initial funding of $100,000 from the appropriation of March 1895. Costs were saved using a thinner, 5-ft. concrete thickness for protective walls. The heavy forest was cleared on a new site considerably to the north of the old fortress, not far from the shore. Actual construction began in April 1896, being completed except for sand embankment by July 1897. Transfer to troops was made on December 3, 1898 for a building cost of $194,680. In August 1896 the fort was informed that they would be sent a mix of four early Model 1891 carriages and twelve Model 1896 carriages. The original armament consisted of two mortars Model 1890 (#4 and #7) and fourteen Model 1890M1 12-inch Bethlehem mortars (#1, #2, #4, #5, #6, #10, #12, #17, #18, #20, #21, #24, #25 and #26) on four Model 1891 carriages (#77, #78, #84 and #85) and twelve Model 1896 carriages (#1, #2, #3, #4, #5, #6, #7, #8, #9, #10, #11, and #12). Thus, two types of carriages were used in the emplacement. The northernmost pit (Pit "D") was armed with the four older Model 1891 mortar carriages. They held their mortars more offset to the front than the later type, requiring circular cut-outs in the adjacent pit walls for clearance of the mortars when lowered horizontally. The other three pits had the later Model 1896 mortar carriage designed for the Model 1890 mortar, and no cut-outs were required. In the fall of 1901, the older type Model 1891 carriages were removed and sent to Watertown for storage and then replaced with newly made standard Model 1896 types (#303. #304, #305, and #306). It was named for Brigadier General Robert Anderson of Fort Sumter Civil War fame by General Orders No. 105 on October 9, 1902. In 1906 by General Orders No. 20 the two northern pits were tactically split off and named for Brigadier General George D. Ruggles of Civil War service. Early in service problems were encountered with water seepage into the magazines, such that temporary wooden magazines had to be built nearby. After several solutions failed, the problems were eventually solved and the ammunition moved back into the internal magazines. The battery was modified with the addition of a new battery commander's station and plotting rooms in 1914 for $6924. In 1920 two mortars and carriages per pit were removed, going to rearm Battery Butler-Capron at Fort Moultrie, Charleston South Carolina. The final eight mortars served until removed under authority of December 29, 1942. Though somewhat damaged, the emplacement still exists under the control of the National Park Service. The battery is closed to the public, except for one pit.

- **BOMFORD:** This battery for two 10-inch guns on disappearing carriages was the first modern, Endicott period work for Fort Monroe. Known as the Redoubt A battery during construction. it was located in the old infantry redoubt outside the northeast bastion of Fort Monroe across the moat. This redoubt had carried six 15-inch Rodman guns since the 1870s, but this old work was built over for the new construction. From the February 24, 1891 Fortification Act $158,848 was provided for this construction. A design plan, albeit with details and even gun sizes lacking pending availability of new carriage types, was submitted by engineer P.C. Haines on May 27, 1891. New concrete work was underway in February of 1892, but completion was delayed until new carriages were developed and their mounting specifics known. The design was for two adjacent, internal oriented platforms with adjacent. left-side magazines. Additional completion plans were submitted on July 20, 1895 to finish platforms for two 10-inch Model 1888 guns on Model 1894 disappearing carriages. More funds were even needed in September 1896 for temporarily filling in the moat around the redoubt to allow a pathway to tow the heavy guns to the battery. Armament was in fact not mounted until 1897. It then received two 10-inch Model 1888 guns (Bethlehem Steel #2 and #3 received on March 22, 1897) on disappearing carriages Model 1894M1 #12 and #13. The battery was transferred to service troops on January 19, 1897 at a cost of $154,379.99.

The battery was named on General Orders No. 78 of 25 May 1903, for Colonel George Bomford, former Chief of Ordnance. By 1913 it was modernized with new loading platform extensions, a battery commander's station, plotting room and electric ammunition hoists. The battery served for a considerably period of time, being designated a school training unit by a board in 1932. It was finally deleted and the armament removed shortly before the start of World War II in 1940. Most of the emplacement was destroyed by army engineers in March of 1951. A few traces of the concrete foundation are all that remains today.

- **HUMPHREYS**: A single emplacement for a 10-inch disappearing gun located in the re-entering place of arms (salient) opposite old masonry front no. 5. Funding originated in 1891, but work was suspended after construction of the magazine in 1892 pending availability of mounting details. A plan submission was made on July 20, 1895 for just over $5000 for the finalization of the pit and platform for a Model 1894 disappearing carriage. Transfer was made on March 19, 1898 for a cost of $59,964.50. It held a single 10-inch Model 1888 Bethlehem Steel gun (#1) on front-pintle disappearing carriage LF Model 1894 (#11). It was named for Lt. Colonel Charles Humphreys of Civil War service on General Orders No. 20 of January 25, 1906. The battery served only a short while, and was out of commission by 1910. The tube was eventually removed and used to rearm a carriage at nearby Battery Eustis. The emplacement itself was destroyed in the early 1960s; no trace remains of the battery today.

- **CHURCH**: A battery for two 10-inch guns, though it was started for just a single emplacement. Appropriations of March 3, 1897 called for an emplacement for a single gun to use the first, experimental prototype disappearing carriage (slightly modified and later produced as the Model 1894). Engineers submitted a plan on May 7, 1897 for a new location, known at the time as the Redoubt B battery, on the beach some 800-feet south of the new mortar battery. Considerable concern and discussion ensued about the suitability of the local foundation, as well as how close the battery would be to the shore. The intended carriage required certain detailed changes, like a deeper counterweight well, longer hold-down bolts, and modified fittings. Structural work was done that following winter. It was armed with 10-inch Model 1888 Bethlehem gun #31 on the experimental carriage (not assigned a serial number). A year later with National Defense Act, funding a second emplacement was authorized. Plans were submitted on March 23, 1898. This was on the right flank; the No. 1 emplacement. Finished in the fall of 1899 it carried 10-inch gun Model 1888 Bethlehem gun #25 on Model 1896 disappearing carriage #33. The battery was transferred to troops on January 3, 1901 for a cost of $90,473. It was named on General Orders No. 105 in 1902 for Albert E. Church, professor of mathematics of West Point. The experimental carriage never proved adequate for service and was changed out for a standard Model 1896 LF carriage moved from the Northeast Bastion emplacement about 1903. Modernization of the emplacement's loading platforms, hoists, and a battery commander station took place through 1912. The original guns were removed in 1917 for service on railroad carriages. The battery was rearmed in 1919 when it received Model 1895 guns for use on the existing carriages (Watervliet Model 1895M1 guns #32 and #33 on disappearing carriages Model 1896 #33 and #51 respectively). This armament was ordered removed under authority of the 1932 Board, but not actually accomplished until April 30, 1942. The emplacement was then used as a radar station for the rest of the war. The emplacement still exists under the control of the National Park Service. The battery is closed to the public.

- **EUSTIS**: Another dual 10-inch battery built for Fort Monroe. Eustis was authorized through the funding of the National Defense Act of 1898. Money was available for a battery of three 10-inch guns, but it was decided to instead build the second emplacement of Battery Church and a new two-gun battery near the shore. Initial plans were submitted by engineer Captain Thomas Casey on March 23, 1898. These plans closely followed approved Mimeograph No. 22. The location selected was a site located northeast of Redoubt A along the shoreline, part way between Battery Church and old Fort Monroe. Work was begun in March of 1898, and it was complete except for mounting of guns and some electrical installations by the end of 1899. The battery was transferred on January 3, 1901 for a cost of $80,766.67. It was armed with 10-inch Model 1888MII Bethlehem guns on Model LF 1896 disappearing carriages (gun #20 on carriage #49 and gun #21 on carriage #63). It was referred to as Redoubt C while being constructed until named in General Orders No. 105 of October 1902 for Colonel Abraham Eustis, first commandant of the Artillery School. The battery was modernized in 1908 with platforms extended, a BC station added, and hoists installed. Mechanical range transmission devices were experimentally tried at the battery in 1909. The original armament was removed in 1917 but replaced in 1919 with Model 1888 Bethlehem gun #1 (from Battery Humphreys) and gun #24 from the Northeast Bastion. These served until removed on April 30, 1942. The battery had served as a school instructional unit during its last ten years. The emplacement was destroyed in March 1959 to make room for new post housing.

- **Northeast Bastion**: An emplacement for a single 10-inch disappearing gun also approved with funding from the Act of July 7, 1898 ($38,000 initially). Plans were submitted even before funding, on July 2, 1898, for an estimated total cost of $79,425. It was built into the form of the fifth bastion, with the parapet crest level at the same height of the original masonry work. Building it required removal of one 15-inch and two 10-inch Rodman guns already on the bastion parapet. Because of its elevation, concrete and building materials had to be carried with wheel barrels to the site from the railway siding adjacent. Eventually it received and mounted 10-inch Model 1888M1 Bethlehem Steel gun #23 on disappearing carriage Model 1896 #51. The restricted emplacement was apparently never considered acceptable. The battery was not physically transferred, nor ever given an official name. By 1902 it was slated for deletion, the armament to be transferred where it could be more useful. By 1903 the armament was dismounted, the carriage being allocated to replace the experimental carriage at Battery Church at Fort Monroe, the tube to be held temporarily as a spare unit. The site was used as a foundation for several subsequent fire control structures—modified as such it still exists under the control of the Fort Monroe Authority. The battery site is open, but the interior is closed to the public.

- **Parapet Battery**: Over time a variety of guns had been emplaced on Fort Monroe's southern and eastern front parapets. In 1896 the south bastion had one 15-inch and three 10-inch Rodmans along with five 8-inch converted rifles. The southeast bastion had one 15-inch, and nine 8-inch converted rifles. The east bastion had a single 15-inch and four 10-inch Rodmans. Funding from the National Defense Act of March 1898 allowed this to be augmented with a few new 8-inch breechloading guns on modern Model 1892 barbette carriages. On April 16, 1898 the local engineers were instructed to prepare platforms for three 8-inch Model 1888 guns on barbette carriages along with a $6000 allocation of funds. Eventually four of these mounts were emplaced that summer and in service by the end of 1898. One went to the southern bastion, two to curtain wall No. 3, and the final one at the point of the southeast bastion. The Model 1888 gun tubes used were Watervliet #6 (having been received at Fort Monroe in 1894), West Point Foundry #8 and #9, and Bethlehem Steel #24. Model 1892 barbette carriages were Watertown Arsenal #1, #2, #4, and #8. The emplacements were simple concrete gun blocks containing the required hold-down bolts. They were situated on the ramparts, using only the original fort parapet wall for cover, with no dedicated magazines. The battery served only a short while, two guns and carriages were dismounted and sent off for use in San Francisco by June 30, 1899. The last two mounts were authorized for removal in 1912. These two carriages were returned to Watertown. One of the carriages was in fact damaged while trying to move it in a vertical position through the main gate of old Fort Monroe. While never formally named, the original battery was transferred on November 1, 1898 for an engineering cost of $1611.22. After removal the mounting blocks were sodded over, the only trace today being the open, missing spots of original 1830s gun blocks on the parapet.

- **Water Battery**: The old, casemated water battery outside Fort Monroe continued to hold armament up to the turn of the century. In 1896 it held three 15-inch and thirty-eight 10-inch Rodman smoothbore cannons. Since 1895 the extension to the battery across the moat to the north was also used. It carried four 8-inch converted rifles and nearby a single modern 12-inch mortar for use as a practice mount for the Coast Artillery School located at Fort Monroe. This model 1890 mortar and model 1891 carriage was removed by 1898 and remounted completing mortar Battery Anderson. Also a modern Model 1888 8-inch gun and Model 1892 barbette carriage on an open emplacement was at the northern end of the water battery which was also used by the school for target practice firings. The battery was armed with 8-inch Model 1888 West Point Foundry gun #11 on Model 1892 barbette #7. This mount was also removed in 1898 and transferred to Battery Barber. The areas for these school batteries were extensively reworked with the advent of other Endicott batteries, no trace of their emplacement blocks remain today.

- **BARBER**: On February 18, 1898 the engineers at Fort Monroe suggested that the 8-inch modern Model 1888 gun on Model 1892 barbette carriage that had been used for practice firing near the old Water Battery be relocated to a better position at the southern end of the cover face in front of the new mine casemate on the north side of the fort. Here it could still be used for training practice but also in case of hostilities. The National Defense Act of 1898 appropriated money for a new emplacement with a single 8-inch gun on barbette carriage at Fort Monroe. A position was built near the mining casemate not far from Battery Bomford from April to June of 1898. It consisted of a single concrete platform at a base ring height of 16.5-feet above mean low water. No magazine was needed—the ammunition could be stored at nearby Battery Bomford and moved by hand and trucks to the emplacement. It was transferred on November 1, 1898 for a construction cost of $1663.39. The battery was armed with 8-inch Model 1888 West Point Foundry gun #11 on Model 1892 barbette #7. The battery was named for Brigadier General Thomas H. Barber, U.S. Volunteers, in General Orders No. 20 of January 25, 1906. The unit did not serve long, and the battery was disarmed by 1915. The emplacement was destroyed in March 1951, no traces of it remain today.

- **MONTGOMERY**: The last Endicott period battery for Fort Monroe, Montgomery was designed for two 6-inch pedestal type guns. It was located between batteries Church and DeRussy. Plans were submitted on July 22, 1901 by engineer Major James Quinn. Work was done using funds appropriated on November 9, 1900 in the latter half of 1901 and early 1902. The armament was mounted in May of 1905. Transfer to service troops came on June 30, 1904 for a cost of $24,000. The battery was the first of the Model 1900 6-inch emplacement type. It was named on General Orders No. 78 of May 25, 1903 for Major Lemuel Montgomery, who was killed in action against the Creek Indians at Horseshoe Bend, AL in 1814. The original armament was two 6-inch Model 1900 guns on Model 1900 pedestal mounts (#15/#33 and #42/#34). Both the guns and carriages were removed on February 6, 1917 and emplaced in a temporary battery at Fort Story. In July 1919 Montgomery was rearmed with two guns transferred from the Delaware Capes (6-inch Model 1900 on Model 1900 pedestals #39/#39 and #41/#38). These then served during the interwar years, until subsequently removed in September 1941. Then in an interesting coincidence, the battery was again rearmed, this time with its two original guns that were returned here. This final armament served until removed when Montgomery went out of service at the very end of the Coast Artillery in May of 1948. The emplacement was itself destroyed in the early 1950s for a new housing construction program.

- **GATEWOOD**: A Spanish-American War emergency battery for four 4.7-inch rapid-fire guns. With National Defense Act funds, two-gun emplacements were authorized on April 16, 1898, and just a week later on April 13th an additional two were allocated. While originally it was thought to place such guns in the old bastion points, eventually all four were placed linearly in a single battery on the curtain wall parapet of front No. 4. It consisted of simple gun blocks with concrete containing the hold-down bolts and a single, small shared concrete magazine between guns No. 2 and 3. Transfer had been made on September 24, 1898 for a construction cost of $2955.36. The armament consisted of four Armstrong (British-built) 4.72-inch guns and pedestals. These were from a batch of 34 such guns purchased as an emergency measure at the start of the war and used to arm RF batteries in multiple ports. Actual serial numbers used (gun/pedestal) were: #9837/#10847, #9856/#11008, #9850/#11020 and #9853/#10848. The guns were all of the 50-caliber type. The battery was named on General Orders No. 105 on October 9, 1902 for 1st Lieutenant Charles Gatewood of Indian War service. The battery did not serve long, the guns being sent to August Arsenal by 1914. Eventually several of the guns and mounts were emplaced in Puerto Rico at the military post at El Moro in San Juan harbor. Only part of the magazine is left of the battery on the parapet of old Fort Monroe.

- **IRWIN**: A light battery for four 3-inch, 15-pounder guns emplaced in a site outside the southeast bastion of the old fort near the beach, and subsequently Battery Parrott. Plans were submitted on December 1, 1898. A considerable debate took place within the engineers about the location of this battery, some arguing strongly to put it up on the Fort Monroe curtain wall. Almost two years elapsed before approval for the beach site, still allowing the road along the shore and avoiding the lighthouse property, was finalized. It was built from early spring 1900 to 1901. To provide a wide coverage, the usual traverse mounds over the magazines were eliminated in favor a ¼-inch boiler plate coverage in addition to the usual concrete. It was transferred on May 28, 1903 for a cost of $12,500. Naming came on General Orders No. 78 of May 25, 1903 for 1st Lieutenant Douglas Irwin who was killed in action at Monterrey Mexico in 1846. It was armed with four 3-inch Model 1898 guns on M1898 masking parapet mounts. The assigned guns with pillars were Driggs-Seabury #85/#85, #86/#86, #92/#92, and #93/#93. Due to the shifting of the mine field, a proposal was made in 1906 to enhance the battery's coverage by removing guns No. 2 and 3 to new positions beyond each flank. Approval for accomplishing this was denied due to lack of funds. The armament was removed after World War I, in 1920. One pit was subsequently modified to take a 3-inch Model 1917 anti-aircraft piece (Watervliet #57) as part of a three-gun battery of that type of gun. Then after World War II, two pedestal 3-inch guns Model 1902 were re-located to Irwin from Fort Wool on May 31, 1946 to serve as saluting guns. These guns were Bethlehem M1902 guns #6 and #7 on M1902MI mounts. Long since abandoned, the two saluting guns and their late war box-type gun shield remain emplaced under the control of the Fort Monroe Authority. The battery is open to the public.

- Two 6-pound (2.24-inch bore) coast defense guns were emplaced on the Fort Monroe parapet in 1901. This gun (there were three different models, but with generally the same characteristics) was a light, wheeled field piece intended to give boast defense posts the capability of mobile light fire to shift as needed to cover minefields, obstacles, landing beaches and anti-personnel support. Often a section of two guns was assigned, but not permanently emplaced, to most major posts. However, at Fort Monroe an experimental emplacement of two concrete platforms and command station was built by Captain Edmund Zalinski as a non-official expedient in 1901. These were located over old rampart gun platforms Nos. 33 and 34. Transfer was made on May 31, 1902 for just $277.28. It was assigned Model 1900 guns and carriage (#24/#2 and #31/#9). Before long it fell into disuse and the platforms were removed, no trace of them remains today.

- *Battery #124* (planned): A 1940 Program projected battery for two 16-inch casemated barbette guns at Fort Monroe. It was to have been one of five such batteries for the defenses of Chesapeake Bay; but was never actually funded or built. It was to be emplaced at the far northern end of the Fort Monroe reservation. The project was assigned a very low priority of just #35 in the early schedule (September 11, 1940) for such batteries. No work was ever undertaken beyond site surveys, and the project was officially cancelled on November 23, 1943.

- **AMTB Tac-23**: A 1943 Program anti-torpedo boat battery for Fort Monroe. It consisted of two 90mm mobile and two fixed guns of the same size. Work on the emplacements for the fixed guns was done in the summer of 1943. It transferred on November 1, 1943 for a cost of $17,289.08. Two fixed platforms were emplaced in the gun pits of former Battery Parrott, which had just recently been disarmed of its 12-inch guns. The battery also got a new battery commander station but used the magazines of the older work. The two 90mm guns on M3 fixed, shielded carriages, served until removed in December of 1950. A similar gun and carriage were moved here from a battery at Fisherman Island for display purposes in 1976 by the U.S. Army. With this gun in place, the battery still exists under the control of the Fort Monroe Authority. The battery is open to the public.

Fort Wool (1819-1946) is located on an artificial island known as the "Rip Raps" on the south side of the channel from Chesapeake Bay into Hampton Roads. Fort Wool was constructed to provide crossfire with Fort Monroe on the north side of the channel. It was designed as tower battery with three tiers of casemates to mount a total of 216 guns. This Third System fort was never finished due to the difficult construction problems presented by the fort's foundation. After originally being named Fort Calhoun, it was officially renamed in 1862 for Maj. Gen. John E. Wool, U.S. Army. While construction started in 1819, in 1830 work on the fort stopped and it was not until 1858 did work start again. The start of the Civil War found the most of first tier complete and a few cannons mounted. The fort did take part in the Battle of Sewell's Point and naval engagement between the ironclads in 1862. The fort was rearmed during the Endicott Program with five small caliber concrete batteries constructed into the walls of the Third System fort, providing protection to the mine fields controlled by Fort Monroe. During World War II, one of the 6-inch disappearing gun batteries was converted into a uniquely designed #200 Series 6-inch shielded battery with its battery commander tower still standing (the only one in the USA). While its coast artillery role ended in 1946, it was not until 1967 that it was turned over to the Commonwealth of Virginia. Starting in 1985, the City of Hampton leased the island as a historic site with tour boat landing visitors for daily guided tours. In 2020, Fort Wool was turned into a temporary seasonal seabird nesting habitat by the Commonwealth of Virginia with no public access due to the expansion of Hampton Road Bridge Tunnel. It is hoped that once the tunnel construction is completed the island fort will be returned to being a historic site open to the public.

FORT WOOL

HAMPTON ROADS, VA.

SERIAL NUMBER

124

LEGEND
7. BARRACKS.
10. MESS HALL.
11. COAL SHED.
12. BATH HOUSE.

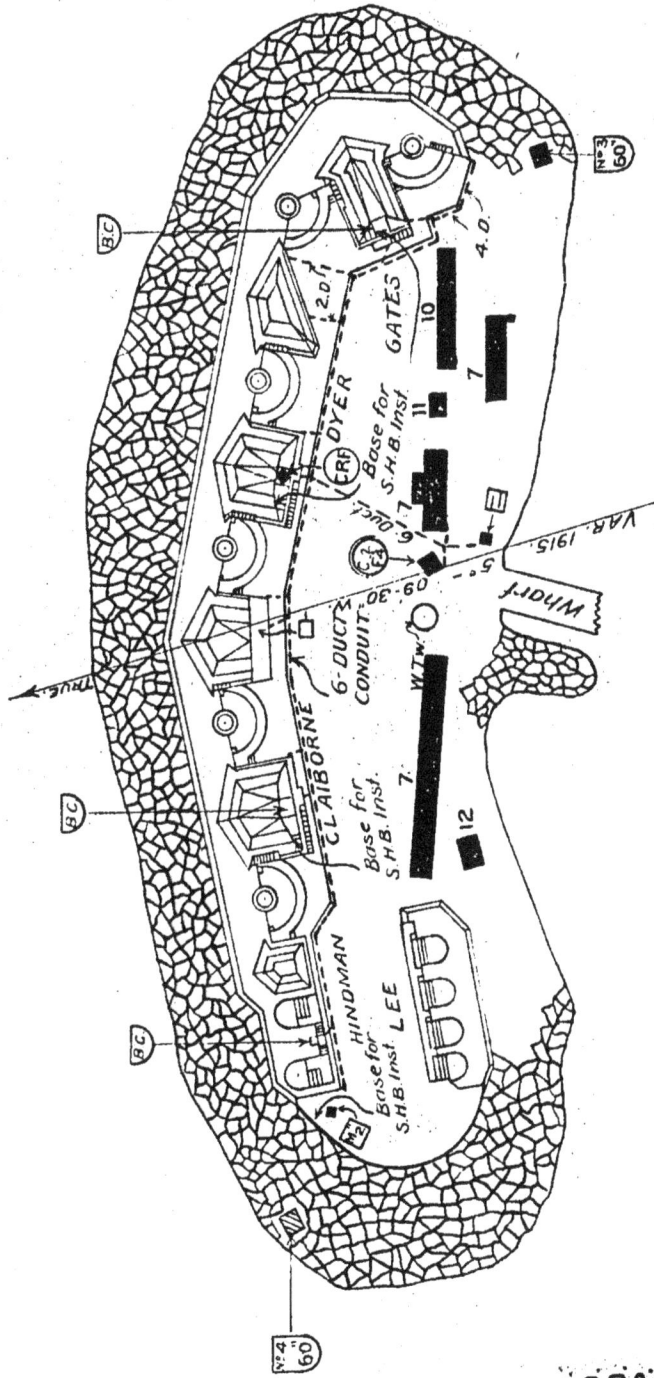

EDITION OF JUNE 9, 1916.
REVISIONS: APR. 1, 1921.

CLAIBORNE
Base for S.H.B. Inst.

DYER
E.R.F.
Base for S.H.B. Inst.

GATES

HINDMAN
Base for S.H.B. Inst.

LEE

6-DUCT CONDUIT.

Wharf

VAR. 1915.

THE PLANE OF REFERENCE IS MEAN LOW WATER

500 ft.

BATTERIES.
* CLAIBORNE .... 2-6" D.
* DYER .... 2-6" D.
  GATES .... 2-6" D.
  LEE .... 4-3" P.
  HINDMAN .... 2-3" P.
* Guns Dismounted.

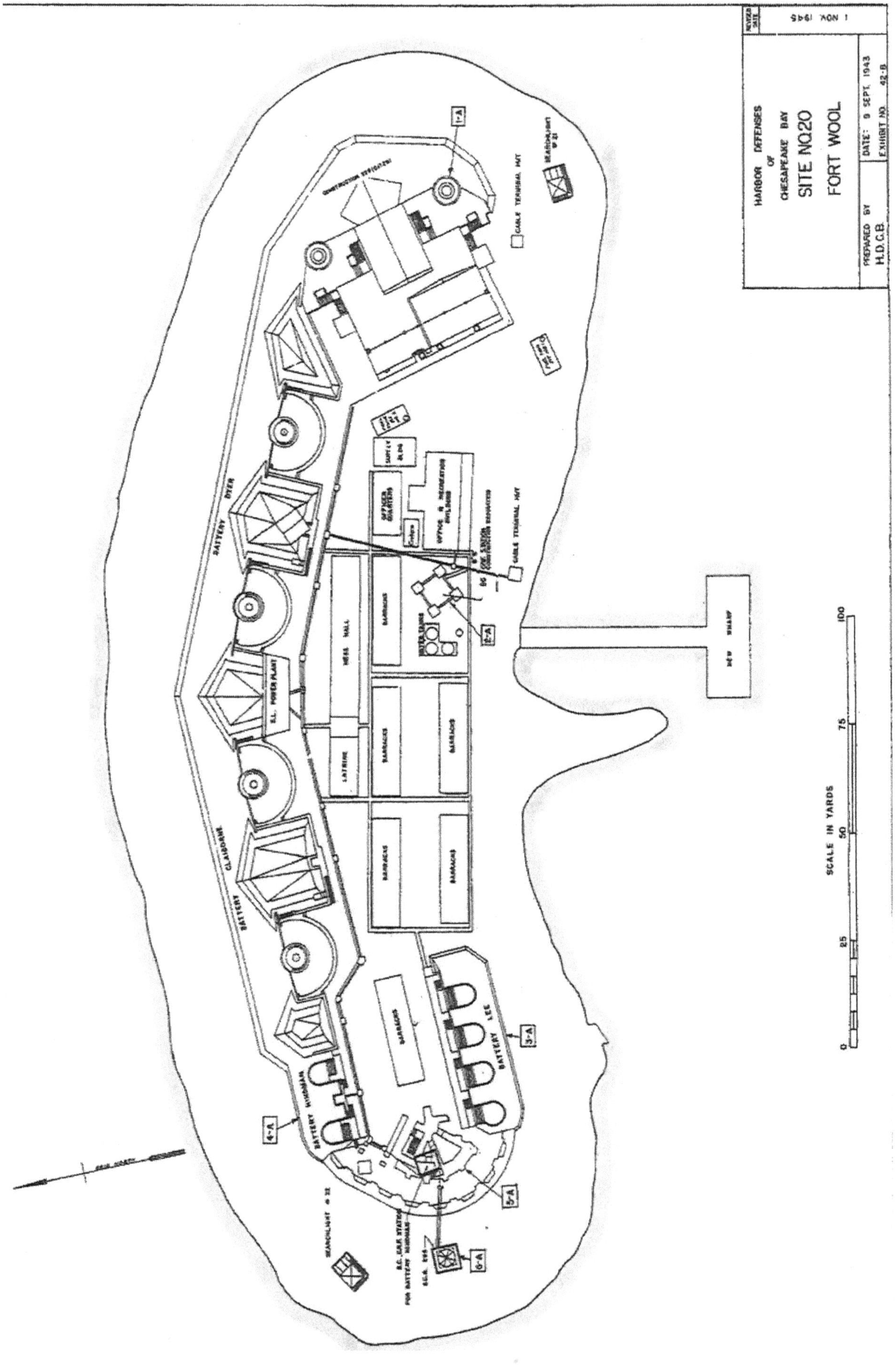
HARBOR DEFENSES
OF
CHESAPEAKE BAY
SITE NO20
FORT WOOL
PREPARED BY
H.D.C.B.
DATE: 9 SEPT. 1943
EXHIBIT NO. 42-B
1 NOV. 1945
SCALE IN YARDS
0
25
50
75
100
BATTERY DYER
BATTERY CLAIBORNE
BATTERY HINDMAN
BATTERY LEE
S.L. POWER PLANT
MESS HALL
LATRINE
BARRACKS
NEW WHARF
1-A
2-A
3-A
4-A
5-A
6-A

Fort Wool 1920s (NARA)

Fort Wool 1920 (NARA)

Fort Wool (Terry McGovern)

Fort Wool (Terry McGovern)

Fort Wool Gun Batteries

- **LEE:** The first modern battery for Fort Wool was this emplacement for four 3-inch guns on pedestals mounted to fire to the southwest. Original submission by engineer James Quinn called for four 15-pounder guns on masking parapet mounts to defend the channel between Fort Wool and Willoughby Spit. It was thought that as the granite fort had now been here for over thirty years without subsidence, the foundation was firm. Magazines were to be built with floors ten feet over mean low water as protection for occasional storm surges. After a change in plans to substitute pedestal guns for masking parapet types, concrete work was conducted in 1902. It was transferred on November 15, 1905 for a cost of $40,000. It was armed originally with four 3-inch Model 1902M1 Bethlehem Steel guns on Model 1902 pedestal mounts (#8/#8, #9/#9, #10/#10 and #11/#11). The battery was named on General Orders No. 194 on December 27, 1904 for Major General Henry "Light Horse" Lee of Revolutionary War service. This battery served well into World War II. Two guns and pedestals were transferred to a temporary site on Fisherman Island on May 28, 1942 (#8 and #9). Records are unclear, but these two guns may have been subsequently replaced, as in 1944 four guns are still carried at Battery Lee. The remaining armament was removed in 1946-47. The emplacement still exists. The battery is open, but the island is closed to the public.

- **HINDMAN:** A battery for two 3-inch pedestal guns mounted on the opposite side of the island from Lee, on the old fort parapet firing to the north. Plans were submitted by Captain James Quinn on November 20, 1902. Planned from the start for 3-inch pedestal guns, it closely followed the standard plans included in Mimeograph No. 30. In common with the other batteries at Fort Wool, the parapet of the battery was entirely of concrete, it being felt that it would be impossible to maintain sand or sod in such a location. It was built from January 1903 to late 1904, being transferred on November 15, 1905 for a construction cost of $13,444. It was armed with two 3-inch Model 1902M1 guns on Model 1902 pedestals ($6/#6 and #7/#7). The battery was named on General Orders No. 194 on December 27, 1904 for Brevet Colonel Jacob Hindman of the War of 1812 service. This armament was active well into the start of World War II, though there may have been different guns moved here about 1943-44 (records are unclear). A battery commanders station with a CRF on the roof was built on the old casemates in 1943. In any event the final armament was removed about 1946. The final two guns and pedestals were moved to Fort Monroe and re-emplaced at Battery Irwin as saluting pieces. The emplacement still exists. The battery is open, but the island is closed to the public.

- **CLAIBORNE:** A battery for two 6-inch disappearing guns. On July 10, 1901 a submission was made for an initial battery of two 6-inch RF guns on pedestal mounts for old Fort Wool. Type plans were found inappropriate to this site, rather a modification of a type being built at Fort Standish, Boston was consulted. Due to the constrained dimensions of available space at Fort Wool, and the impending adoption of the new Model 1903 disappearing carriage, plans were reworked over the next couple of years. Finally on June 19, 1903 Captain E.E. Winslow submitted new plans for six disappearing guns in three two-gun batteries for the site. Gun centers were set at 127-feet, the rearward rooms were shifted into the more numerous traverses between each gun. In November 1903 the sum of $165,000 was allocated for the construction of the six emplacements, they were organized into three two-gun batteries—Claiborne, Dyer, and Gates. Work was done from December 1903 until late 1904. Transfer was made on August 18, 1908 for a construction cost of $55,000. Claiborne occupied the south end of the sequence, between batteries Dyer and Hindman. It was armed with two Model 1903 6-inch guns on Model 1903 disappearing carriages (#23/#47 and

#24/#46). Claiborne was named on General Orders No. 194 on December 27, 1904 for Brigader General Ferdinand L. Claiborne, a War of 1812 American Army officer. This armament was removed about 1917, though eventually the gun tubes were used to replace those removed from Battery Gates in 1919. The emplacement still exists. The battery is open, but the island is closed to the public.

- **DYER**: The second pair of 6-inch disappearing guns in the sequence of six such guns at Fort Wool. It was positioned between Claiborne and Gates. Transfer was made on August 18, 1908 for an allocated cost of $55,000. It was armed with two 6-inch Model 1903 guns on Model 1903 disappearing carriages (#18/#49 and #22/#48). It was also named on General Orders No. 194 on December 27, 1904, for ordnance officer Brevet Major General Alexander B. Dyer. The battery itself was disarmed in 1917, and was not rearmed. The emplacement still exists. The battery is open, but the island is closed to the public.

- **GATES**: The final pair of 6-inch disappearing guns at the eastern end of Fort Wool. It was built with the other 6-inch batteries in 1904-05 and transferred with Claiborne and Dyer on August 18, 1908 for a cost of $55,000. Itws armed with two Model 1903 6-inch guns on Model 1903 disappearing carriages (#3/#50 and #1/#51). It was named on General Orders No. 194 of December 27, 1904 for Revolutionary War officer Major General Horatio Gates. Because of its advantageous position on the eastern end of the island, it was retained longer than its sisters. After its tubes were removed in 1917 for field mounts, they were soon replaced with those originally taken from Battery Claiborne in 1919. Thus, it would up with guns/carriages #23/#50 and #24/#51. The position was selected for a new 1940 Program 6-inch battery. Consequently, it was disarmed and the original emplacement was extensively modified in 1942-43.

- **Battery #229**: A 1940 Program dual 6-inch barbette battery. It was approved in September 1940 and built with FY 1943 appropriations. It was designed as one of the "inner" defense batteries for the Chesapeake, to cover minefields and nets of Hampton Roads. Because of the site size restriction of the island, it was constructed to a modified plan, unique in comparison to other 200-series 1940 batteries. The magazines were at ground level, the two-gun platforms were carried on the upper, exposed parapet. Work was done from March 31, 1943 to January 31, 1944. It involved extensively modifiying much of the old Endicott Battery Gates. Transfer was made on April 17, 1944 for a cost of $262,330.45. The name "Gates" appears to have informally adopted by the new 6-inch emplacement, but it was not an official designation. The battery was never completed and the designated armament of two 6-inch M1 guns on Model M4 carriages (carriages #45 and #46 assigned) was not mounted. It was left suspended after the war until finally abandoned. The emplacement still exists. The battery is open, but the island is closed to the public.

Fort John Curtis (1940-1946) is located at Cape Charles, the southern end of the Delmarva peninsula, near the town of Kiptopeke, Virginia, off State Highway 13. Fort John Curtis and its nearby Fisherman Island military reservation were primarily developed during World War II to defend the entrance to Chesapeake Bay, though a temporary World War I 5-inch battery was located on Fisherman Island. The 1940 Program resulted in the construction of a #100 Series 16-inch casemated battery and two #200 Series 6-inch shielded batteries. Supporting these main batteries were mine casemate, two railway batteries with 8-inch guns, a 155mm GPF battery, several large control towers, and two ATMB batteries mounting 90mm and 3-inch guns. Originally named Fort Winslow in 1941, it was renamed in General Orders 9 of 1942 for Capt. John Parke Custis, U.S. Army, George Washington's stepson and aide-de-camp who died of illness in 1781. After World War II, Fort John Curtis was used as a U.S. Air Force radar base, while the U.S. Navy used Fisherman Island. The Cape Charles Air Force Station and Fisherman Island reservation were declared surplus in

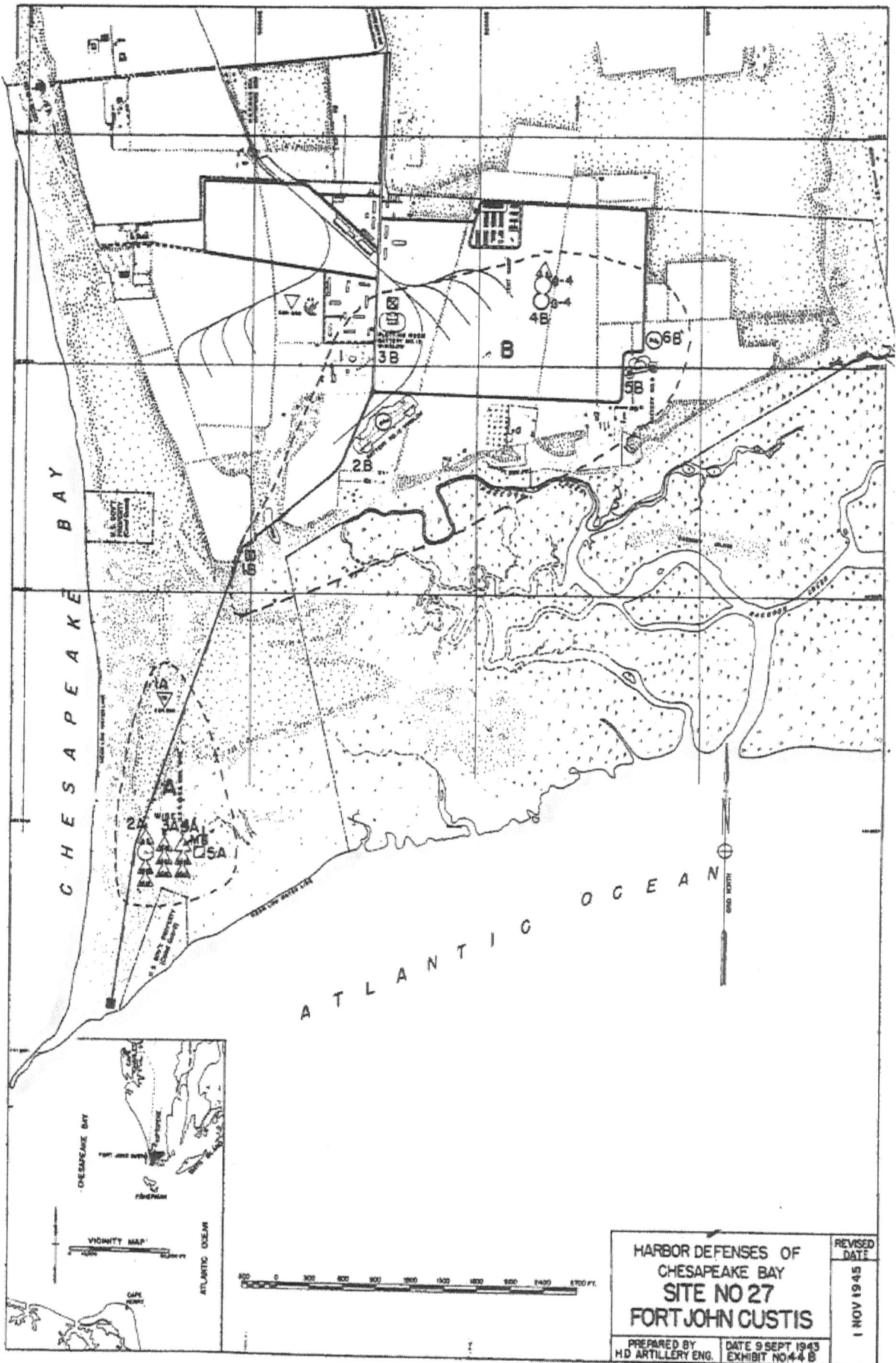

CHESAPEAKE BAY

ATLANTIC OCEAN

B

4B

3B

2B

6B

5B

7A

A

2A

3A 4A

4B

5A

VICINITY MAP

CHESAPEAKE BAY

FORT JOHN CUSTIS

FISHERMAN

ATLANTIC OCEAN

CAPE HENRY

HARBOR DEFENSES OF
CHESAPEAKE BAY
SITE NO 27
FORT JOHN CUSTIS

REVISED DATE
1 NOV 1945

PREPARED BY H.D. ARTILLERY ENG.	DATE 9 SEPT 1943 EXHIBIT NO.44 B

Battery Winslow, Fort John Custis, Eastern Shore of Virginia National Wildlife Refuge (Terry McGovern)

Battery #228 Fisherman Island, Eastern Shore of Virginia National Wildlife Refuge (Terry McGovern)

1981 and most of the reservation became part of the Eastern Shore of Virginia National Wildlife Refuge. Of special note was the batteries on Fisherman Island which retain their armament until 1966 when the two 6-inch guns on shielded barbette carriages were removed (now located at Fort Pickens, Florida) and the two 90mm guns on fixed mounts (now located at Fort Monroe and Fort MacArthur) were also removed. Battery Winslow received a navy 16-inch MkVII barrel installed in one of the open casemates in 2012, while the interior of the battery is closed to the public. Battery #227 on Fisherman Island still retains its power room equipment. Today, Fort John Curtis and Fisherman Island remain part of the wildlife refuge and open for seasonal day use, though most structures have been demolished or buried.

Fort Custis Gun Batteries

- **WINSLOW**: A battery for two 16-inch barbette guns erected on the Cape Charles reservation of Fort Custis. This site had long been planned for heavy guns to secure the entrance to Chesapeake Bay in conjunction with similar heavy batteries at Cape Henry. The project was one of the first five new dual casemated batteries approved on April 20, 1934 prior to the 1940 Program. The design was consistent with these other early plans, being of the recessed magazine of the "straight" design for the main internal gallery. Appropriations were delayed until FY-1941 ($30,000 beginning funds that year, followed with the balance in FY-1942). Actual work was done from August 18, 1941 until January 1, 1943 for transfer on November 12, 1943 at a building cost of $1,554,866. When the 1940 Program was formalized this work was given Battery Construction Number 122, with a priority of #7 on September 11, 1940. A plotting PSR room was to the north and eight base-end stations were also built serving the battery. It was named on General Orders No. 46 of September 17, 1942 for Coast Artillery engineer Brigadier General E. Eveleth Winslow. It was armed upon completion with two 16-inch MkIIM1 guns on Model 1919M4 barbette carriages (#108/#25 and #69/#26). The battery served out the war, being finally disarmed in 1949 and shortly thereafter abandoned. The emplacement still exists on the Eastern Shore of Virginia National Wildlife Refuge (the closed Cape Charles Air Station). Today Battery Winslow has a 16-inch navy MkVII gun barrel on display in one of the casemates. The battery site is open, but the interior is closed to the public.

- *Battery #123* (planned): A 1940 Program dual 16-inch battery planned for Fort Custis. It was to be emplaced to the east-northeast of Battery Winslow and had a field of fire to the east. As it was the last of the four heavy batteries planned for the entrance to the bay, it was assigned a relatively low priority for construction of #26 on the September 1940 priority list and then it was #27 in the list of August 1941. No work was ever accomplished on this project, it being entirely deleted from plans on authority of December 13, 1942.

- **Battery #228**: A standard 1940 Program dual 6-inch battery built-up to the east of Battery Winslow. Of low priority, it was not funded until the FY-1943 budget. Construction work was done from September 1, 1942 until November 30, 1943. Transfer was made on November 12, 1943 for a building cost of $230,937.68. While the emplacement was basically finished by early 1944, the designed armament of two 6-inch M1(T2) guns on M4 barbette carriages was never mounted. However, it does appear that the M4 carriages (#19 and #20) may have been physically delivered to the site. Work was indefinitely suspended after the war, and finally the emplacement was abandoned. It was never named. The emplacement still exists on the property of the National Wildlife Refuge; however, all entrances are covered with earth and debris.

Fisherman Island Military Reservation (1917-1919, 1942-1949/1969) is located on the southernmost island on the Delmarva Peninsula chain of barrier islands. Originally here was a U.S. Marine Hospital / Public Health Service Quarantine Station (1890-1919). Two two-gun 5-inch gun batteries were built 1917-1919. The U.S. Army post was then called Fisherman's Island Camp. The military reservation was officially established in 1919 in a land swap with the U.S. Treasury Department (for Craney Island, Portsmouth). The reservation later became a sub-unit of Fort John Custis from 1942-1949. New batteries built were Battery 227, Anti Motor Torpedo Boat Battery 20 (aka New Battery Lee), and Anti Motor Torpedo Boat Battery 24. A mine casemate was also here, now buried. The U.S. Navy took control of the island from 1949-1969. The remaining abandoned guns were finally removed in 1966, and the steel-frame BC stations and searchlight towers were knocked down in place in 1986. The two six-inch guns from Battery #227 were sent to Fort Pickens, FL, one 90mm AMTB gun went to Battery Parrott at Fort Monroe, and one 90mm AMTB gun went to Fort Moultrie, SC. The site became part of a National Wildlife Refuge in 1969. Public access to the island, including beach access, is restricted.

Fisherman Island M.R. Gun Batteries

- Two batteries, each for two 5-inch pedestal mounted guns were erected on Fisherman Island in 1917 as part of World War I emergency gun relocations. Moved to this site was the armament previously held at Battery Ritchie at Fort DuPont and Battery Fraser at Fort Slocum. The guns were Model 1900 5-inch guns emplaced on Model 1903 pedestal mounts (Watervliet Arsenal #8, #9, #10, and #11 on pedestals #11, #12, #18, and #19). The exact location on the island reservation is currently unknown. They were probably removed in 1919-1920. No remains appears to be left.

- **Battery #227**: A standard 1940 Program dual 6-inch battery built on Fisherman's Island; it was given the relatively high priority of #11 in September 1940, and funded with the FY-1942 Budget. It was of standard 200-series design. Building took place between April 15, 1942 and October 31, 1943. Transfer was made on November 12, 1943 for a cost of $271,352. It was armed with two 6-inch guns Model 1905A2 on Model M1 carriages (#16/#58 and #21/#59). The battery was never named. The armament was initially abandoned in place when the reservation was eliminated early postwar. Then in July 1966 the armament was removed for historic preservation. Eventually the guns and barbette carriages were restored and displayed by the National Park Service at Fort Pickens in 1976. The battery retains is three motor generator and switchboard in abandoned condition. The emplacement still exists at the Fisherman Island National Wildlife Refuge. The battery is closed to the public and access to the island is restricted.

- **AMTB Tac-20**: A temporary emplacement for two 3-inch guns built on the island to provide minefield coverage in 1942. The site was built from May 25, 1942 to August 31, 1942 and transferred on November 20, 1942 for a cost of $16,465. It consisted of gun blocks and a covered concrete magazine, The two 3-inch Model 1902M1 guns and carriages (Bethlehem Steel #8 and #9) were moved here from Battery Lee at Fort Wool. The role of the battery was replaced by Tac-24 of 90mm guns later in 1943. The guns were removed from their blocks on January 15, 1944 and one subsequently remounted at Tac-19 at Fort Story. The emplacement still exists at the Fisherman Island National Wildlife Refuge. The battery is open to the public, but access to the island is restricted.

- **AMTB Tac-24**: A 1943 Program 90mm AMTB battery built at Fisherman Island. It was constructed from December 7, 1942 to November 26, 1943, being transferred on February 12, 1944 for a cost of $6,936.56. The basic construction consisted of a BC station and two permanent gun blocks. It was abandoned postwar, but the guns were not physically removed until the 1976, currently being

Harbor Defenses of Chesapeake Bay
SITE NO. 28
FISHERMAN ISLAND

DATE IO SEPT 1943   EXHIBIT NO 45 B
PREPARED BY H.D. ARTILLERY ENG.

REVISED DATE 1 NOV 1945

used for display purposes at Battery Tac-22 at Fort Monroe and at Fort MacArthur. The emplacement still exists at the Fisherman Island section of the National Wildlife Refuge. The battery is open to the public, but access to the island is restricted.

Fort Story (1914-1946) is located at Cape Henry, 18 miles east of Norfolk and six miles north of Virginia Beach, Virginia. The active base covers 1,451 acres and nearly four miles of seashore at the tip of Cape Henry. Fort Story was acquired in 1914 to upgrade the defenses at the entrance to Chesapeake Bay. It was named in General Orders 31 of 1916 for Maj. Gen. John Patten Story, U.S. Army. The fort's first coast artillery defenses were two temporary World War I batteries. The fort's primary armament during this period was a unique battery of four 16-inch M1920 howitzers. The 16-inch Model 1920 was the U.S. Army's follow-on program to the 12-inch mortars, but only Fort Story's Battery Pennington was constructed. The fort also received railway artillery, 155mm mobile artillery, a controlled mine casemate, and 3-inch anti-aircraft guns. The coming of World War II and growing importance of Norfolk as a naval base caused the 1940 Program to warrant the construction of two #100 Series 16-inch casemated batteries and three #200 Series 6-inch shielded batteries. Supporting these main batteries were a new mine casemate (supporting the new mine facility at Little Creek), a railway battery with 8-inch guns, several large steel fire control towers, and several AMTB batteries mounting 90mm and 3-inch guns. These new batteries, along with the 16-inch howitzers (which received armored shields to provide some protection from aerial bombing), resulted in Fort Story (with eight 16-inch guns and six 6-inch guns) being the most heavily armed coast artillery fort in the United States. Near the end World War II, Fort Story military functions switch to being a convalescent hospital and an amphibious training center. After the end of its use as a seacoast artillery fort in 1948, it was home to a Nike site and several other Army functions during the Cold War. Fort Story is currently part of the Joint Expeditionary Base Little Creek-Fort Story and continues to be used by both the U.S. Army and U.S. Navy for training, testing and recreation. The several major batteries have been repurposed for other uses, while some are abandoned. Currently access to the site is very limited, except for visiting the historic Old Cape Henry Lighthouse which has a small visitors center and gift shop. Recently, a 16-inch MKVII barrel was placed on display next to the lighthouse to honor military history of Fort Story as the 16-inch casemated Battery Ketcham is located near-by.

Fort Story Gun Batteries

- **PENNINGTON – WALKE:** This was the first permanent battery to be installed at Cape Henry military reservation of Fort Story. A number of projects for heavy 14-inch and 16-inch guns and howitzers had emerged from the recommendations of the 1915 Board of Review. This battery was designed for four 16-inch mortars (howitzers) on open, barbette carriages on the central part of the reservation. The plans called for open, simple circular concrete platforms for the four howitzers, separated from each other by about 1000-feet. Guns were connected by a standard-gauge rail system delivering ammunition from dispersed, unprotected magazines. Work was done from July 1921 until the following summer. They were transferred on October 21, 1922 for a cost of $332,174.23. Initially the emplacements were organized as one battery, named Battery Pennington in General Orders No. 13 of March 27, 1922 for Brigadier General Cummings McWhorter Pennington Jr., an artilleryman of Civil War service. In 1940 the two northern batteries were separated administratively and named Battery Walke for Brigadier General Willoughby Walke. The battery was armed with four 16-inch howitzers Model 1920 and Model 1920 carriages (#2/#4, #3/#2, #4/#3, and #5/#1). A considerable number of support buildings were associated with the plan. A major, buried plotting room was built in the hills behind the battery, being transferred on May 5, 1923 for a cost of $28,656. Power plants were also required for gun and railroad operation. The battery served

ENTRANCE TO CHESAPEAKE BAY

FORT STORY

GENERAL MAP

5000 Ft.

3000

1000

0

SERIAL NUMBER 124

EDITION OF JAN. 14, 1915.
REVISIONS: NOV. 8, 1916.
MAR. 21, 1919; JAN. 1, 1921.

Abandoned

2-6" BATTERY
2-5" BATTERY
NO ARMAMENT

3-Anti-Aircraft Guns.

Norfolk Southern R.R.

Platform for
12" Barbette Mount
Platform for
14" R.R. Mount

L.S. St.

Old L.H.

L.H.

Artillery Observation
Stations No.

PARCEL A.

PARCEL B.

PARCEL C.

PARCEL D.

PARCEL E.

GRANITE

EMERSON

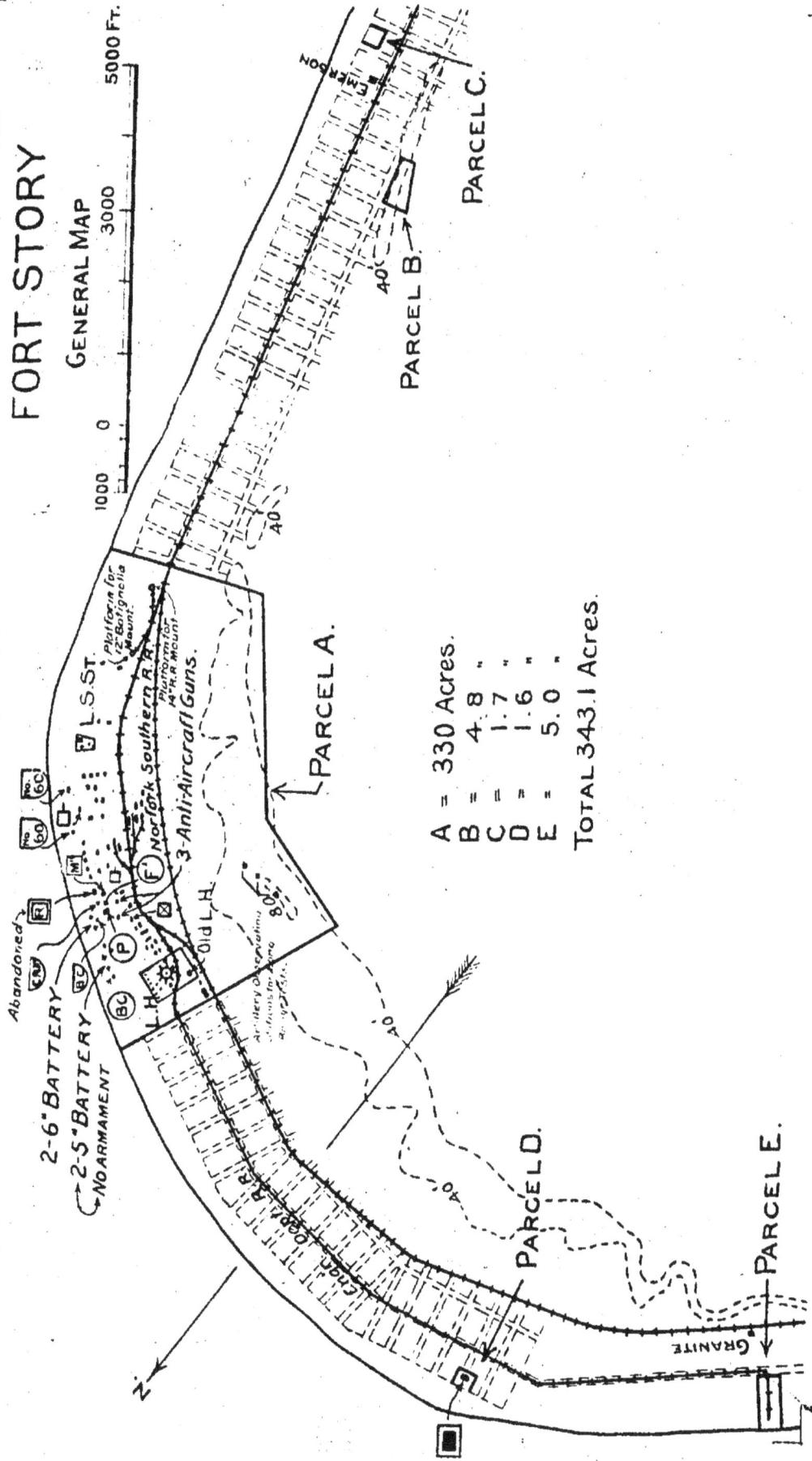

A = 330 Acres.
B = 4.8 "
C = 1.7 "
D = 1.6 "
E = 5.0 "

TOTAL 343.1 Acres.

ENTRANCE TO CHESAPEAKE BAY

FORT STORY D-1.

PARCEL "A"

SERIAL NUMBER 124

EDITION OF JAN. 1, 1921.
REVISIONS JUNE 9. 1922.

BATTERIES

PENNINGTON—4-16"H.

A-Anti-Aircraft gun 3-3"

LEGEND

1. ADMINISTRATION BLDG.
2. COMM. OFFICERS QRS.
3. OFFICERS QRS.
4. HOSPITAL.
6. N.C. OFFICERS QRS.
7. BARRACKS.
8. GUARD HOUSE.
9. POST EXCHANGE.
10. MESS.
11. POST BUILDING.
12.
13. OIL HOUSE.
20. Q.M. QUARTERS.
21. Q.M. WATER TOWER.
22. Q.M. STABLE.
23. Q.M. FIRE HOSE HOUSE.
24. Q.M. STORE HOUSE.
25. Q.M. LAVATORY.
26. Q.M. BATH.
27. Q.M. COMMISSARY.
28. Q.M. CARPENTER SHOP.
29. Q.M. COAL BIN.
30. ORDNANCE QRS.
40. E.D. OFFICE.
41. E.D. CEMENT HOUSE.
42. E.D. SHOP.
43. E.D. & Q.M. WATER TOWER.
44. E.D. BOILER ROOM.
45. E.D. LOCOMOTIVE HO.
46. E.D. LUMBER SHED.
47. E.D. COAL BIN.
48. E.D. QUARTERS.
49. E.D. OIL HOUSE.
49a. E.D. FIRE HOSE HOUSE.
100. E.D. MESS.
101. E.D. STORE HOUSE.
70. SERVICE CLUB.
90. N.&S.R.R. PASS. SHED.
91. N.&S.R.R. FREIGHT PLAT.
92. N.&S.R.R. DEPOT AT CAPE HENRY.

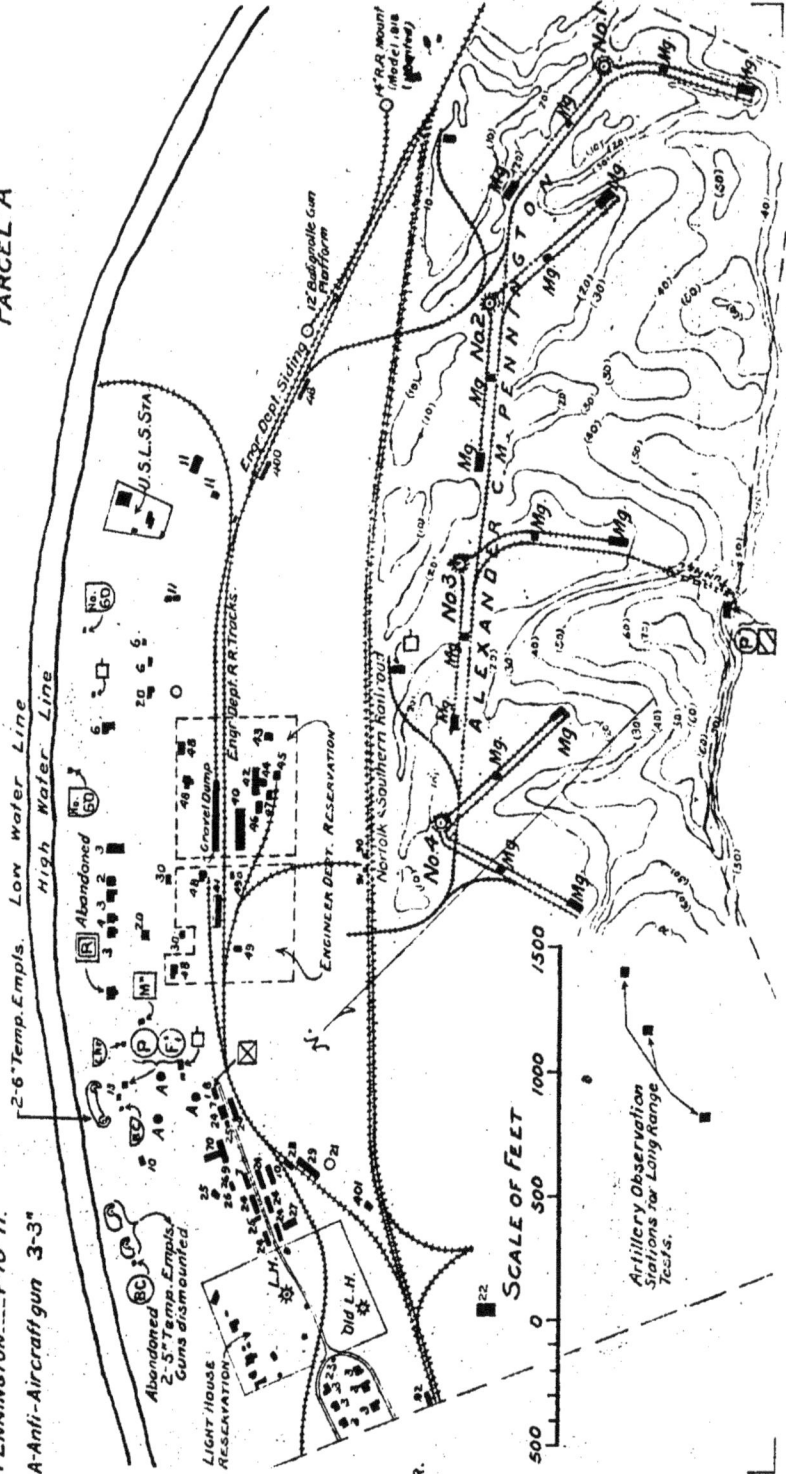

2-6"Temp. Empls. Low Water Line

High Water Line

Abandoned

U.S.L.S. STA.

Abandoned
2-5"Temp. Empls.
Guns dismounted.

Light House
Reservation

Old L.H.

SCALE OF FEET
500   0   500   1000   1500

Artillery Observation
Stations for Long Range
Tests.

Engr. Dept. Siding

12 Railgroune Gun Platform

Engr. Dept. R.R. Tracks

ENGINEER DEPT. RESERVATION

Norfolk & Southern Railroad

ALEXANDER   CAMP   PENNINGTON

No.2

No.3

No.4

Gravel Dump

FORT STORY, VA
EXISTING AND PROPOSED
INSTALLATIONS
5/11/41

EXHIBIT 2-A

ATLANTIC OCEAN

SMALL ARMS RANGE

ATHLETIC FIELD

BCN 226

Pennington

BCN 225

BCN 222

WALKE

PARADE

BCN 121

BCN 120

SCALE OF YARDS

FORT STORY, VIRGINIA

Battery #121, Joint Expeditionary Base Little Creek-Fort Story (Terry McGovern)

Cape Henry, Joint Expeditionary Base Little Creek-Fort Story (Terry McGovern)

up through World War II as the only completed 16-inch howitzer battery ever built. In 1944 the guns were given some degree of protection by installing light steel shields over the carriages. The battery was finally abandoned and the armament removed under authority of April 29, 1947. Parts of several of the emplacements and some magazines and the plotting room still exist on the Joint Expeditionary Base Fort Story. The battery site is partly open, but access to the base is restricted.

- A temporary 5-inch battery was erected on the Cape Henry beach in 1917. In common with many other such sites, a pair of 5-inch Model 1900 guns and carriages were relocated here during the summer of 1917. Guns Model 1900 on carriages Model 1903 (#3/#1 and #5/#2) were moved here in July 1917 from Battery Rice, Fort Andrews, Boston. Construction consisted of simple concrete gun blocks. The guns and carriages were removed after World War I in early 1920 and shipped off for storage at Watertown Arsenal on December 7, 1920. Some partial remains of the blocks and bolt circles still remain on the beach at JEB Fort Story. Access to the base is restricted.

- A temporary 6-inch battery erected on the beach at Fort Story, not far from the 5-inch battery in 1917. Two 6-inch Model 1900 guns and pedestal mounts were moved from Battery Montgomery at Fort Monroe for this emplacement. They were Model 1900 #15/#31 and #42/#34. Like the sister 5-inch emplacements, the only permanent work was the pair of concrete gun blocks. The armament was authorized for removal on July 2, 1919. While there was some delay in removal, shore erosion of the battery necessitated the actual removal in 1922. They were dismantled, slushed with oil, and stored on cribbing near the old emplacement for several more years. The broken blocks still remain on the beach at JEB Fort Story. Access to the base is restricted.

- **KETCHAM:** A battery for two 16-inch barbette, casemated guns located on the highlands at the western side of the Cape Henry reservation. This project was the third of the new 1934 approved batteries (following the two for approved for Narragansett Bay). It had the early casemate plan with a recessed gun room having two separate entrances (approved on April 20, 1934) but not funded until Fiscal Years 1941 and 1942 ($625,000 appropriated each year). Work was done from April 12, 1941 until November 12, 1943 for transfer on November 12, 1943 for $1,553,592. It was later assigned Battery Construction Number 120 for reference and assigned priority #5 on September 11, 1940 and #2 on August 11, 1941. It was named on General Orders No. 7 on January 24, 1942 for Brigadier General Daniel Ketcham, U.S. Army. Itwas armed with two 16-inch navy Mk II on Model M1919M4 barbette carriages (#109/#21 and #106/#22). Also built was a separate plotting room and numerous base-end stations. Ketcham was probably the single most important battery for the defenses of the Chesapeake. The battery was completed in 1942 and served until deleted under authority of May 5, 1948—though this was extended slightly later until January 20, 1949. The emplacement still exists and is currently abandoned at the JEB Fort Story. The battery is closed to the public and access to the base is restricted.

- **Battery #121:** A second pair of 16-inch casemated guns for Fort Story originating in the 1940 Program. It was located to the south of Ketcham, though with about the same field of fire and being on the same ridge line. The design was consistent with other 1940 Program batteries, with the later "straight" design for the main internal gallery, and thus slightly different from Ketcham. The battery was built from January 6, 1942 until January 31, 1943 for transfer on November 12, 1943 for a cost of $1,881,850.47. It had been funded under the FY-1942 and 1943 budgets, original priority of #19 on September 11, 1940 and #11 on August 11, 1941. This increase was motivated by taking advantage of the existing construction plant on site from the previous projects. The battery was armed with two 16-inch MkII guns on Model 1919M4 barbette carriages (#52/#32 and

#60/#41). The battery was never named. Like Ketcham it was deleted on January 20, 1949. The emplacement still exists and has been incorporated into US Navy SEAL Training facility at JEB Fort Story. The battery is closed to the public and access to the base is restricted.

- **WORCESTER**: A dual 6-inch modern battery built at Fort Story, near the beach on the eastern side of the reservation. It pre-dates the 1940 Program slightly and was the prototype example of the standardized 6-inch battery design. As such it differs considerably from the plan finally adopted for that generation of works. It was funded under the FY-1941 budget and was actually constructed from November 14, 1940 to April 1, 1941. Transfer was made on April 23, 1941 for a cost of $76,237. It was named on General Orders No. 5 of January 20, 1942 for Colonel Philip Worcester. The new 6-inch barbette guns were not yet available at the time of completion, so it was armed with older Model 1900 guns and M1900 pedestal mounts. These were obtained from surplus stocks held in the Chesapeake defenses (#1/#33 and #5/#34). This armament served the battery until removed under authority of April 29. 1947. During its final period of completion, it was assigned Battery Construction No. 224 to fit into the 1940 Program numbering sequence. The emplacement still exists, and the structure is still in use for military purposes at JEB Fort Story. The battery is closed to the public and access to the base is restricted.

- **CRAMER:** A second dual 6-inch battery built on the Fort Story reservation, south of Worcester near the beach. It was a standard 1940 Program battery in design and construction and was given construction number #225. It was funded under the 1942 Fiscal Year budget and at the time given a high priority of #6. Construction was done from October 8, 1941 to May 15, 1942 for transfer on November 23, 1943 at a cost of $175,832. It was named on General Orders No. 63 of November 14, 1942 for Lieutenant Colonel Raymond V. Cramer, of the Coast Artillery Corps. The battery was armed with two 6-inch model 1903A2 guns on M2 barbette carriages (#18/#2 and #62/#1). The battery served until the final authority to stand down the Chesapeake Defenses in early 1949. The emplacement still exists and is used for storage at JEB Fort Story. The battery is closed to the public and access to the base is restricted.

- **Battery #226**: The final dual 6-inch barbette battery for Fort Story. It was located on the rise to the west of Worcester, firing to the east. Built as part of the 1940 Program, it was funded in the FY-1943 budget and was of standard two-gun emplacement design. Work was done from September 1, 1942 to October 3, 1943 for transfer on November 12, 1943 at a cost of $234,602.06. It was never named, just carrying the No. 226 Battery Construction Number designation. The battery was armed with two 6-inch T2 (M1) guns on Model M4 barbette carriages (#8/#2 and #9/#1). It was retained until 1948 or 1949 in the harbor defenses before being disarmed. The emplacement still exists and is abandoned at JEB Fort Story. The battery is closed to the public and access to the base is restricted.

- **AMTB Tac-21**: A new AMTB battery built as a prototype for the dual 90mm fixed and shielded M3 mount. It was completed to a relatively complex design, which due to its cost and resources consumed was never duplicated. It had separate gun blocks for fixed guns and an intervening re-inforce concrete magazine and battery commander's station. Work was done from September 25, 1942 to October 31, 1942 for transfer on August 2, 1943, the combined magazine/plot had a cost of $33,622, while the gun blocks and ready ammo boxes cost $5844. It was armed with its two emplaced 90mm guns on M3 barbette carriages and two mobile 90mm M1 guns usually placed in open field positions on either flank. It served until disarmed on April 26, 1949. The emplacement still exists and is in use by the military at JEB Fort Story. The battery site is open to the public, but access to the base is restricted.

- **AMTB Tac-22**: A 1943 Program AMTB battery consisting of two 90mm fixed gun and two 90mm mobile mounts emplaced on the beach to the south of Tac-21 (at the end of 21st Street). It was built from February 16, 1943 to April 13, 1943 for transfer on December 30, 1943 at a cost of $5,085.76. It was removed under authority of April 26, 1949. Parts of the gun blocks still exist but are rapidly being undermined by the beach surf at JEB Fort Story. Access to the base is restricted.

Fort Story (NARA)

Power room Battery #228 Fisherman Island (Mark Berhow)

3-inch guns on a pedestal mount in Battery Irwin, Fort Monroe (Mark Berhow)

16-inch MkVIII gun barrel in Battery Winslow, East Virginia National Wildlife Refuge (Mark Berhow)

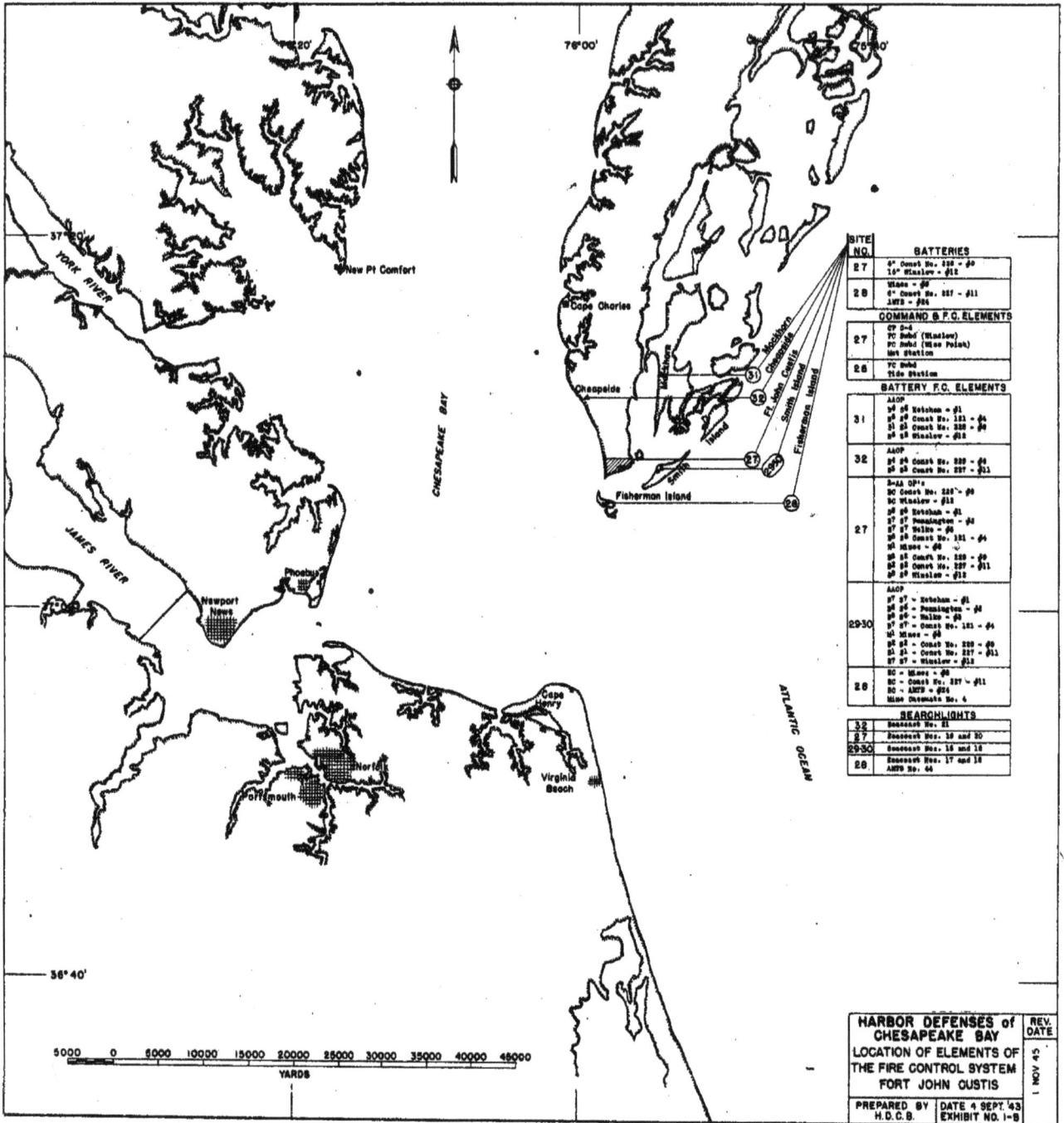

HARBOR DEFENSES of CHESAPEAKE BAY
LOCATION OF ELEMENTS OF THE FIRE CONTROL SYSTEM FORT JOHN CUSTIS

REV. DATE 1 NOV 45

PREPARED BY H.D.C.B.   DATE 4 SEPT. '43   EXHIBIT NO. 1-B

PAGE 3

76°00'

YORK RIVER

New Pt Comfort

Cape Charles

CHESAPEAKE BAY

Smith Island

Fisherman Island

JAMES RIVER

Phoebus

Newport News

NORTH BUCKROE   (21)

FORT MONROE   (19)

FORT WOOL   (20)

OCEAN VIEW   (11)

ATLANTIC OCEAN

Cape Henry

Norfolk

Portsmouth

Virginia Beach

37° 20'

36° 40'

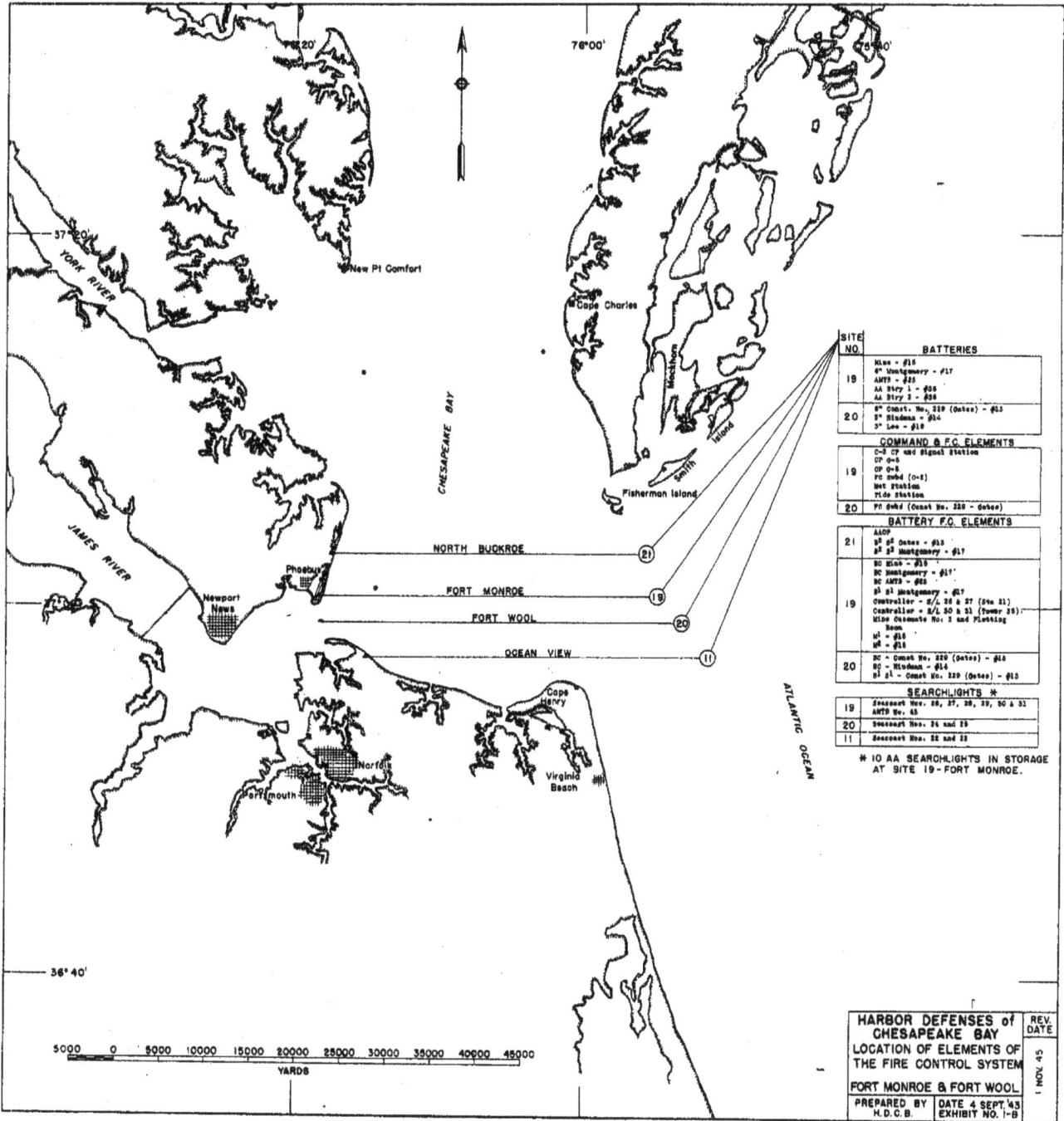

SITE NO.	BATTERIES
19	Mine - #16 6" Montgomery - #17 AMTB - #23 AA Btry 1 - #36 AA Btry 2 - #36
20	8" Cont. No. 229 (Gates) - #13 3" Hindman - #14 3" Lee - #15

	COMMAND & F.C. ELEMENTS
19	C-8 CP and Signal Station CP O-5 OP O-8 FC swbd (O-8) Met Station Tide Station
20	FC swbd (Const No. 229 - Gates)

	BATTERY F.C. ELEMENTS
21	AADP B² B² Gates - #13 B² Montgomery - #17
19	BC Mine - #16 BC Montgomery - #17 BC AMTB - #23 B² B² Montgomery - #17 Controller - S/L 26 & 27 (Sta 21) Controller - S/L 30 & 31 (Tower 29) Mine Casemate No. 2 and Plotting Room B¹ - #16 B² - #16
20	BC - Const No. 229 (Gates) - #13 BC - Hindman - #14 B² B¹ - Const No. 229 (Gates) - #13

	SEARCHLIGHTS ✱
19	Seacoast Nos. 26, 27, 28, 29, 30 & 31 AMTB No. 45
20	Seacoast Nos. 24 and 25
11	Seacoast Nos. 22 and 23

✱ 10 AA SEARCHLIGHTS IN STORAGE AT SITE 19-FORT MONROE.

| 5000 | 0 | 5000 | 10000 | 15000 | 20000 | 25000 | 30000 | 35000 | 40000 | 45000 |

YARDS

HARBOR DEFENSES of CHESAPEAKE BAY	REV. DATE
LOCATION OF ELEMENTS OF THE FIRE CONTROL SYSTEM	1 NOV 45
FORT MONROE & FORT WOOL	
PREPARED BY H.D.C.B.	DATE 4 SEPT. '43 EXHIBIT NO. 1-8

PAGE 2

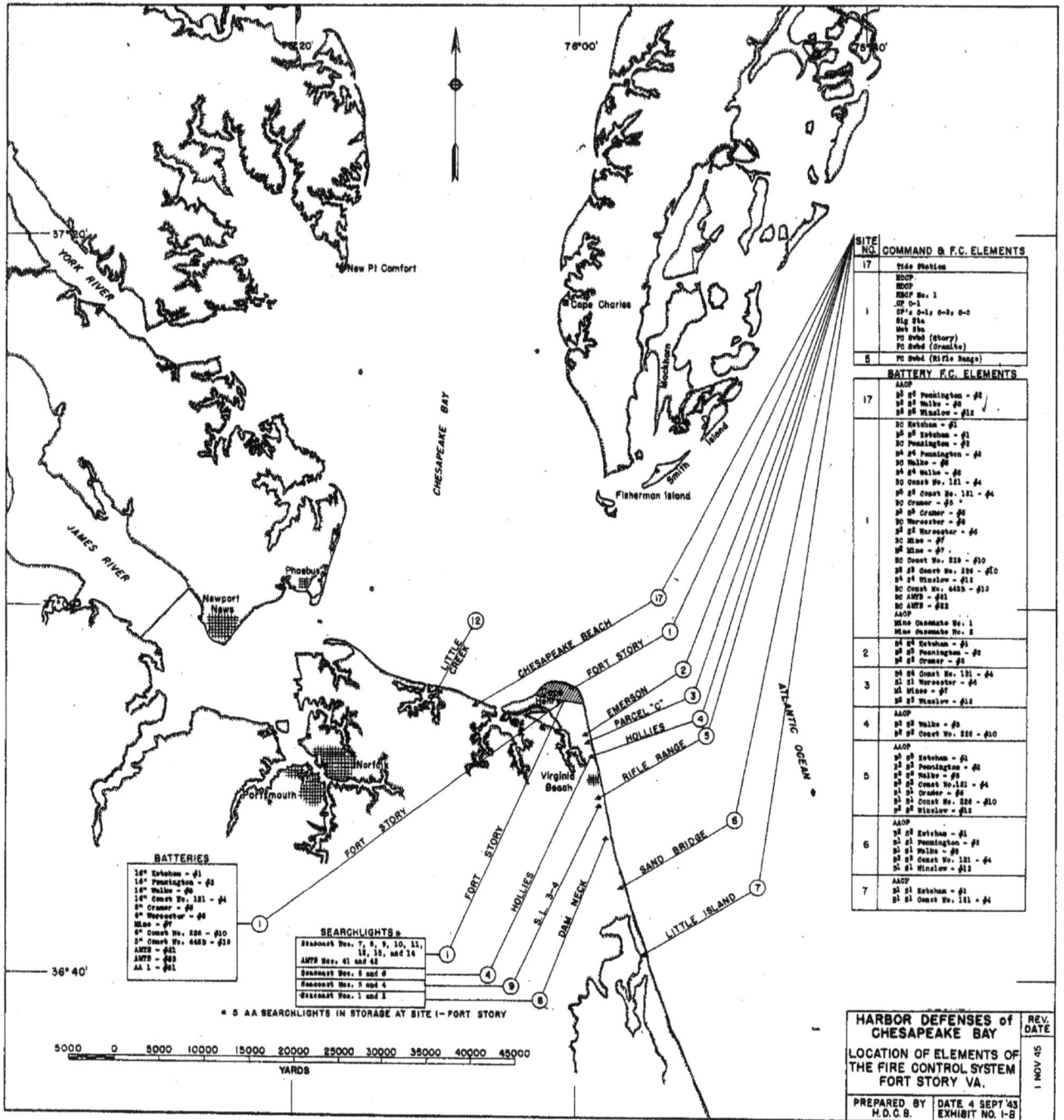

Weinert, Richard P. and Robert Arthur. *Defender of the Chesapeake, the Story of Fort Monroe.* White Mane Publishing Co. Shippensburg, PA, 1989.

McGovern, Terrance, *The Chesapeake Bay at War! – The Coast Defenses of Chesapeake Bay during World War Two*, Three Sisters Press, McLean, VA 20

Chesapeake Bay World War II-era Site Locations. Stations housed in a single structure are connected by dashes (-)

location	Loc#	Purpose
Fort Story	1	Batt. Tact. #1 Ketcham, Batt. Tact. #2 Penningon, Batt. Tact. #3 Walke, Batt. Tact. #4 BCN 121, Batt. Tact. #5 Cramer, Batt. Tact. #6 Worchester, Batt. Tact. #7 mine, Batt. Tact. #21 AMTB, Batt. Tact. #22 AMTB, HECP-HEDP, HDOP1, HDOP2, G1, G2, G3, T, Met, SBR, BC/Ketcham, BC/Pennington, BC Walke, BC 121 (lighthouse), BC/Cramer, BC/Worchester
Fort Story Granite	15	Batt. Tact. #19, BS5/Ketcham, BS4/Pennington, BS4/Walke, BS5/121, BS3/Cramer, BS4/Winslow, SBR
Fort Story Casemate		Batt. Tact. #10 BCN 226, BS2/Worchester, M1-BC, M1/2, BC/226, BS3/226
Emerson	2	BS4/Ketcham, BS3/Pennington, BS2/Cramer
Parcel C	3	BS4/121, BS1/Worchester, M1/1, BS3 Winslow
Hollies	4	BS3/Walke, BS2/226
Rifle Range Camp	5	BS3/Ketcham, BS2/Pennington, BS2/Walke, BS1/Cramer, BS1/226, BS2/Winslow, SBR
Sand Bridge	6	BS2/Ketcham, BS1/Pennington, BS1/Walke, BS2/121, BS1/Winslow
Little Island	7	BS1/Ketcham, BS1/121
Naval Res	8	Searchlight
	9	Searchlight
	10	SCR Radar Searchlight
	11	Searchlight
Little Creek	12	mine facility
120th Street	15	SCR Radar
	16	SCR Radar
Chesapeake Beach	17	BS5/Pennington, BS5/Walke, BS5/Winslow
Fort Monroe	19	Batt. Tact. #17 Montgomery, Batt. Tact. #16 Lee, Batt. Tact. #15 mines, Batt. Tact. #23 AMTB, G5, G6, T, Met, SBR, M2-BC, M2/1, M2/2, BC/Montgomery, B1/Montgomery, BC/Lee
Fort Wool	20	Batt. Tact. #14 Hindmann, Batt. Tact. #13 BCN 229 Gates, BC-BS1/229, BC/Hindmann
North Buckroe	21	BS2/229, BS2/Montgomery
Fort John Custis	27	Batt. Tact. #12 Winslow, Batt. Tact. #9 BCN 228, G4, T, Met, SBR, BC/Winslow, BC/228
Wise Point	27	SBR, BS8/Ketcham, BS7/Pennington, BS7/Walke, BS8/121, M3/1, BS3/228, BS2/227, BS8/Winslow
Fisherman Island	28	Batt. Tact. #8 mine, Batt. Tact. #11 Batt. Tact. #24 AMTB, BCN 227, SBR, M3-BC, BC/227
Smith Island	29&30	BS7/Ketcham, BS6/Pennington, BS6/Walke, BS7/121, M3/2, BS2/228, BS1/227, BS7/Winslow
Mockhorn Island	31	BS6/Ketcham, BS6/121, BS1/228, BS6/Winslow
Cheapside	32	BS4/228, BS3/227

THE HARBOR DEFENSES OF THE CAPE FEAR RIVER – NORTH CAROLINA

The port of Wilmington, founded in 1730 on banks of the Cape Fear River, grew into an important commercial and trade center by the time of the Revolutionary War. This key port received a Third System fort. During the Civil War the very large Fort Fisher was constructed using earthworks along with other supporting defenses to allow the Confederates to control this harbor until early 1865 as it was key port for Confederate blockade runners. The US Army built new concrete batteries and garrison structures on Oak Island during the Endicott Program.

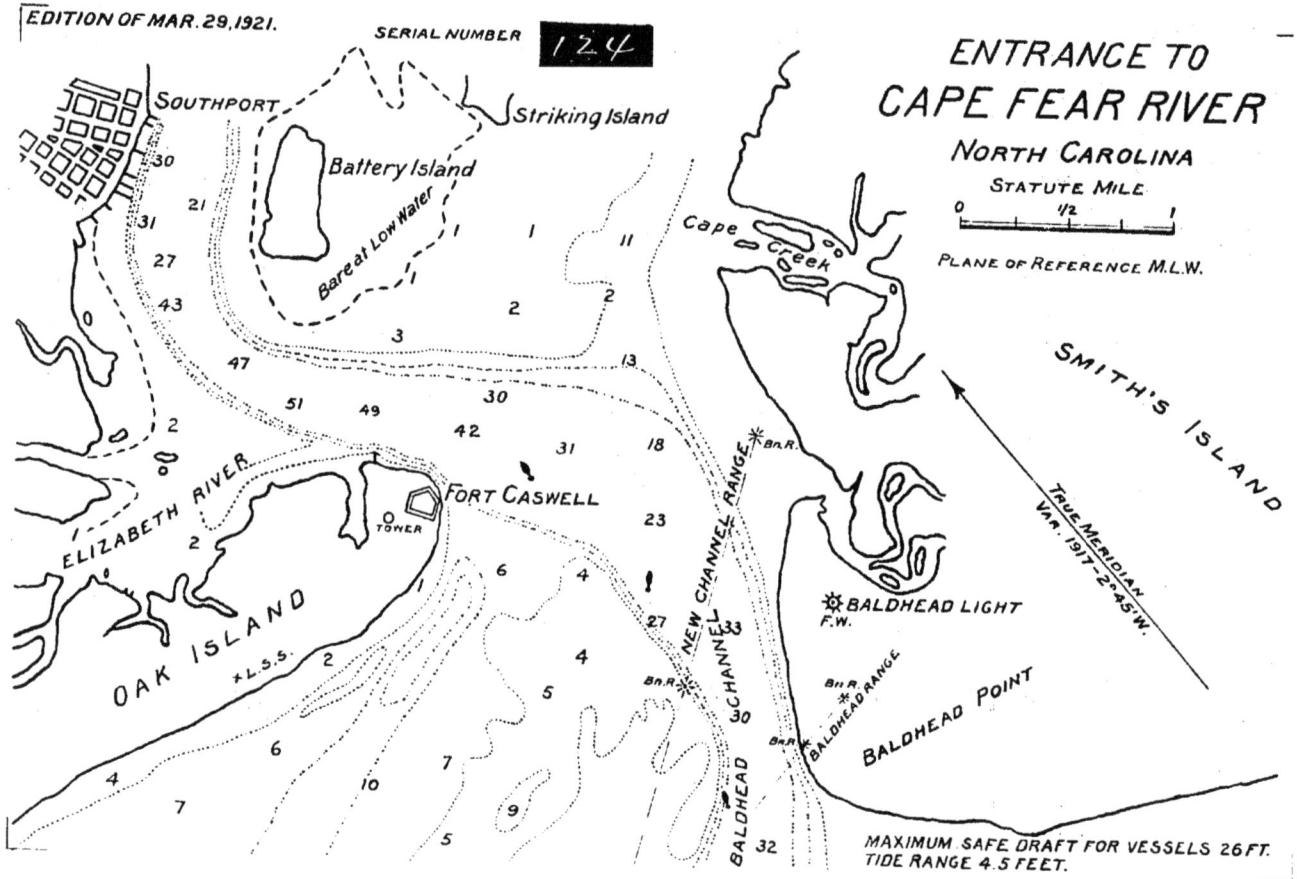

Fort Caswell (1826-1925) is located on Oak Island at the mouth of the Cape Fear River, just two miles south of the town of Southport, North Carolina. The partly existing Third System fort, called Fort Caswell, was built from 1826 to 1838. It was named in General Orders 32 of 1832 for Maj. Gen. Richard Caswell of the Continental Army. The casemate fort of stone and brick served as the primary defense of Wilmington, North Carolina via the Cape Fear River up to and through the Civil War. The fort was held by Confederate forces until the fall of Fort Fisher in 1865 when the fort's magazine with 100,000 pounds of gunpowder was burned resulting in destruction of part of the fort. Placed on caretaker status after the Civil War, Fort Caswell's reservation was used during the Endicott Program for nine concrete batteries, large sea wall, and controlled submarine mine facilities. Installed into the walls of the old fort was the fort's primary battery (later used as a swimming pool) of two 12-inch guns on barbette carriages. In 1925, the coast artillery activities at Fort Caswell ceased and the fort was sold and converted into a summer resort. In 1941, the U.S. Navy purchased the property inside the seawall to be used as a support base for minesweepers and inshore patrol boats. Declared surplus in 1949, the reservation was sold to the Baptist State Convention of North Carolina to be used as a retreat facility. Known today as the Fort Caswell Coastal Retreat and Conference Center is run by the Baptist State Convention of North Carolina, Fort Caswell remains one of the best-preserved examples of an Endicott Program fort, as most of the fort's buildings and batteries have not been altered since their construction in 1895-1904. Visitation outside of the retreat and conference center use must be arranged in advance as the site is not open to the general public.

Fort Caswell Gun Batteries

- **CASWELL**: The primary battery for Fort Caswell of two 12-inch barbette guns. It was emplaced physically within the perimeter of the old Third System masonry fort and required its partial destruction. Plans were submitted on May 3, 1898. It was funded with the National Defense Act of March 1898. While generally following type plans, there were some significant changes to allow the battery to fit the perimeter of the old work. The right-hand gun (emplacement No. 1) was withdrawn in echelon to allow it to fit and for better coverage to the west. The old covered way in front of the fort was left to provide better coverage for the modern emplacement. Also, the battery had a larger than normal power plant in order to serve the new mortar battery and outlying searchlight emplacements. Work was done rapidly (and the wartime urgency caused a serious cost overrun). One emplacement was declared ready for armament as soon as May 20, 1898. It received two 12-inch Model 1888 Watervliet guns on barbettes Model 1892 (#27/#15 and #29/#16). Transfer was made on May 22, 1899 for a cost of $125,993.19. The battery was named in General Orders No. 134 of July 22, 1899 for Richard Caswell, former Governor of North Carolina. For a short while the 4.7-inch emplacement on the right flank of the battery was also considered a part of Battery Caswell but was soon named separately as Battery Madison. After World War I, the defenses of Cape Fear were abandoned. The armament of two 12-inch guns remained until removed in 1926. The emplacement still exists, though somewhat obscured by a later recreational swimming pool built on it, at the Fort Caswell Coastal Retreat and Conference Center run by the North Carolina Baptist Convention. The battery is open, but access to the retreat center is restricted.

- **BAGLEY**: A mortar battery for the Caswell reservation, emplaced on the western end of the reservation, firing to the southwest. Plans were submitted on June 20, 1898. It was 200-yards beyond the closest flank emplacement of Battery Swift. There were foundation problems at this site, subsidence occurring. In May 1899 work had to stop pending placement of new pilings under the gun platforms in order to ensure stability for firing. The design had some unique aspects, part of the traverse wall in each pit being oblique rather than straight. Work was finished in late 1899, and transfer was made on March 6, 1903 for a cost of $115,787.50. The considerable delay in transferring was due to various structural faults and the continual collapse of the earthen parapet. It received eight 12-inch Model 1890M1 mortars on Model 1896 carriages Watervliet tube #3/#216, Niles tube #7/#215, Niles tube #8/#213, Niles tube #9/#212, Builders tube #37/#214, Builders tube #39/#217, Builders tube #41/#218 and Bethlehem tube #37/#219. The battery was named on General Orders No. 138 on July 27, 1899 for Ensign Worth Bagley, U.S. Navy, who was killed in action at Cardenas Cuba in 1898. In 1910 it was decided to transfer the armament to Boston for use in the remodeled Battery Lincoln. Bagley was to receive in return the old cast-iron Model 1886 mortars originally at Lincoln. Local officers protested against the exchange as a lessening of the defenses of Cape Fear, but to no avail. In 1911 the exchange was made. Battery Bagley was then armed with Model 1886 mortars on Model 1891 carriages (#9/#30, #16/#28, #26/#27, #23/#29, #63/#17, #41/#19, #72/#20, and #61/#18). These mortars were removed and scrapped

CAPE FEAR RIVER, N.C.

FORT CASWELL

OAK ISLAND

SERIAL NUMBER. 124.

EDITION OF MAR. 27, 1919.
REVISIONS: MAR. 25, 1921.

Azimuth 90°-39'-30'; distance 5813.400'
" 105° 44'-22"; " 11756.134'

From M¹(M¹) to M²(M²)
From F₁(B₁) to F₂(B₂)

Scale of Feet

LEGEND

1. ADMINISTRATION BLDG.
2. COMMANDING OFF. QRS.
3. OFFICER'S QRS.
4. HOSPITAL
5. HOSPITAL STEW.D
5a. HOSPITAL BARRACKS.
6. N.C.OFFICER'S QRS.
7. BARRACKS.
7t. "
7a. QUARTERS.
8. GUARD HOUSE.
9. POST EXCHANGE.
9t. "
10. GYMNASIUM
11. BAKERY.
11a. NEW BAKERY.
12. FIRE APPARATUS HO.
13. ICE HOUSE.
14. STORE HOUSE.
15. MESS HALL.
15a. NEW MESS HALL.
16t. LAVATORY.
16a. NEW LAVATORY.
17. WORK SHOP.
18. PAINT SHOP.
19. TAILOR SHOP.
100. COAL SHED.
101. WAGON SHED.
102. STABLE.
102t. "
103. COAST ARTILLERY
       MILITIA ST. ROOM.
104. TELAUTOGRAPH
       BOOTHS
20. CONST. Q.M.OFFICE.
20a. Q.M.OFFICE & ST.HO.
21. Q.M.PLUMBER'S SHOP.
22. Q.M.STOREHOUSE.
22t. "
23. Q.M.TRAMWAY.
24. COMMISSARY ST.HO.
31. ORDNANCE. BLDG.
41. ENGR.CARPENTER
       SHOP & OFFICE.
42. ENGR.BLACKSMITH
       SHOP.
43. ENGR.CEMENT SHED
61. NURSE'S. QRS.
61t. "
62t. OFFICER'S MESS HALL.
71. Y.M.C.A. BUILDING.
72. "
73. BOWLING ALLEY
301. CONTRACTOR'S OFF.
91t. " MESS HALL.

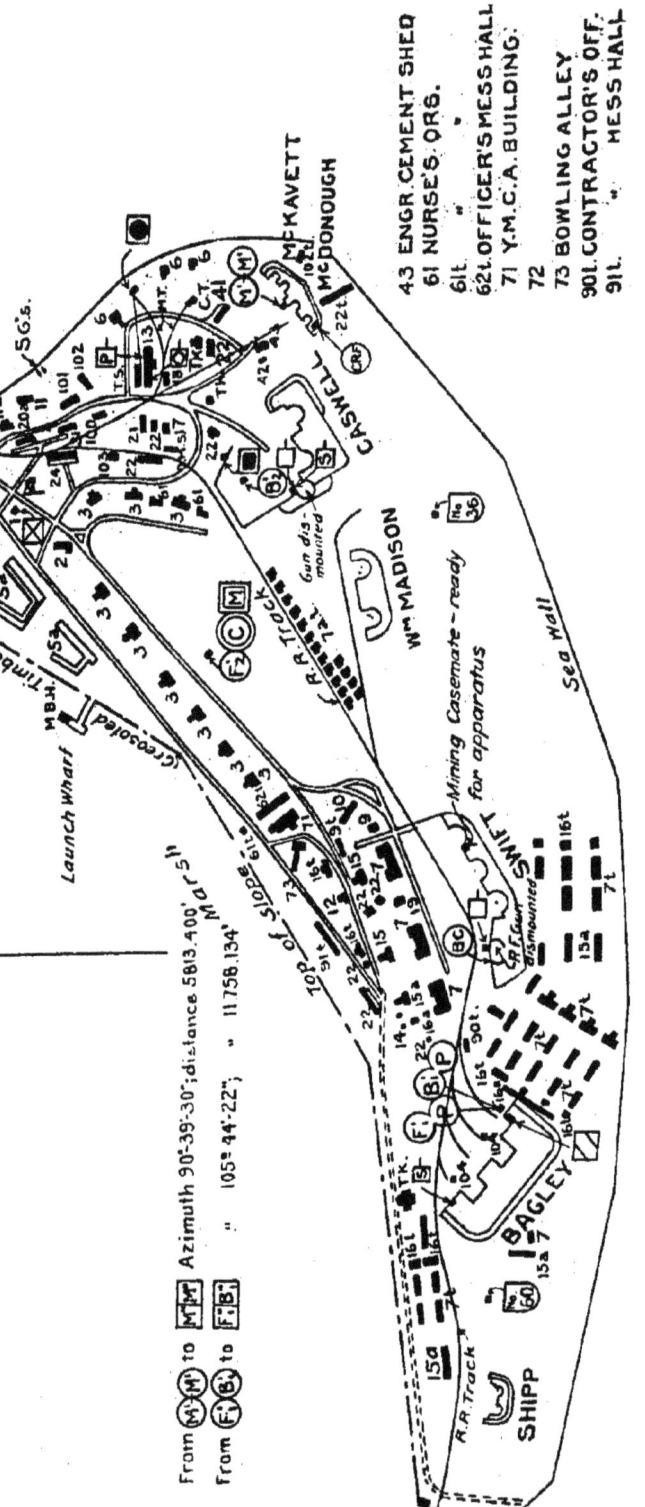

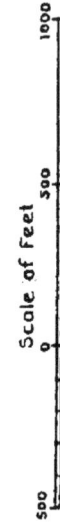

BATTERIES

BAGLEY   4-12"M.
CASWELL   2-12" N.Dis.
*SWIFT   4-8" Dis.
*MADISON   2-6" "
*SHIPP   2-5" B.P.
McKAVETT 2-3" B.P.
McDONOUGH 2-3" P.

*Dismounted

Mining Casemate — ready
for apparatus

Gun dis-mounted

Wm MADISON

Sea wall

CASWELL

McKAVETT

McDONOUGH

SWIFT

BAGLEY

SHIPP

in September 1919. A final set of four mortars (two per pit) of the Model 1890M1 on Model 1896 were sent from New York's Fort Tilden under authority of November 11, 1919. They were present until the discontinuance of the Cape Fear harbor defenses in 1926 after which they were removed. The emplacement still exists at the FCCRCC run by the North Carolina Baptist Convention. The battery is open, but access to the retreat center is restricted.

- **SWIFT**: A large battery eventually of four 8-inch disappearing guns. It was emplaced on a location west, southwest of the old masonry fort. On October 15, 1896 submission was made for a two-gun section of this battery, soon authorized to be increased to three emplacements in line. Work began in late 1896 and continued through 1897. These were all for disappearing carriages Model 1894. This series of guns were of standard type plan. It was decided to collect the rainfall that fell onto the battery as a source of fresh water, which was not otherwise obtainable at the post. Three large cylindrical cisterns, each of 16,500-gallon capacity, were placed vertically in front of the battery and connected by gutters to the battery. These were built with the battery in late 1897. These emplacements received three 8-inch Model 1888 Watervliet guns on Model LF Model 1894 carriages (#25/#8, #27/#7, and #28/#6). These three emplacements were transferred in March of 1898. A fourth gun emplacement, with an M1896 disappearing carriage, was submitted on March 21, 1898 for the right flank of the structure. This final position was angled to the right and slightly withdrawn in echelon. The space that opened up between guns No. 1 and No. 2 was used for a position for a single 5-inch gun. Work on the final emplacement was done in 1898-1899. It was armed with one 8-inch Model 1888 Watervliet gun on LF Model 1896 disappearing carriage (#34/#12). The last emplacement was transferred on December 6, 1898. All four emplacements together cost $121,170.86. It was named on General Orders No. 134 of July 22, 1899 for Captain Alexander Swift, Corps of Engineers officer. With some modifications and repairs to heavy storm damage Battery Swift served during the pre-war years. The gun tubes were removed in 1917 for use on railway carriages, the disappearing carriages were scrapped in the early 1920s. The emplacement still exists at the FCCRCC run by the North Carolina Baptist State Convention. The battery is open, but access to the retreat center is restricted.

- **MADISON (1)**: A battery for a single 4.7-inch pedestal gun erected adjacent to Battery Caswell atop the parapet of old masonry Fort Caswell. The plan for this emplacement was submitted on April 6, 1898, following authorization with funds from the National Defense Act. The gun was one of the thirty-four 4.7-inch guns purchased from Armstrong just at the start of the Spanish American War. It was located on the rampart of old Fort Caswell, in its own emplacement withdrawn in echelon to the right, rear of Battery Caswell. It was a simple single gun block with front parapet and a 260-square foot magazine directly below. Work was done in mid-1898 and transferred on May 33, 1899 for a cost of $6,510.28. The battery was armed with one 4.7-inch gun and pedestal (#11858/#10980). This armament was mounted in July of 1898. It was named on General Orders No. 78 of May 25, 1903 for Surgeon William Madison who died in action against hostile Indians in Wisconsin in 1821. As proposed in a letter of August 1904 and implemented in October, this gun was dismounted and sent to Battery Backus at Fort Screven (Savannah, Georgia). The emplacement was not subsequently used for armament. The name Madison was soon also transferred to the 6-inch disappearing battery recently completed at Fort Caswell. The emplacement still exists atop Fort Caswell at the FCCRCC run by the North Carolina Baptist Convention. The battery is open, but access to the retreat center is restricted.

- **MADISON (2):** A battery for two 6-inch disappearing guns emplaced between Battery Caswell and Swift in the center of the Fort Caswell reservation. It closely followed the standard mimeograph No. 59 plan for a modern 6-inch disappearing gun battery for the Model 1903 carriage. The appropriation of April 21, 1904 funded this construction. It was submitted on July 26, 1904. Work started that October and continued into 1905. Transfer was made on April 24, 1907 for a construction cost of $44,972.31. It was armed with two 6-inch guns Model 1903 on Model 1903 disappearing carriages (#20/#79 and #25/#80). The General Orders No. 194 of December 27, 1904 moved the name from the 4.7-inch battery to this emplacement (for Surgeon William Madison who died in the Indian Wars). The guns were removed in the Spring of 1917 for use on field mounts, the carriages were left in place but scrapped a couple of years later. The emplacement still exists at the FCCRCC run by the North Carolina Baptist Convention. The battery is open, but access to the retreat center is restricted.

- **McDONOUGH (1):** An emplacement for a single 5-inch balanced pillar located between emplacements No. 1 and No. 2 of 8-inch Battery Swift. When the final emplacement on the right flank of the 8-inch battery was added, it was in a slightly canted, withdrawn location. A position for a 5-inch balanced pillar gun was placed on the left flank of the added gun. It was submitted on the same plan as that emplacement on March 21, 1898. Concrete work was done with the larger battery's construction in 1899-1900. The balanced pillar was placed at the appropriate height to fire over the parapet with an entrance to an adjacent magazine. It received 5-inch Model 1897 Bethlehem gun on Model 1896 balanced pillar (#16/#1) mounted in 1901. This armament had previously been on display at the World's Exposition at Buffalo, New York. The emplacement was transferred on April 14, 1903 for a cost of just $187.70. It was named on General Orders No. 78 of May 25, 1903 for 1st Lieutenant Patrick McDonough of the Corps of Artillery killed in action in at Fort Erie in 1814. Very soon, in fact still in 1904, it was decided to move this gun to a new emplacement adjacent to the post's other 5-inch balanced pillar gun at Battery Shipp. Under authority of August 8, 1904 this was accomplished—though there is some evidence that the gun tube itself was not physically shifted until 1907. The battery's name was later transferred to the new 3-inch pedestal emplacement just being finished at Fort Caswell. The original emplacement itself still exists within Battery Swift, at the FCCRCC run by the North Carolina Baptist Convention. The battery is open, but access to the retreat center is restricted.

- **SHIPP:** A battery for 5-inch rapid-fire guns emplaced at the western extreme of the Oak Island's Fort Caswell reservation. Originally a single 5-inch balanced pillar emplacement plan was submitted on August 9, 1898. Work was done in late 1898 on a simple platform for the pillar and adjacent flank magazine. It was transferred on August 15, 1901 for a cost of $10,500. It was armed with one 5-inch Model 1897 Bethlehem gun on a Model 1896 pillar (#20/#15). The battery was named on General Orders No. 134 of July 22, 1899 for 1st Lt. William E. Shipp killed in action at San Juan Hill Cuba in 1898. In 1904 it was decided to move the other single 5-inch balanced pillar gun at Battery McDonough to the Battery Shipp location as a second gun. That plan was submitted on July 26, 1904. The new emplacement was on the right flank of the existing one and canted more to the north to increase the field of fire on that flank. As the balanced pillar was anchored in concrete, the original carriage could not be reused, and the new emplacment had to have a new one. That acquisition delayed the transfer of the gun, which as late as 1907 was still mounted at its old emplacement. Transfer was finally done on April 24, 1907 for a cost of $11,000 for a 5-inch Model 1897 Bethlehem gun on Model 1896 balanced pillar (#16/#1). The two guns then served until dismounted in 1917, the carriages were scrapped in place (in 1920. The emplacement, somewhat damaged and overgrown, still exists at the FCCRCC run by the North Carolina Baptist Convention. The battery is open, but access to the retreat center is restricted.

- **McKAVETT**: A battery for two 3-inch masking parapet guns emplaced near the shore and mine facilities on the eastern end of Oak Island. It was submitted on September 5, 1900 along with an adjacent 5-inch gun, and then again on April 15, 1901 with a modified site and just the two 3-inch masking parapet guns. Work was done in 1901-1902, generally conforming to standard plans for a two gun battery. It was armed with two 3-inch, 15-pounder Model 1898 guns on Model 1898 balanced pillar mounts (#108/#108 and #109/#109). Transfer was made on October 2, 1903 at a cost of $14,489.25. The battery was named on General Orders No. 78 of May 25, 1903 for Captain Henry McKavett, killed at Monterrey in 1847 during the Mexican War. It was later flanked by the new Battery McDonough. In 1916 the balanced pillar mounts were modified to M1898M1 pedestal mounts (losing their retraction ability). They were removed and scrapped in 1920 with the discontinuance of this type of armament in army inventory. The emplacement still exists at the FCCRCC run by the North Carolina Baptist Convention. The battery is open, but access to the retreat center is restricted.

- **McDONOUGH (2)**: A battery for two 3-inch pedestal guns emplaced adjacent to Battery McKavett on the eastern end of Oak Island. Funding came from the Act of June 6, 1902. It was submitted on April 5, 1903. They were on the right flank of McKavett, but the pair was canted 30-degrees further to the east. Their position well covered the minefield just offshore. Work was done in 1903-1904. It was transferred on February 20, 1906 for a cost of $14,450. The battery was armed in late 1905 and carried two 3-inch Model 1902 Bethlehem guns and M1902 pedestal mounts (#14/#14 and #15/#15). The name McDonough was transferred here from the recently relocated 5-inch battery at the post. The armament served until dismounted in 1926 with the abandonment of the defenses here at Cape Fear and the abandonment of the post. The emplacement still exists at the FCCRCC run by the North Carolina Baptist Convention. The battery is open, but access to the retreat center is restricted.

Fort Caswell (Glen Williford)

Fort Caswell 1922 (NARA)

Fort Casewell Coastal Retreat and Conference Center. (http://www.starforts.com/caswell.html)

Battery Bagley Fort Caswell (Glen Williford)

Main gun battery line Fort Caswell (Glen Williford)

Herring, Ethel, and Carolee Williams. *Fort Caswell in War and Peace.* Broadsfoot's Bookmark. Wendell, NC, 1983.

Gaines, William C. "Defending the Cape Fear River, North Carolina, 1803-1945." *CDSG Journal* Vol. 11, No. 4, Nov. 1997, p. 15

THE HARBOR DEFENSES OF CHARLESTON – SOUTH CAROLINA

The Carolina Province was chartered by Charles II of England in 1663. In 1670 settlers arrived and established Charles Town on the west bank of the Ashley River near its junction with the Cooper River. The town soon grew into a wealthy commercial center for shipping of agricultural goods and import to/and from England. Initially siding with the revolutionaries in war of independence, the harbor was attacked repeatedly by the British until it fell after a siege in 1780. Returned to American control in 1782, the town experienced an economic boom that made it high on priority for fortification efforts, with the old colonial fort on Sullivans Island being rebuilt during the First, Second, and Third Systems of defense construction along with a new fort on shoal at the southern edge of the harbor entrance. The Charleston forts were where the first shots of the Civil War were fired and remained in Confederate hands until 1865. Rearmed during the 1870s, and again after the establishment of the Charleston Navy yard in 1901, with 9 new concrete batteries during the Endicott Program, the defenses remained in place with an upgrade of four new batteries during World War II, before being disarmed and declared surplus in 1947.

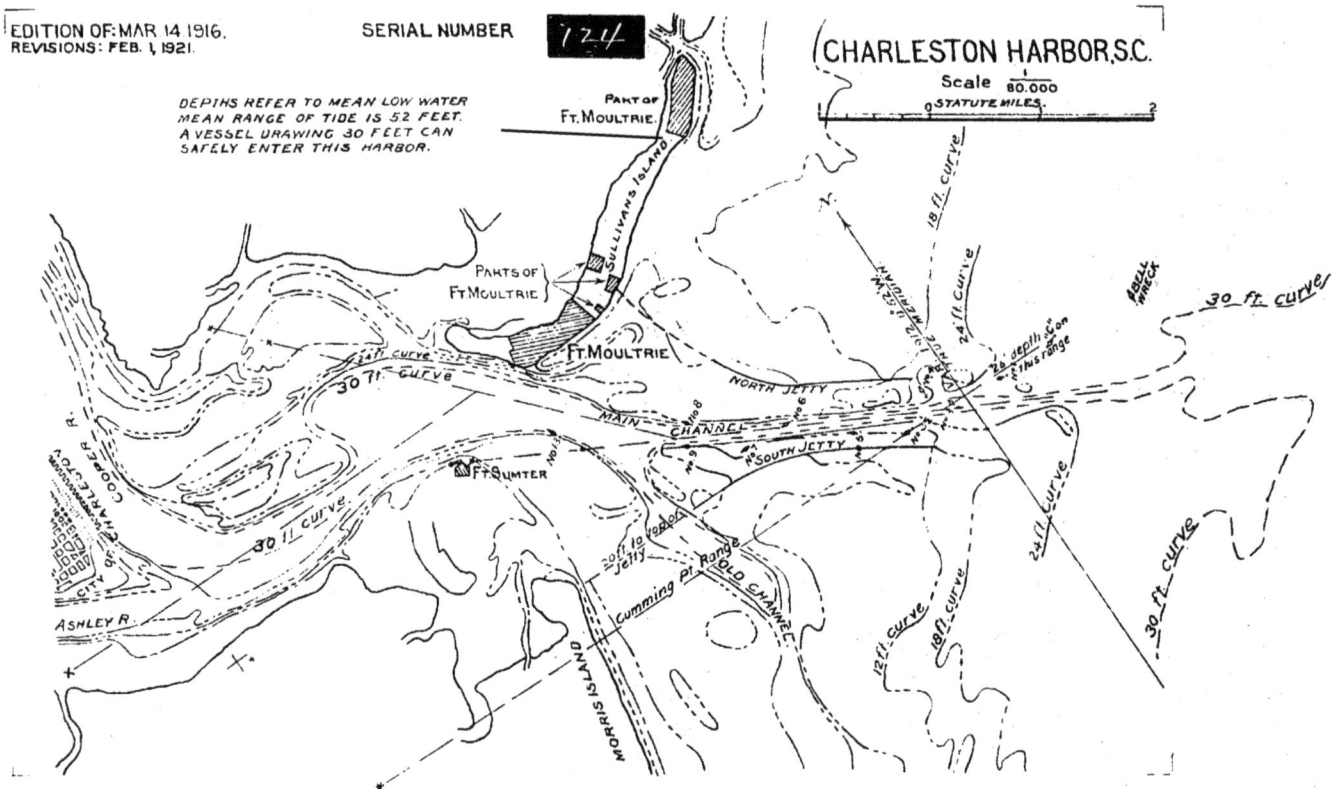

Fort Sumter (1829-1947) is located on shoal at the entrance of Charleston's Harbor, just southeast of Fort Moultrie. Construction of this five-sided, three-tiered, Third System fort was begun in 1828 by building a granite foundation. It was named in General Orders 32 of 1833 for Brig. Gen. Thomas Sumter, Continental Army. By 1860, the fort was still not complete and only a few of planned 135 cannons had been mounted. In April 1861, the famous two-day bombardment of Union forces in Fort Sumter by Confederate forces took place. The fort was then in Confederated hands until 1865 though attacked many times by the Union forces. At the end of the war, the fort was reduced from a masonry fort to an earthen fort by two years of bombardment. The U.S. Army worked to restore it as a useful military installation during the 1870s Period. The damaged walls were re-leveled to a lower height and partially rebuilt. The third tier of gun emplacements was removed. Eleven of the original first-tier gun rooms were restored with 100-pounder Parrott rifles. Fort Sumter greatly altered by the construction of an Endicott Program work, Battery Huger, across the center of the fort in 1897-99. Battery Huger mounted two 12-inch guns, one on a disappearing carriage and the other on a barbette carriage. During this period, Fort Sumter served as a sub-post of Fort Moultrie. During World War II, Fort Sumter received a 90-mm AMTB battery and two temporary barracks. When declared surplus in 1947, the fort was designated to be the Fort Sumter National Monument. The National Park Service has renovated the fort and displays on its Civil War service. Visitors can reach the site through a ferry service from Patriots Point in Mount Pleasant and Liberty Square in downtown Charleston.

Fort Sumter Gun Batteries

- **HUGER**: A battery for heavy 12-inch guns emplaced at Fort Sumter. Early plans in the mid-1890s called for a very large emplacement holding three 12-inch guns on gun lift carriages. In July 1895 extensive tests were made of the foundation at the fort to see if such a heavy structure could be supported. Using an appropriation of $75,000 included in the 1894 Fortification Act, some clearing and foundation work actually began. However, unease with the plan caused it to be quickly suspended and not taken up again until 1898. With the National Defense Act of March 1898, it was decided to build an emplacement for a 12-inch barbette gun across the interior parade of the old work. Work on the platform was quickly done, being reported ready for carriage on May 14, 1898. On June 11, 1898 local engineers were instructed to design an emplacement adjacent to the building barbette emplacement for a Model 1896 disappearing carriage. Despite the difference in armament, the general emplacement plan follows the mimeo for standard Model 1896 emplacements, with lower-level magazines on the left flanks. Ammunition service was initially with hand-operated lifts, but in 1917 electrical chain hoists were installed. The barbette gun was mounted in March of 1899, and the disappearing mount two months later in May. The battery was transferred on June 15, 1899 for a combined construction cost of $97,200. The work had demanded the destruction of the remaining parts of the second tier of casemates of the old fort, and the complete filling-in of the parade in front of the battery. General clean-up, repairs, a new wharf with new entry, and wall facing was also done in concert with the battery work. It was named on General Orders No. 194 of December 27, 1904 for Brigadier General Isaac Huger of Revolutionary War service. It carried one 12-inch Model 1888MII Watervliet gun on a Model 1892 barbette carriage (#30/#14) and one 12-inch Model 1888MII Watervliet gun ona Model 1896 disappearing carriage (#22/#17). The barbette mount was always seen as a wartime expediency and not suitable for a low-lying, exposed position like at Fort Sumter. As early as 1905 suggestions were made to replace it with another disappearing mount, but without the urgency of conflict that was never done. In 1932 expenditures were made for a new battery commander station and for releveling the platforms. The armament was finally authorized for removal on November 25, 1942. The emplacement still exists, and in fact houses the NPS visitor's center at Fort Sumter. The battery is open at certain times and ferry ride is required to reach the fort.

- **AMTB #1**: A 1943 Program AMTB battery for two 90mm fixed and two 90mm mobile mounts intended for Fort Sumter. The fixed gun blocks were located on the filled slope of the parapet in front of Battery Huger. The two accompanying 37mm guns of this battery were sent to be utilized at Morris Island, to the south of Charleston. Concrete work was done from February 12 to April 3, 1943. Transfer was made on July 2, 1943 at a cost of $13,470.53. The battery was dismounted postwar and at some time the blocks were sodded over, no traces remain today.

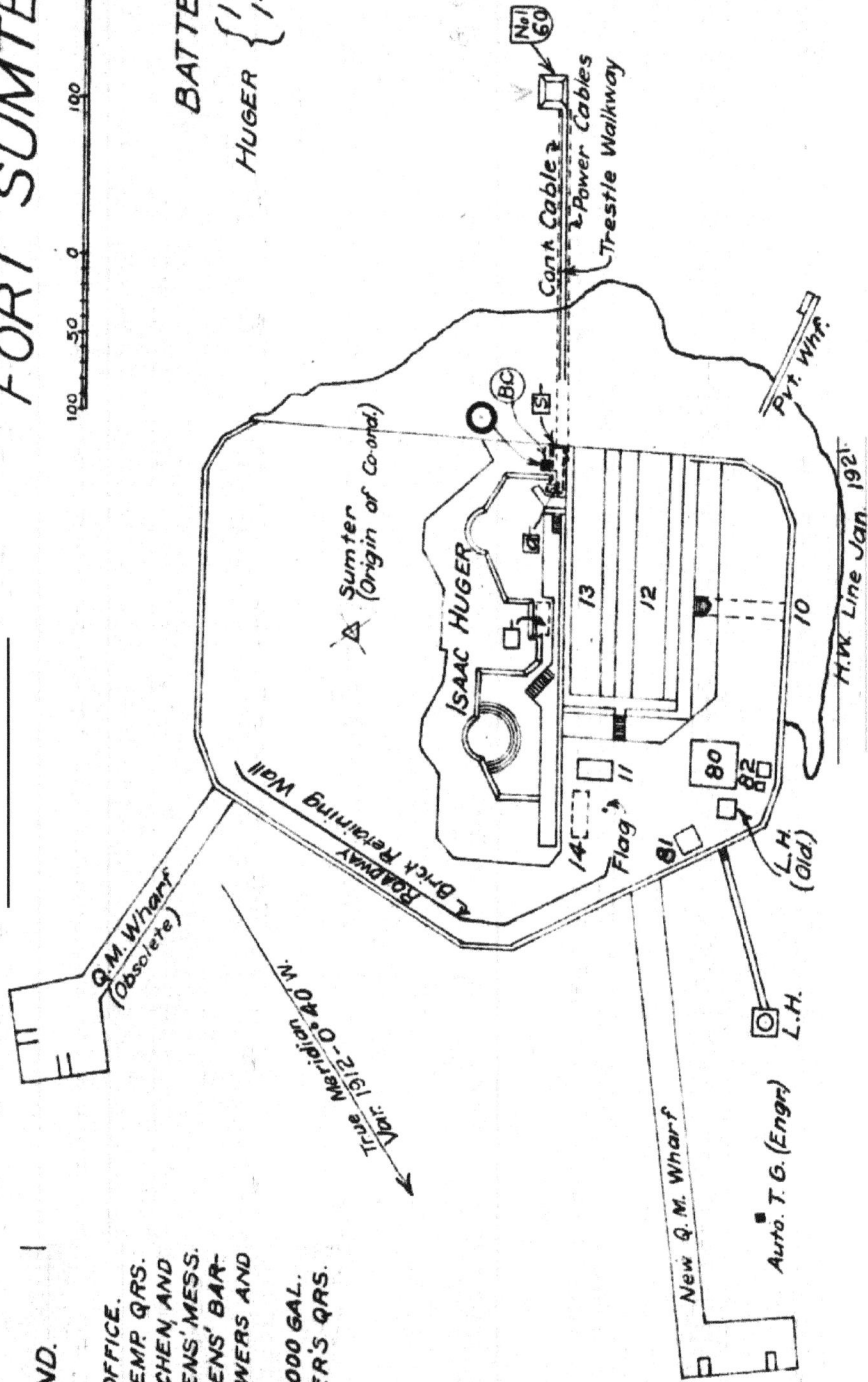

CHARLESTON HARBOR, S.C.

FORT SUMTER

BATTERIES

HUGER { 1-12" Dis.
       { 1-12" N.Dis.

EDITION OF FEB. 1, 1921.

SERIAL NUMBER

LEGEND.

10. SALLY PORT.
11. ARTILLERY OFFICE.
12. OFFICERS' TEMP. QRS.
   & MESS KITCHEN AND
   ENLISTED MENS' MESS.
13. ENLISTED MENS' BAR-
   RACKS, SHOWERS AND
   LATRINE.
14. CISTERN, 20000 GAL.
80. LIGHT-KEEPER'S QRS.
81. FOG BELL.
82. OIL-HOUSE.

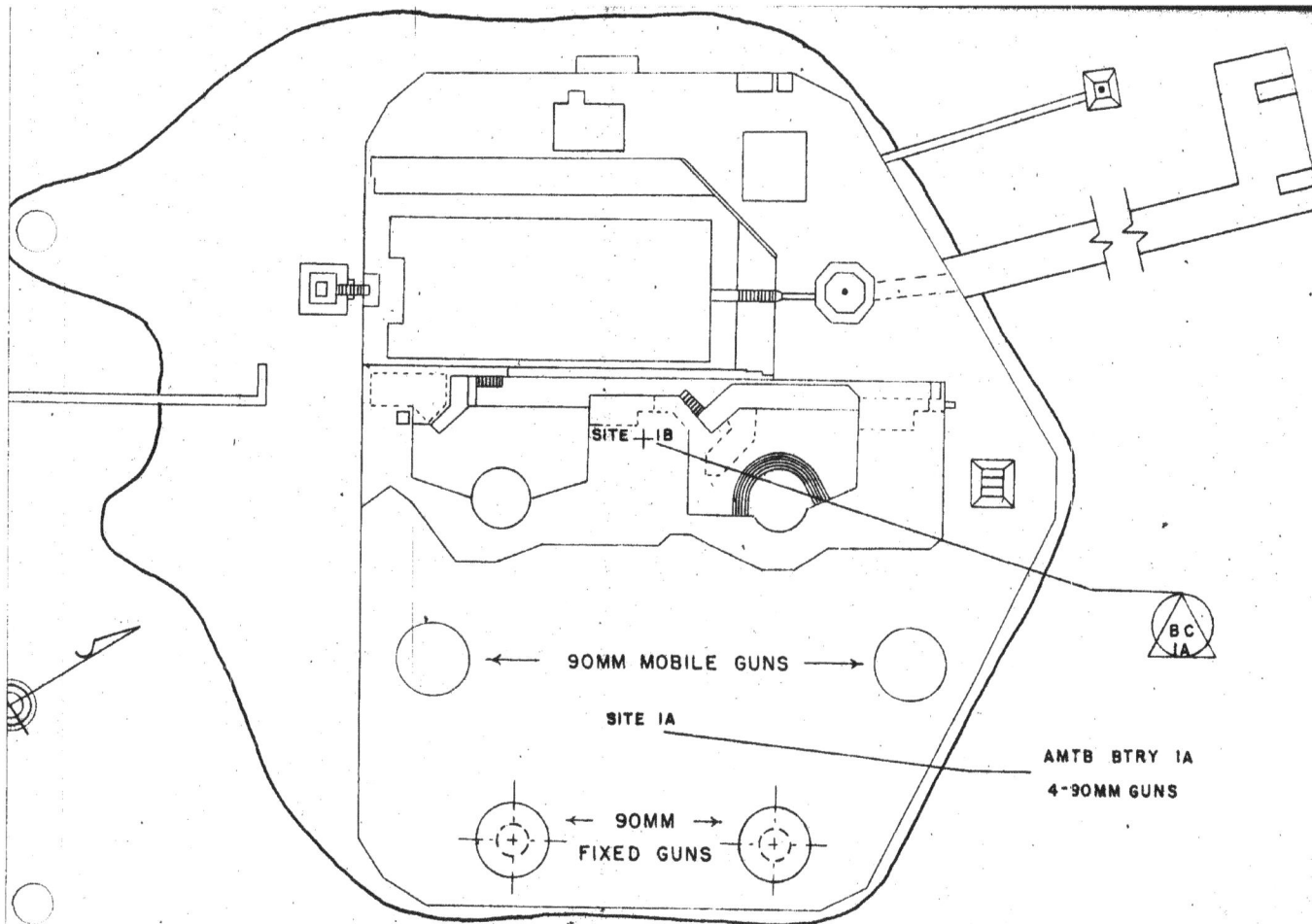

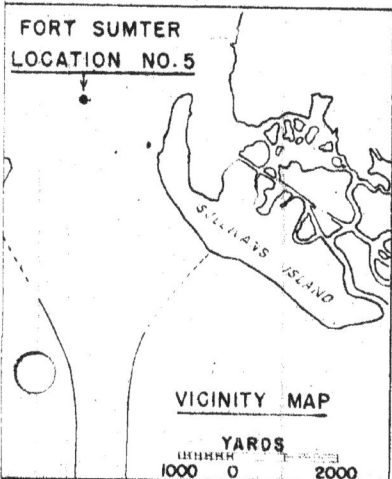

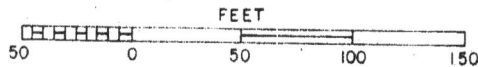

SITE 1B

← 90MM MOBILE GUNS →

SITE 1A

90MM
← FIXED GUNS →

AMTB BTRY 1A
4-90MM GUNS

B C
1A

FEET
50   0   50   100   150

FORT SUMTER
LOCATION NO. 5

SILLIVANS ISLAND

VICINITY MAP

YARDS
1000   0   2000

HARBOR DEFENSES OF
CHARLESTON

LOCATION NO. 5
FORT SUMTER

| REVISION DATE 3/15/45 |

PREPARED BY: HD OF C.   DATE: 3/10/43
APPROVED:
COLONEL CAC   EX. NO.   8B-3

Fort Sumter National Monument (Glen Williford)

Fort Moultrie National Historical Park (Glen Williford)

Fort Moultrie (1808-1947) is located on western end of Sullivans Island at the northern entrance of Charleston's Harbor, just two miles south of the town of Mount Pleasant, South Carolina. The tip of Sullivans Island has been a military site from the colonial period. The first was a palmetto and sand fort built during the American Revolution. The second fort was constructed during 1794-1798 as a First System work. The third fort is the existing Second System fort, which was built from 1808 to 1809. The fort of masonry walls and sand fill served as one of the primary defenses of Charleston, South Carolina up to and through the Civil War. Fort Moultrie played an important role at the start of the Civil War as it participated in the Confederate's bombardment of Fort Sumter. During the war, the fort was involved in repulsing several attempts by the Union Navy to enter Charleston Harbor and was only abandoned by Confederate forces in 1865 as the City of Charleston was seized by the Union Army. It was revamped and reoccupied during the 1870s Period but by 1876, Fort Moultrie had been placed on caretaker status. Fort Moultrie's reservation was used during the Endicott Program for nine concrete batteries and controlled submarine mine facilities. Installed into the walls of the old fort were three rapid-fire batteries, while the fort's primary armament of six 10-inch guns on disappearing carriages and sixteen 12-inch mortars were constructed in several batteries around the reservation. The construction of these batteries from 1898 to 1905 caused Fort Moultrie's military reservation to greatly expand in size and scope. It was finally officially named in General Orders 78 of 1903 for Maj. Gen. William Moultrie, Continental Army, commander of the fort during the Revolutionary War. The fort served as a U.S. Army training post during World War I and continued this role until World War II. The 1940 Program saw the construction of a Series #200 battery near the old fort, while a casemated long-range 12-inch battery was built on the eastern end of Sullivans Island. During World War II, a AMTB battery and 155mm GPF battery was also installed. Declared surplus in 1947, the reservation was sold in parcels to private interests or turn over the State of South Carolina. In 1960, the state turned over its parcels to the National Park Service to become part of the Fort Sumter and Fort Moultrie National Historical Park. The entire 2nd System fort has been set up to display the comprehensive history of American seacoast defense. Two batteries within the old pre-Civil War brick fort retain a 4.7-inch gun and a 3-inch gun. Open daily with a visitor center and tours. The remaining garrison structures and four other batteries are privately owned or owned by the local civic authorities. Fort Moultrie serves today as an excellent example of the development of America's coast defenses from the Second System to World War II.

Fort Moultrie Gun Batteries

- **BUTLER – CAPRON**: The battery for sixteen mortars assigned to the Fort Moultrie reservation on Sullivan's Island north of the main entry to Charleston Harbor. Located northeast of old Fort Moultrie, plans were submitted on August 24, 1896. The plan was the older 2 x 2 quadrangular type, with four pits arranged in a rectangle entirely surrounded by parapets. An attempt was made to proportion protection to the expected exposure to incoming fire. The north side was considered relatively safe and was less protected. The northeastern end of the eastern parapet was cut-off obliquely, leaving it open. The spacing was reduced, the rear set of pits were moved 40-feet closer to the others versus the type plans. This meant that the pit slopes were steeper than normal, and they had considerable erosion problems over time. The mortars were at just a 10-foot elevation over mean low water level. The pits were of the older, narrow type (40 x 60-feet). The battery had a relocator room, as well as firing battery and dynamo rooms. Ammunition service was by same-level overhead trolley to trucks. Work was done from 1896 to mid-March 1897 except for mounting base rings. By June 30, 1897 all sixteen carriages and eight of the mortars were mounted. Transfer was made on June 28, 1898 for a cost of $185,357.71. It was named on General Orders No. 134 of July 22, 1899 for Allyn K. Capron, 7th U.S. Cavalry who was killed in the Battle of Las Guasimas in Cuba in 1898. It was armed with eight model 1886 cast-iron, steel hooped mortars on Model 1891 car-

CHARLESTON HARBOR, S.C.

FORT MOULTRIE

SULLIVAN'S ISLAND
GENERAL MAP.

BATTERIES.

PIERCE BUTLER 4 –12" M.
CAPRON __ 4 –12" M.
JASPER __ 4–10" Dis.
THOMSON __ 2 –10" " "
LOGAN ____ 1– 6" " "
GADSDEN __
BINGHAM __
MCCORKLE.
LORD __ P.

A. Anti β Guns

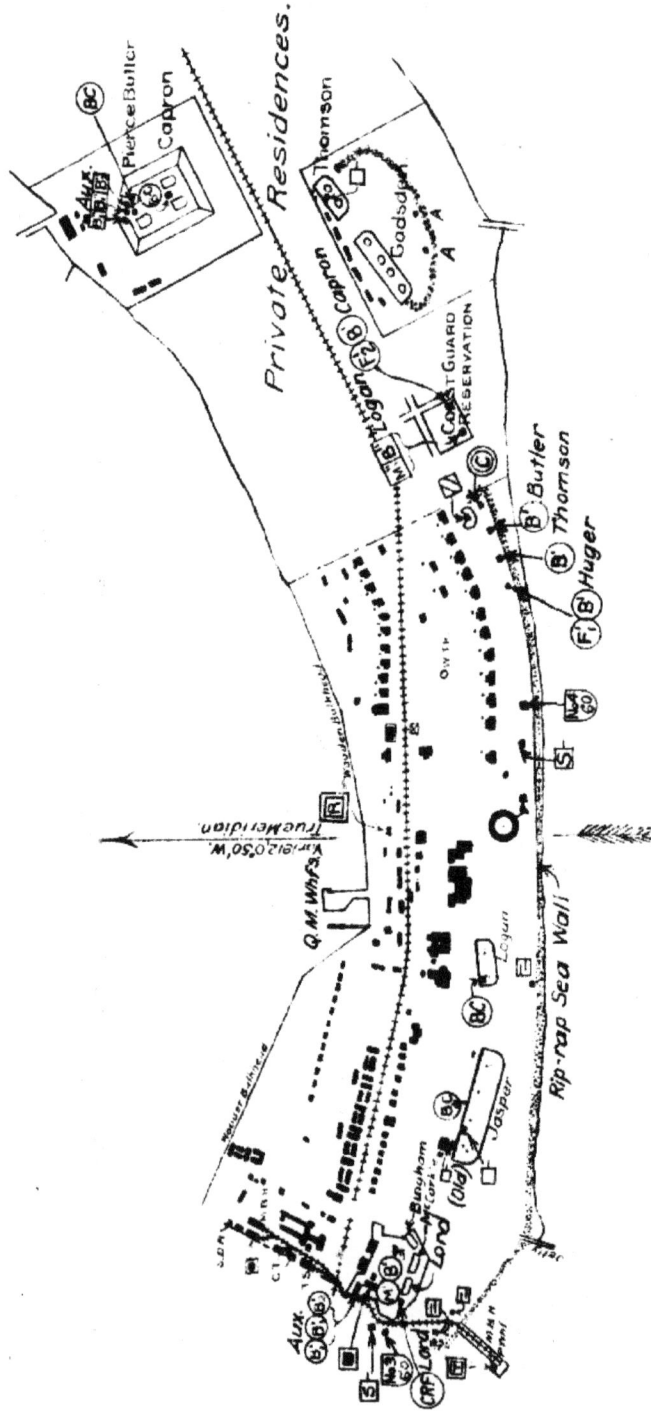

SERIAL NUMBER

EDITION OF APR. 23, 1915.
REVISIONS: DEC. 7, 1915, MAR. 14, 1916.
NOV. 8, 1916; FEB. 1, 1921.

Scale of Feet.

4000 Ft.
3000
2000
1000
500 0 500 1000

CHARLESTON HARBOR, S.C.
FORT MOULTRIE - D1.
SULLIVAN'S ISLAND, S.C.
WESTERN PORTION.

SERIAL NUMBER

EDITION OF APR.23,1915.
REVISIONS: DEC.7,1915; MAR.14,1916.
NOV.8,1916; FEB.1,1921.

LEGEND.

1 ADMINISTRATION BLDG.
2 COMMANDING OFFICERS QRS.
3 OFFICERS QUARTERS.
3a BACH.OFFICERS QRS.
4 HOSPITAL.
4a DISPENSARY.
4b DEAD HOUSE.
5 HOSPITAL STEWARD'S QRS.
6 N.C.OFFICERS QRS.
7 BARRACKS.
8 GUARD HOUSE.
9 POST EXCHANGE.
10 ENGR.OFFICE.(ARTY.)
11 COAL SHED.
12 OIL HOUSE.
13 STABLE.
14 WORK SHOP.
15 RESERVOIR.
16 FIRE HOUSE.
17 BAKE HOUSE.
18 LAUNDRY.
19 BAND STAND.
20 Q.M.STOREHOUSE.
21 COM. STOREHOUSE.
22
23 GARAGE (Q.M.).
24 WOOD SAW (ELEC.).
30 ORD.STOREHOUSE.
31 ORD.WORK SHOP.
41 ENGR.STOREHOUSE.
70 CHURCH.
71 SERVICE CLUB.
72
100 WAGON SHED.
101 OLD BAKE HOUSE.
102 BAND BARRACKS.
104 MARRIED ENLISTED
    MEN.
105 CISTERNS.
106 FOWL HOUSE.
107 OUT HOUSE.
5a NURSES QUARTERS.
103 SCALES.
108 WAR GAME HOUSE.
109 POWERHOUSE (ABANDOND)

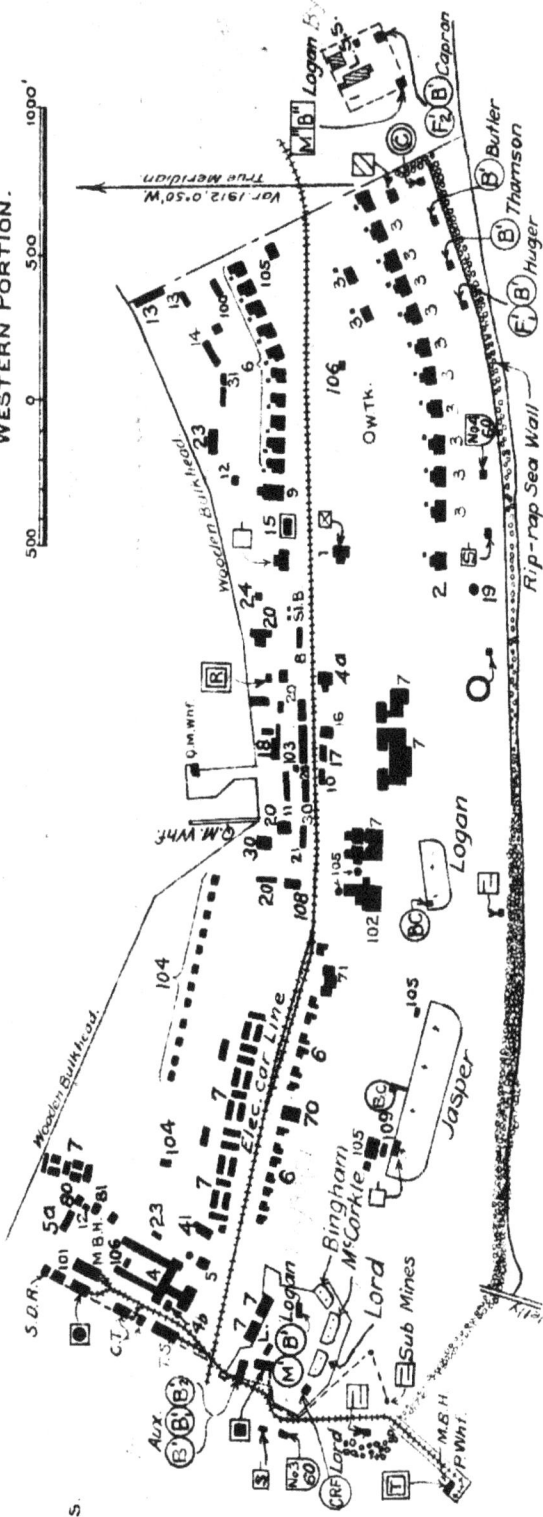

BATTERIES.

JASPER......4-10"DIS.
LOGAN.......1-6" "
BINGHAM.
McCORKLE.
LORD.......2-3" P.

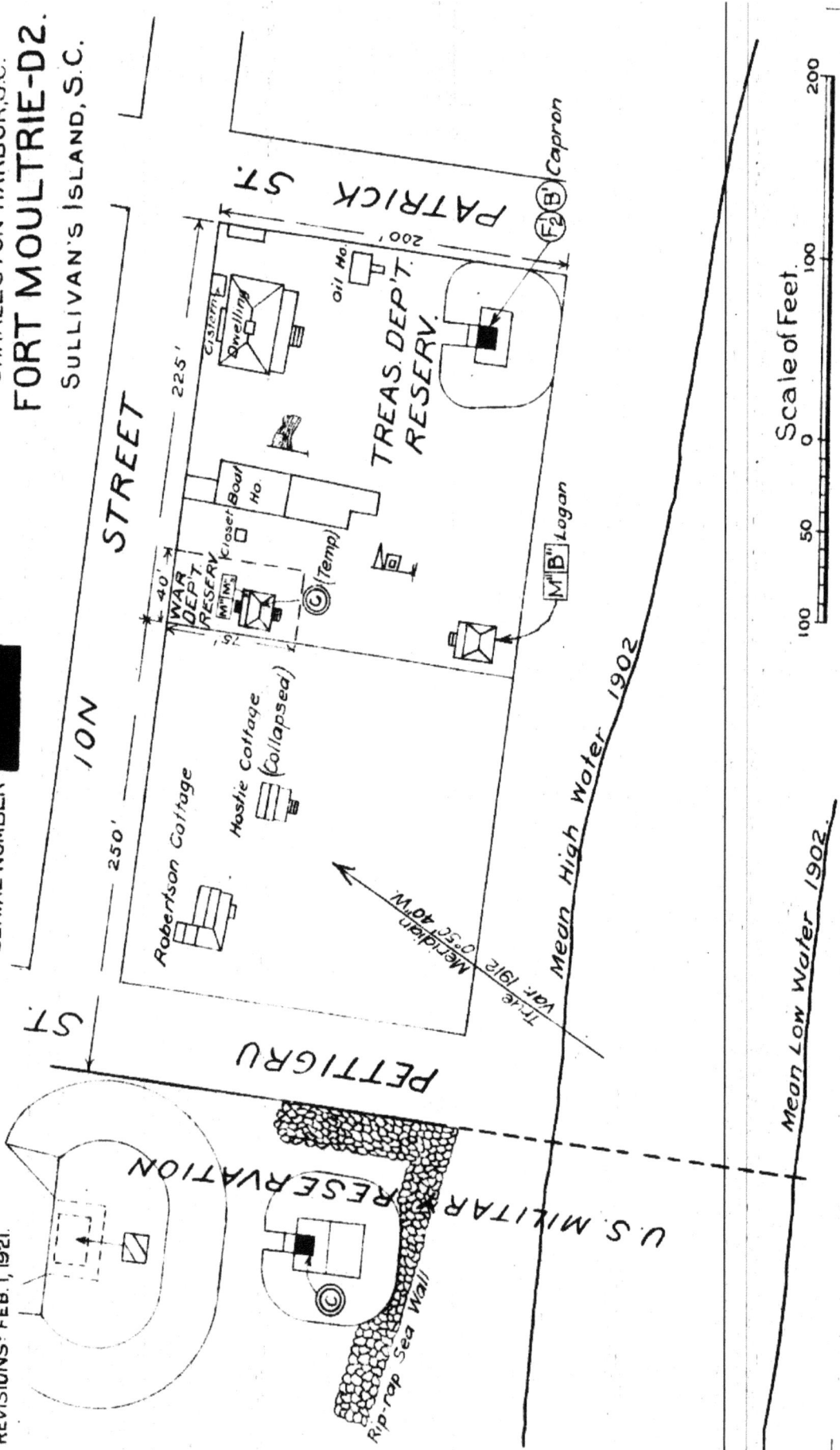

CHARLESTON HARBOR, S.C.
FORT MOULTRIE-D2.
SULLIVAN'S ISLAND, S.C.

SERIAL NUMBER

TREAS. DEP'T. RESERV.

WAR DEP'T RESERV.

Dwelling

Cistern

Oil Ho.

Boat Ho.

Kitchen

(Temp.)

(Temp.)

Robertson Cottage

Hostie Cottage
(Collapsed)

PATRICK ST.

PETTIGRU ST.

STREET

ION

ST.

U.S. MILITARY RESERVATION

Rip-rap Sea Wall

Mean High Water 1902

Mean Low Water 1902

True Meridian 1912
Var. 1912 0°5C'40"W

200'

225'

250'

40'

F₂ B' Capron

M B' Logan

Scale of Feet.
100   50   0   100   200

EDITION OF APR. 23, 1915.
REVISIONS: FEB. 1, 1921.

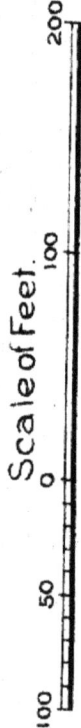

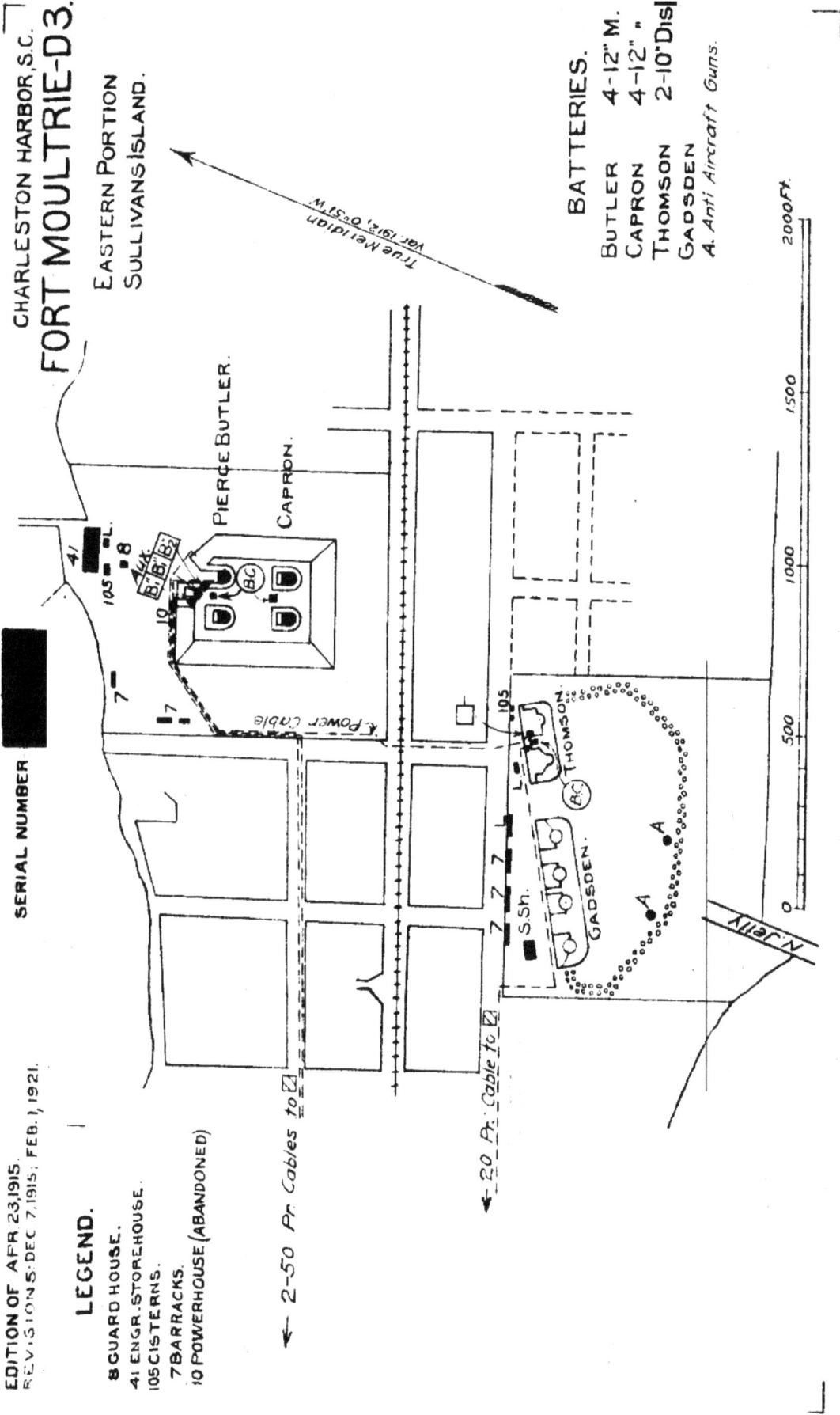

CHARLESTON HARBOR, S.C.
FORT MOULTRIE-D3.

EASTERN PORTION
SULLIVANS ISLAND.

BATTERIES.

BUTLER   4-12" M.
CAPRON   4-12" "
THOMSON   2-10"Dis|
GADSDEN
A. Anti Aircraft Guns.

True Meridian
Var. 1912, 0°51'W.

SERIAL NUMBER

EDITION OF APR 23,1915.
REVISIONS: DEC 7,1915; FEB. 1, 1921.

LEGEND.

8 GUARD HOUSE.
41 ENGR. STOREHOUSE.
105 CISTERNS.
7 BARRACKS.
10 POWERHOUSE (ABANDONED)

PIERCE BUTLER.

CAPRON.

Power Cable

2-50 Pr. Cables to

20 Pr. Cable to

THOMSON

GADSDEN.

S.Sh.

JETTY

SS

H

HE
CP   H

BTRY
LORD
2-3" RF
GUNS

M

BC
2A

SCR 296
FOR 230

T   P

IA

IE   IC

IP

+ ID

+ IJ

+ IH

IL

+ IN

IF +
IB

+ IK

CH-1

IM +

IG

CH-2

CONST. 230
2-6" BC GUNS

BC₁
CRF

BC₂

B S³ ³
³ ³

AMTB BTRY 2A
4-90MM GUNS

B S³ ³
2 2

SCR 582

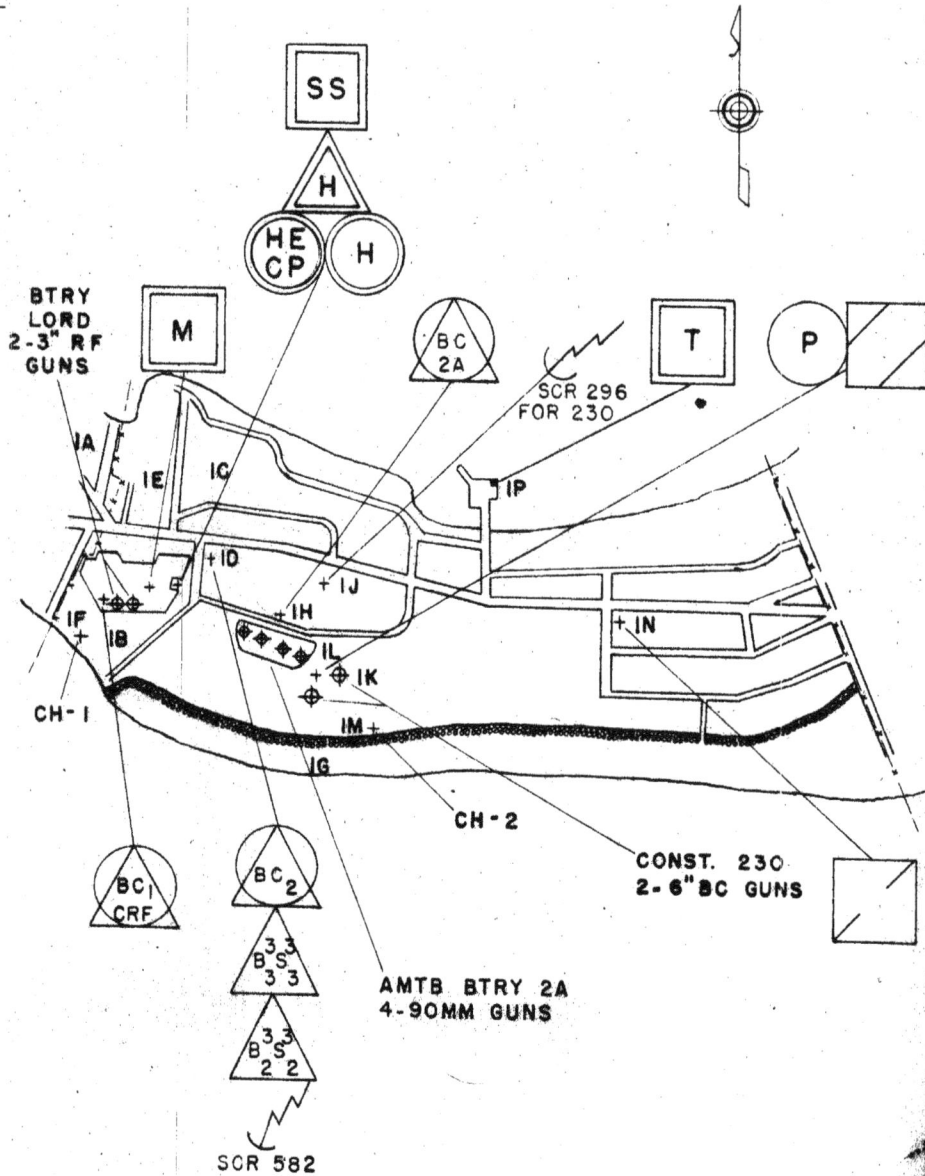

NOTE: SYMBOLS IA, IK ETC.
REPRESENT SITE NUMBERS.

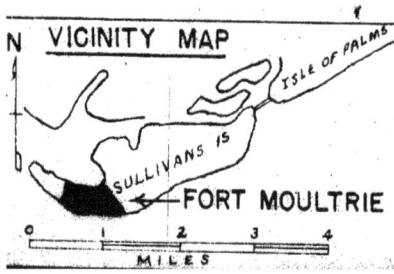

FEET
200 0 200 400 600 800 1000 1200 1400

VICINITY MAP

N

ISLE OF PALMS

SULLIVANS IS

FORT MOULTRIE

0   1   2   3   4
MILES

HARBOR DEFENSES OF
CHARLESTON
LOCATION NO.6
FORT MOULTRIE

REVISION
DATE
3/15/43

PREPARED BY HD OF C DATE 3/18/43
APPROV _____ EX. NO. BB-6
COL. CAC

Fort Moultrie 1929 (NARA)

Fort Moultrie 1920s (NARA)

Fort Moultrie 1920s (NARA)

Fort Sumter 1920s (NARA)

riages (#62/#58, #51/#60, #28/#65, #32/#61, #33/#62, #45/#36, #57/#37 , and #2/#35) and eight specially-ordered Model 1886-1890 all-steel mortars on Model 1891 carriages (#7/#74, #3/#75, #2/#76, #8/#79, #5/#81, #6/#82, #4/#83, and #1/#80). In 1906 the northern two pits were split off administratively to form Battery Butler. It was named in General Orders No. 20 of January 25, 1905 for Colonel Pierce Butler, South Carolina Volunteers, who was killed at Churubusco Mexico in 1847. For several years problems with the design and construction plagued the work, In July 1901 the retaining walls on the slopes began disintegrating with firing, and their Rosendale cement had to be replaced with Portland cement in select locations, which cost $2016.00 In 1904 a more extensive rebuilding was required for over $22,000. Due to settling and the ramps then required for ammunition trucks, doors had to be adjusted, lifts and tables installed for shell and shot, and telephone booths erected in each pit. Also, a new plotting room was built into the place of the old power room of the battery. The armament was changed in 1920. All sixteen old-type mortars and carriages were removed and the battery re-armed with eight Model 1890M1 mortars and Model 1896 carriages—but with just two per pit. Five of the tubes and all eight of the carriages came from Battery Anderson-Ruggles at Fort Monroe. The new armament was mortars Bethlehem #1, #21, #24, and #25, Watervliet #7, #21, #24, and #128 on carriages #1, #2, #5, #7, #12, #11, #305, and #304. This armament served until removed during World War II under authority of November 25, 1942. For years after 1947 the emplacement existed in a generally abandoned state, but it was filled-in for public safety in the early 2000s. The site is controlled by the Town of Sullivan's Island as a town park.

- **JASPER**: A battery for four 10-inch disappearing guns emplaced on the old Fort Moultrie reservation. It was placed about 600-yards to the east of the masonry fort, generally in a west-east straight line. Additional property had to be purchased to use this space. Plans were submitted for the first two emplacement using funds from the Act of June 6, 1896 on September 21, 1896. The recommendation was to start the eastern two emplacements first. Later in late 1896 and then early 1897 the final two emplacements were added. The plan followed type plans in most regards. Noting the danger from hurricane flood surges, the front slope was exchanged for a vertical retaining wall. Admitting that it would be exposed to enemy fire, the protection of concrete and earth depth in front was increased. Work was done in 1897, and except for a few details, it was completed by July 1, 1898. It was transferred on May 4, 1898 for a cost of $238,100. It was named on General Orders No. 23 of April 8, 1898 for Sergeant William Jasper who died defending Charleston in 1776. The guns were received in November 1897 and soon mounted. The battery carried four 10-inch Model 1888 Watervliet guns on Model 1896 disappearing carriages (#33/#30, #58/#26, #60/#24, and #62/#25). Minor modifications were made for telephone niches in 1903, range finding stations in 1907, wooden extensions to loading platforms in 1917 and a new battery commander's station in 1931. In 1910, a new power plant was erected in an adjacent brick building replacing the power room in the battery. Chain hoists were installed in 1904. In 1915 the battery was administratively split, simply becoming Jasper I and Jasper II. Jasper I was disarmed in 1917 and then rearmed with new tubes in 1920— Model 1888 Watervliet tubes #9 and #20. The final set of four guns was not removed until early in World War II authority coming in a command of November 25, 1942. The emplacement still exists at the Fort Sumter and Fort Moultrie National Historical Park. It has been partly restored by the NPS with working shell lifts, shell magazines, and static display power room are on public display.

- **LOGAN**: A battery for two 6-inch guns emplaced during the Spanish American War. It was located to the east of Battery Jasper and fired to the south across the main channel. In April of 1898 local engineers were asked to build an emplacement for one of the eight 6-inch British guns purchased during the Spanish American War emergency. The platform was reported ready on August 19, 1898, but they were awaiting the gun. On February 4, 1899 plans were submitted for a single 6-inch Model 1897 gun on disappearing carriage for a second emplacement at this battery, and with two types of guns the battery design was a hybrid type. The first (No. 2) emplacement was a simple concrete platform with a single, lower left flank magazine room. The second (No.1) emplacement was an interior emplacement with a lower magazine using a back-delivery hoist and usual store and support rooms. Work was done in 1898 to August 1899. Both the Vickers and the disappearing gun unit were transferred on September 25, 1899 for a total cost of $34,618. It was named on General Orders No, 78 of May 25, 1903 for Captain William Logan killed in action against the Nez Perce Indians at Big Hole Montana in 1877. It was armed with one 6-inch Vickers gun and pedestal (#12137/#11158) and one 6-inch Model 1897M1 gun on Model 1898 disappearing carriage (#3/#4). The pedestal gun was removed early, being shipped out on December 28, 1904 for eventual use in Battery Bankhead at Fort Adams. The disappearing unit remained active into World War II. It was finally deleted using authority of February 17, 1944. The emplacement still exists, though heavily overgrown. The Town of Sullivan's Island owns the battery, but in an abandoned condition. A local effort is underway to remove vegetation from the battery. The battery is open to the public.

- **BINGHAM:** A battery for two 4.7-inch, rapid-fire guns emplaced inside old Fort Moultrie during the emergency of the Spanish American War. This was emplacement for two of the 34 Armstrong 4.7-inch guns purchased in April of 1898. Simple separate platforms for each gun were erected along the ramparts of the old fort, firing to the south. Emplacements were reported ready on July 25, 1898. It utilized an older 1870s brick magazine connected with a new concrete passageway to the guns, as magazine. Otherwise, it in general followed the dimensions for the new 5-inch RF batteries. Concrete work was done from April 12 to May 1, 1898. Transfer was made on November 3, 1899 for a cost of $6,000. It was named on General Orders No. 78 of May 25, 1903 for 2nd Lieutenant Horatio Bingham, who was killed in action against the Sioux near Fort Phil Kearny in 1866. It was armed with two 4.7-inch guns on pedestals (#11687/#11013 and #11008/#11017). This armament served until being dismounted at the start of World War I in 1917. The emplacement was subsequently used for general storage. It still exists today with a 4.7-inch gun and pedestal put on display here by the National Park Service at the Fort Sumter and Fort Moultrie National Historical Park. The battery is open to the public.

- **McCORKLE:** A battery for three 3-inch, 15-pounder masking parapet guns built on the southern parapet of old Fort Moultrie, just to the west of Battery Lord and Bingham. Submission was made on November 9, 1898. The plans followed the design for such batteries in Mimeograph #30, with standard 29-foot spacing between platforms. One change was to incorporate a 1870s brick magazine built for a 15-inch Rodman emplacement into the battery design. Work was done from January 1, 1899 to 1900. Transfer was made to Coast Artillery troops on July 16, 1901 for a cost of $9,390.79. It was named on General Orders No. 78 of May 25, 1903 for 1st Lieutenant Henry L. McCorkle who was killed in Cuba in 1898. It was armed with three 3-inch Model 1898 Driggs-Seabury guns and masking parapet mounts (#6, #13, and #14). The guns were dismounted when all such batteries were declared obsolete in June 1920. One emplacement was later modified to hold a special, hybrid mount. A Model 1898 modified tube (#13) was emplaced on a modified Model 1903 pedestal (#65) on November 19, 1923. There is some evidence that this combination may have been first installed in a mobile emplacement, but most of its service life occurred here at Fort Moultrie. It remained until declared surplus and removed in early 1943 by authorization given on November 12, 1942. The emplacement still exists, with a Model 1911 3-inch practice gun in emplacement No. 1, at the Fort Sumter and Fort Moultrie National Historical Park. The battery is open to the public.

- **LORD**: A battery for two 3-inch pedestal guns also emplaced along the southern rampart wall of old Fort Moultrie. It was in the space between Battery Bingham and Battery McCorkle. Plans were submitted on September 8, 1902. It generally followed type plans, with the two-gun platforms separated by 42-feet as was common with the Model 1902 emplacements. Like McCorkle it incorporated an older 1870s magazine in its design as the traverse ammunition storage. Work was done in 1902-1903, it being reported ready for transfer on December 17, 1903. Transfer was made on December 31, 1903 for a cost of $12,995.85. It was named on General Orders No. 78 of May 25, 1903 for Assistant Surgeon George E. Lord who was killed in 1876 at the Little Big Horn. The battery was armed with two 3-inch Model 1902 guns and pedestal mounts (#26/#26 and #27/#27). There occurred a significant accident here in 1913 when a defect in the breech firing mechanism caused a premature discharge of a round, killing four and wounding three seriously. Following this incident, all Model 1898 and Model 1902 3-inch guns were prohibited from additional firings in the U.S. until all their firing pin guides were inspected. The armament continued to serve as one of the primary light, rapid-fire batteries for Charleston until it was removed in March 1946. The emplacement itself was destroyed in 1975 by the National Park Service in order to reconstruct an 1870s battery for display purposes in the same area.

- **THOMSON**: A new 10-inch disappearing battery emplaced on a separate section of the Fort Moultrie reservation near the North Jetty, to the east of the lighthouse reservation and to the south of the mortar battery. It covered the approaches to the harbor from the east. It fired to the south, southeast. The plans were submitted on April 20, 1903. For a short while this reservation was called Fort Getty but soon consolidated with the rest of Fort Moultrie. It was consistent with type plans with few modifications. It had complete 2-foot-wide air passages around the magazines for ventilation and its own electric plant to serve this and the adjacent 6-inch battery. The No. 1 emplacement was built as an interior emplacement, the No. 2 as a left flank emplacement. During construction the loading platforms were extended by 5-feet by authorization of December 18, 1903. Work was done from 1903-1904. It was transferred on September 10, 1906 for a total cost of $115,000. It was armed with the Model 1900 guns on Model 1901 disappearing carriages (#3/#2 and #7/#3) and thus was among the most modern and powerful works for the defense of Charleston Harbor. The battery was named on General Orders No. 113 of June 23, 1904 for Colonel William Thomson of Revolutionary War service. The battery remained armed through the war years, not being authorized for removal until March 26, 1945. The emplacement still exists, and it is owned and used by the Town of Sullivan's Island as a fire training center. The battery site is open, but the interior is closed to the public.

- **GADSDEN**: The new four-gun emplacement for 6-inch disappearing guns emplaced on the right flank, generally in-line with the two guns of Battery Thomson. Plan submission was made on July 6, 1903. It closely followed the type plans as recommended in Mimeograph #59. It was built with two end, flank emplacements and two interior emplacements. There was a shared magazine between the first two guns and between the third and fourth gun. Work was done in 1904-1905. Transfer was made on September 10, 1906 for a cost of $88,200. It was named on General Orders No. 113 of June 23, 1904 for Brigadier General Christopher Gadsden of Continental Army service during the Revolutionary War. It was armed with four 6-inch guns Model 1903 on Model 1903 disappearing carriages (#15/#13, #62/#14, #63/#16, and #64/#26). Like many other such 6-inch batteries,

Battery 230 and Battery Jasper, Fort Moultrie National Historical Park (Glen Williford)

Battery Butler-Capron, Sullivans Island (Glen Williford)

Battery Gadsden Sullivans Island (Glen Williford)

Battery #520 Sullivans Island (Glen Williford)

it was disarmed in 1917 to provide guns tubes for field artillery mounts. The carriages were then scrapped in place in 1919 or 1920. The emplacement still exists on property owned by the Town of Sullivan's Island. The emplacement is being used for the Poe Public Library, a cultural center, and town storage area. The battery is open, but interior is only open at certain times.

- **Battery #230**: A standard 1940 Program dual 6-inch barbette battery built to the east of Jasper and Logan. It was the only modern 6-inch emplacement allotted to Charleston, being given national priority #22 and funded under the FY-1942 Budget. Original plans called for this unit to be sited at Fort Sumter, but the local board soon transferred it here to Moultrie as a more practical use of space. Construction was done between September 18, 1942 and November 30, 1943. It was transferred on February 29, 1944 for a cost of $297.220. The planned armament was for 6-inch guns M1(T2) on Model M4 barbette carriages. The guns were never delivered, but M4 carriages #43 and #44 were received at the post. The battery was never completed or armed, and eventually indefinitely suspended. The emplacement served a variety of purposes for the military, including the construction of harbor control structure by the U.S. Navy on top of the emplacement during the Cold War. The emplacement and building were transferred to the National Park Service as part of the Fort Sumter and Fort Moultrie National Historical Park. The battery is closed to the public.

- **AMTB #2**: A battery for two 90mm fixed and two 90mm mobile guns emplaced on old Battery Jasper's parapet in late 1943. Two accompanying 37mm guns of the battery were to be located further east on Sullivan's Island. Work was done from February 12 to March 28, 1943. Transfer was made on July 2, 1943 for a cost of $18,521. The simple blocks were set in flush in the earthen parapet atop old 10-inch Battery Jasper, magazines and plotting rooms were in rooms of the Endicott battery. The battery served until removed in 1946. The blocks still remain at the Fort Sumter and Fort Moultrie National Historical Park.

Marshall Military Reservation (1905-1947) is located on the eastern end of Sullivan's Island near the intersection of Jasper Boulevard (State Highway 703) and Middle Street. Batteries built here during World War II were Battery 520 (1944-1947) and a four-gun 155mm battery (1941-1945, two Panama mounts in 1942). Construction #520 was one of the few 12-inch gun batteries (using guns from Battery Kimble, Fort Travis, Texas) that was completely new construction at a new location. Two of the 155mm guns were transferred to Folly Island in 1942. A fire-control tower and a radar tower were also once located here. On 18 May 1944, an SCR-296-A Radar set was installed and accepted for service as a gun laying radar primarily for Battery 520 and secondarily for Battery 230. The gun casemates and plotting room of Battery 520 are now each separate private residential homes on I'on Ave., between Brownell Ave. and Station 30 and Station 31 Streets, known locally as "Fort Marshall". The two Panama mounts are now in the surf behind 2907 and 2917 Marshall Blvd. Now known as the Marshall Reservation sub-division of beach home.

Marshall Military Reservation Gun Batteries

- *Battery #125* (planned): The northern part of Sullivan's Island was considered for heavy batteries for quite some time in the modern period. The reservation itself had been acquired in 1905 for fire control facilities. As early as the 1915 Program the movement here of 12-inch mortars or 10-inch guns from nearby Fort Moultrie had been discussed. As part of the 1940 Program a battery for two modern, casemated 16-inch guns had been planned, and given the designation of Battery Construction No. 125. It was to be one of the standard "100-series" type batteries. However, by 1941 plans had changed, and a new site for this battery was selected to the south of Charleston on

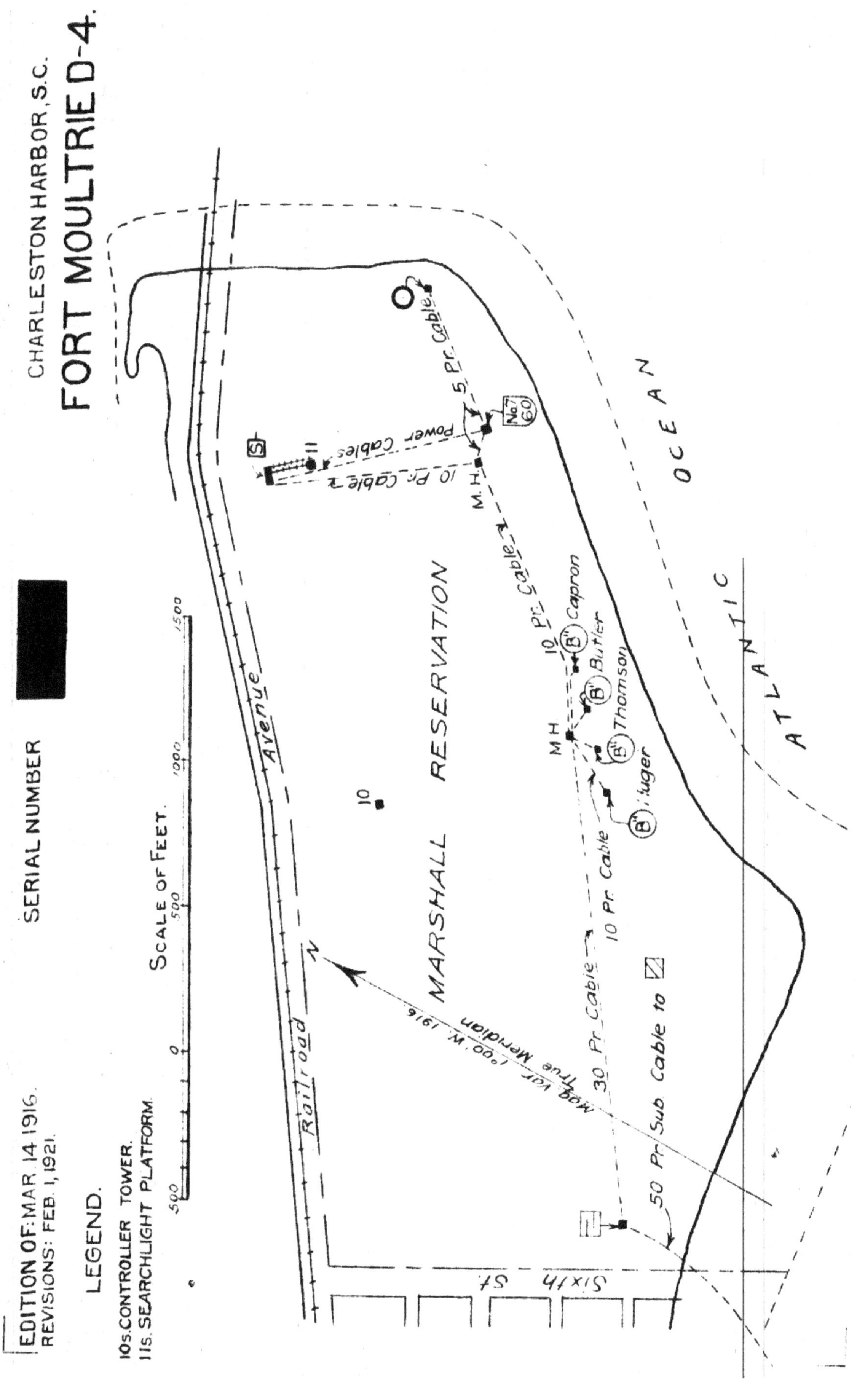

FORT MOULTRIE D-4.

CHARLESTON HARBOR, S.C.

EDITION OF: MAR. 14 1916.
REVISIONS: FEB. 1, 1921.

LEGEND.

105. CONTROLLER TOWER.
115. SEARCHLIGHT PLATFORM.

SERIAL NUMBER

SCALE OF FEET.

MARSHALL RESERVATION

ATLANTIC OCEAN

Avenue

Railroad

Sixth St.

Power Cables
5 Pr. Cable
10 Pr. Cable
No. 60
M.H.
10 Pr. Cable
Capron
Butler
Thomson
M.H.
Luger
30 Pr. Cable
10 Pr. Cable
50 Pr. Sub Cable to
Mag. Var. 1°00' W. 1916
True Meridian
N
10

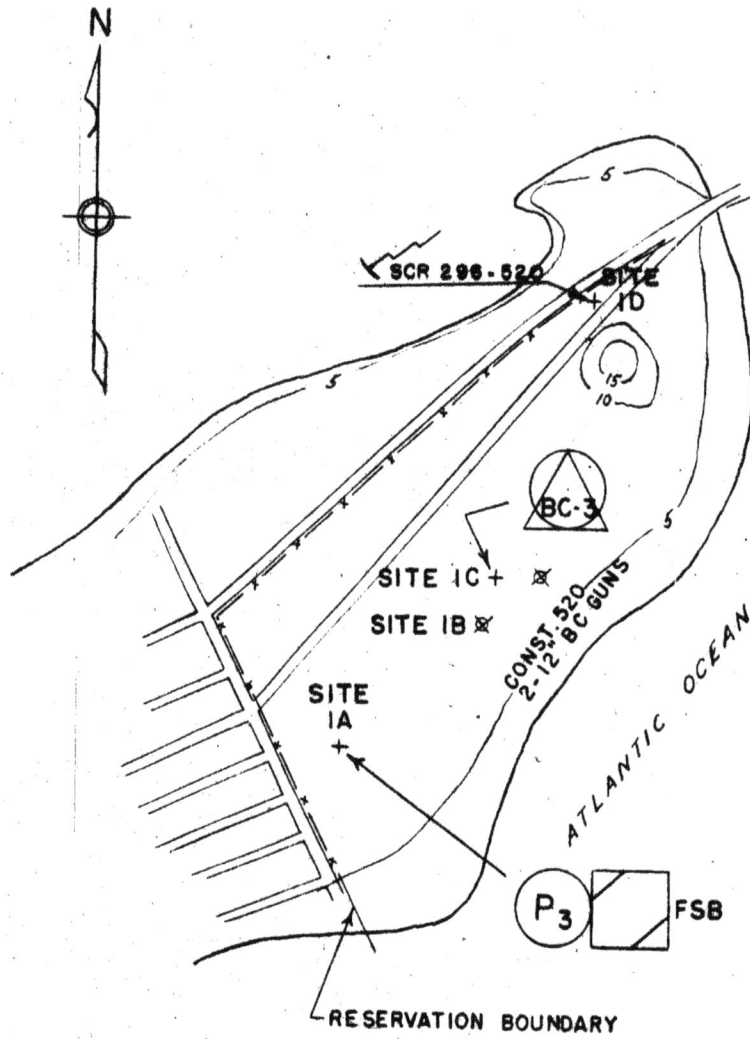

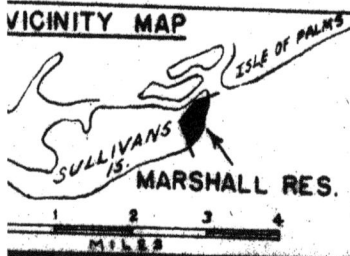

N

SCR 296-520   SITE 1D

15
10

BC-3

SITE 1C + ⊠

SITE 1B ⊠

CONST. 520
2-12" BC GUNS

ATLANTIC OCEAN

SITE
1A
+

P₃ / FSB

RESERVATION BOUNDARY

FEET

200 0 200 400 600 800 1000 1200 1400

VICINITY MAP   ISLE OF PALMS

SULLIVANS IS.

MARSHALL RES.

1   2   3   4
MILES

HARBOR DEFENSES OF
CHARLESTON

LOCATION NO. 9
MARSHALL RESERVATION

PREPARED BY HD OF C | DATE 3/11/43
APPROVED
COL CAC | EX NO. 88-5

James Island. Eventually even this work was cancelled, and Battery #520 became the only large, new work to be built for the Charleston defenses. No work was ever undertaken on project #125.

- **Battery #520:** With the cancellation of the Battery Construction No. 125 16-inch gun project, there was still a need for a heavy gun battery to protect Charleston. Correspondence of November 11, 1942 revealed that two 12-inch guns on Model 1917 barbette carriages were available when the proposed transfer of guns from Battery Haslet in Delaware to Trinidad was cancelled, and it was proposed to move these guns to a new battery at the Marshall Reservation. By December 18, 1942 the move was approved, though the guns from Battery Kimble at Fort Travis were substituted. The James Island site had been eliminated as it was 5000-yards from the shoreline, thus being too restrictive for the range of a 12-inch battery. Thoughthe Marshall reservation was considered to be too far to the north, it was finally selected as the best option available. This was verified by the Board recommendation of February 1, 1943 and by February 10th the construction was assigned as Battery No. 520 and authorization to proceed was given. It was built between May 1st and December 31st of 1943. Transfer to troops was made on February 28, 1944 for a cost of $1,432,443. The guns were received by mid-1944 and they were mounted soon after. It was armed with two 12-inch Model 1895M1A4 Watervliet guns on Model 1917 long-range barbette carriages (#48/#21 and #49/#22). Similar in design to Battery #519 at Fort Miles, #520 was a slightly smaller, more simplified 16-inch casemated design with a gun spacing of 420-feet. It was provided with a separate PSR building and a series of assigned fire control stations. It was located near the northern end of the island and reservation, firing to the southeast. The battery was disarmed shortly after the end of the war in 1947. The reservation was subsequently sold for commercial use, and the emplacement has been modified for use as three private residences. As such it still exists on private property. The battery is closed to the public.

Fort Moultrie (Glen Williford)

Charleston World War II-era Site Locations. Stations housed in a single structure are connected by dashes (-)

location	Loc#	Purpose
Folly South	1	BS1/230-BS1/520
Folly Island	2	SL
Folly North	3	B2/230-BS2/520
Morris Island	4	SL
Fort Sumter	5	Batt. Tact. #1A AMTB 1; SL
Fort Moultrie	6	Batt. Tact. #1 Lord, Batt. Tact. #2 BCN 230, BC/Lord, HECP-HDCP, BC/230, BS3/230, BS3/520
Fort Moultrie	6	Batt. Tact. #2A, AMTB 2, SCR582, SCR296
Fort Moultrie	7	Military Reservation
Fort Moultrie	8	Military Reservation; SL
Marshall Reservation	9	Batt. Tact. #3 BCN 520, BC/520, PSR/530, SCR296/520
Isle of Palms	10	BS4/230-BS4/520; SL
Dewees Island	11	BS5/230-BS5/520

Battery Jasper 25 kW generator Fort Moultrie (Mark Berhow)

Stokely, Jim. *Fort Moultrie, Constant Defender.* National Parks Handbook No. 136, US Dept. of Interior, NPS, 1985.

Gaines, William C. "A History of the Modern Coastal Defenses of Charleston, South Carolina, Part I 1894-1939." *CDSG Journal* Vol. 12, No. 3, Aug. 1998, p. 55; Vol. 12, No. 4, Nov. 1998, p. 35.

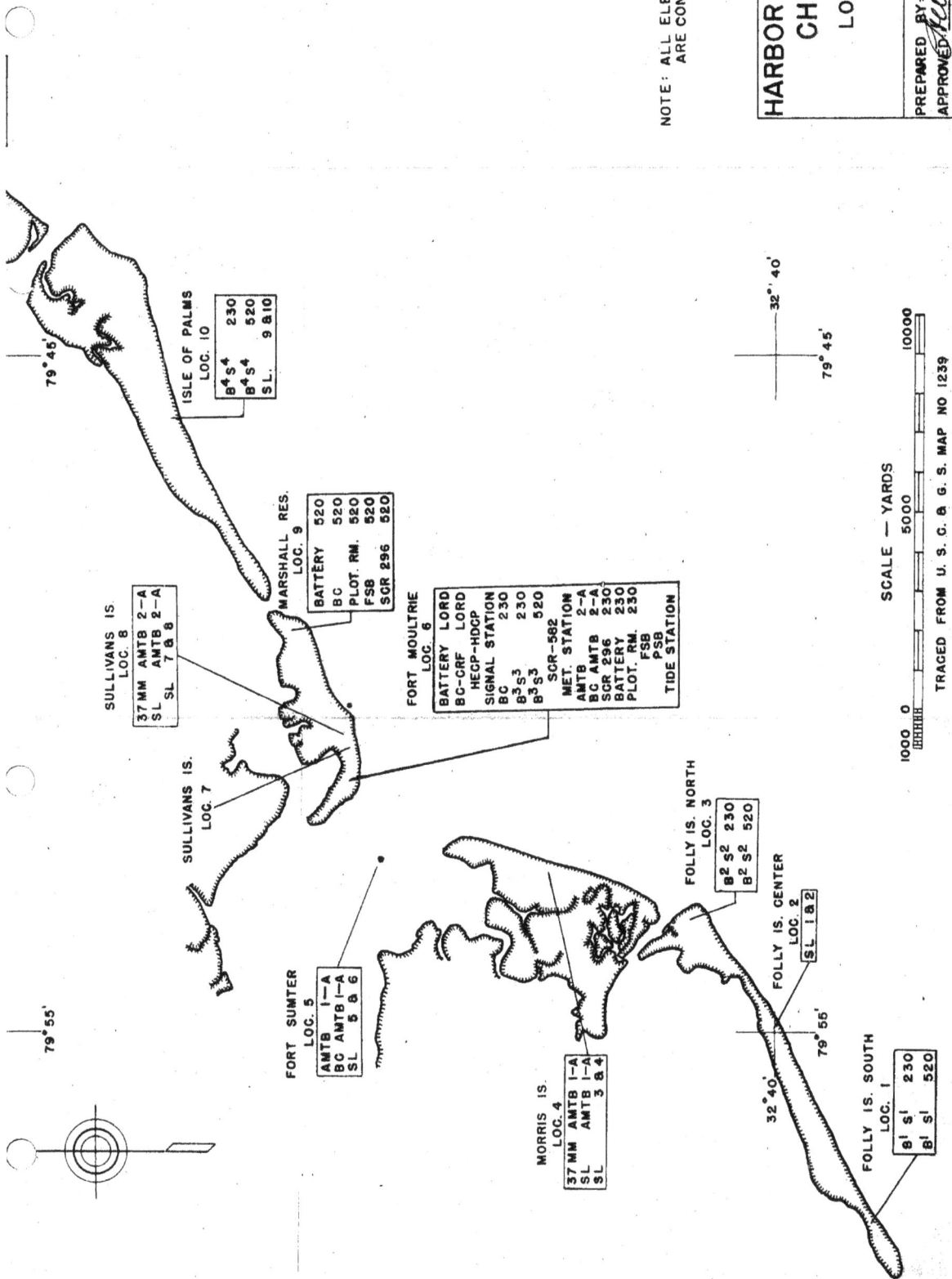

NOTE: ALL ELEMENTS SHOWN
ARE CONSTRUCTED.

HARBOR DEFENSES OF
CHARLESTON

LOCATION OF
ELEMENTS

| PREPARED BY: HD OF C. | DATE: 3/15/43 |
| APPROVED | EX NO. IA |

SCALE — YARDS

TRACED FROM U. S. C. B. G. S. MAP NO 1239

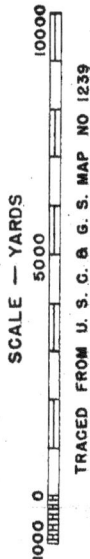

CHARLESTON HARBOR, S.C.

FORT SUMTER

BATTERIE

HUGER---- 1A--- 2-90MM
        --- 2-90MM

On Caretaking Status

SERIAL NUMBER

LEGEND.

7 BARRACKS
9 SALLYPORT
8 MESSHALL
2 LATRINE
4 NEW PUMP HOUSE
15 CISTERN 20000 GAL.
5 BRICK LATRINE & SHOWERS
10 OIL HOUSE
14 CARETAKERS HOUSE
13 LIGHTHOUSE RADIO TOWER
17 MONUMENT & FLAG POLE
18 MONUMENT
19 LIGHT-KEEPERS QRS.

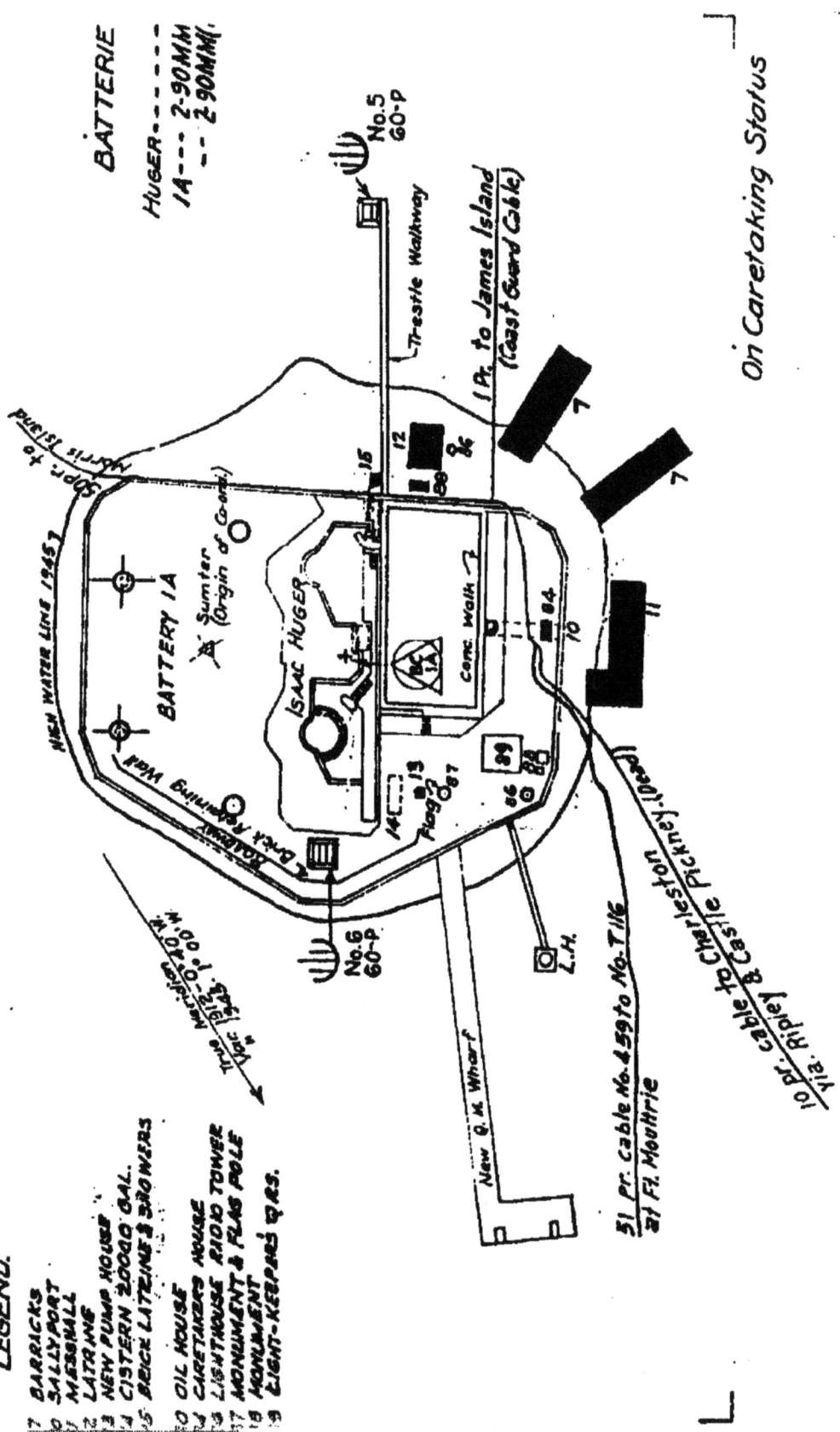

EDITION OF FEB.1, 1921.
REVISIONS: MAR.1, 1925.
OCT. 5, 1928, DEC. 5, 1934
20 NOV. 1945

No.5
60-P

Trestle Walkway

(Pr. to James Island
(Coast Guard Cable)

NEW WATER LINE 1945

Morris Island

SPUR TO

BATTERY 1A

Sumter
(Origin of Co-ord.)

ISAAC HUGER

No.6
60-P

Brick Retaining Wall

Roadway

Meridian 0-0-0.W.
True 0-0-0.W.
Dec. 1945

Conc. Walk

Flag

L.H.

New Q.M. Wharf

51 pr. cable No. 459 to No.716
at Ft. Moultrie

10 pr. cable to Charleston & Castle Pickney, 10600
via. Ripley

CHARLESTON HARBOR, S.C.

FORT MOULTRIE

SULLIVAN'S ISLAND
GENERAL MAP.

BATTERIES.

Pierce Butler 4-12"h
Capron ... 4-12"h
Jasper ... 4-10"Dis
Thomson ... 2-10" "
Logan ... 1-6" "
Gadsden ...
Bingham ...
McCorkle 1-3"P.
Lord ... 2-3" P.

A. Anti Aircraft Gun
Emplacements Only.

6 cottages under lease for 20 yrs.
See Act [Public - No. 782 - 74th Cong] C S. 4432 J

Scale of Feet.

On Caretaking Status

EDITION OF APR.23,1915.
REVISIONS: DEC.7,1915; MAR.14,1916.
NOV.8,1916; FEB. 1, 1921; MAR.1,1925.
OCT. 5,1928, DEC. 5, 1934

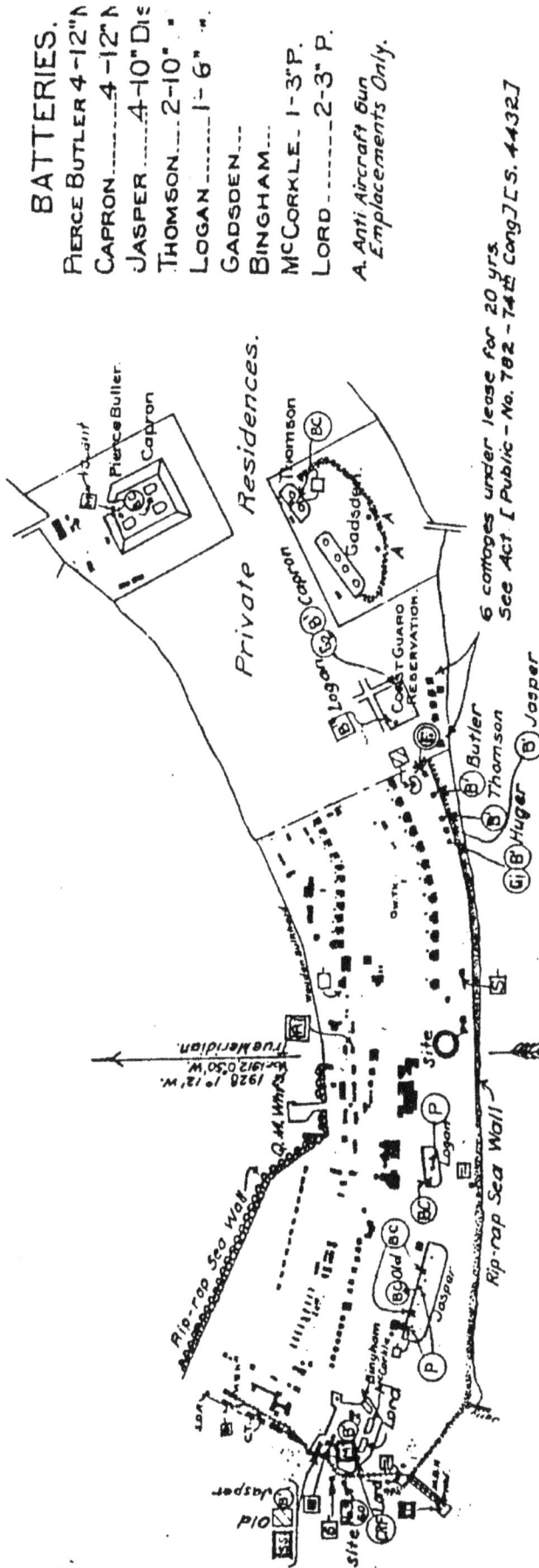

SERIAL NUMBER

Private Residences.

CHARLESTON HARBOR, S.C.
FORT MOULTRIE-DI
SULLIVANS ISLAND, S.C.
WESTERN PORTION

RE-DRAWN 20 NOV. 1945

SERIAL NUMBER

BATTERIES

LORD   2-3"R.F.C.
2A   { 2-90MM (Fixed)
           { 2-90 MM (Mobile)
JASPER
LOGAN
McCORKLE
BINGHAM
CONST. 230
MARINE ONLY.

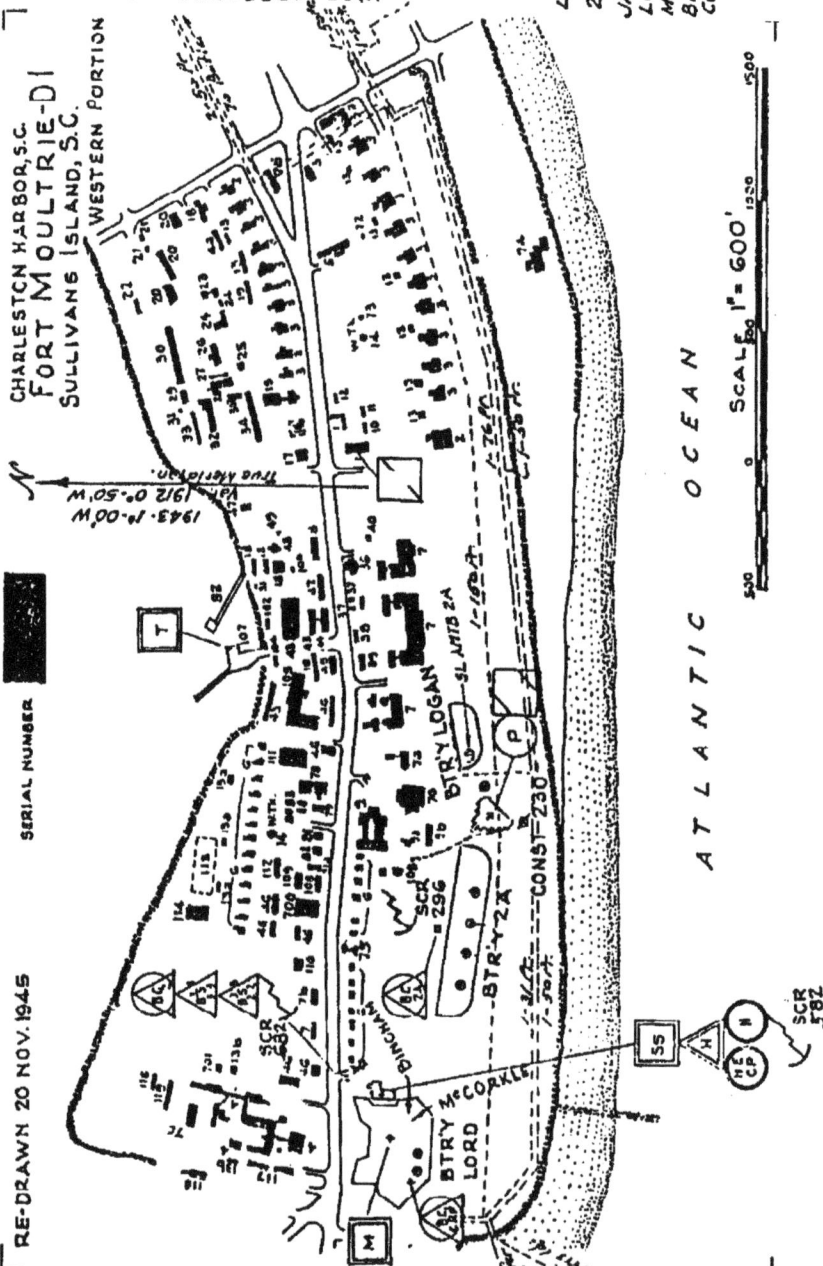

ATLANTIC OCEAN

SCALE 1"= 600'

True Meridian
Var. 1912 0°-50'W
1943- 1°-00'W

BTRY LOGAN
SL. UNTS 2A
BTRY McCORKLE
BTRY LORD
BTRY BINGHAM
CONST-230
SCR 296
SCR 582

LEGEND

1   ADMINISTRATION BUILDING
2   COMMANDING OFFICER'S QRS.
3   OFFICERS QUARTERS
3A   BACHELOR OFFICERS QRS.
4   HOSPITAL
4A   VETERINARIAN HOSPITAL
5   NURSES QUARTERS
6   N.C.O. QUARTERS
7   BARRACKS
7A   BAND BARRACKS
7B   M.P. BARRACKS
7C   HOSPITAL BARRACKS
8   GUARD HOUSE
9   POST EXCHANGE
9A   P.X. SERVICE STATION
9B   P.X. BEER PARLOR
10   MASTER GUNNER
11   CIVILIAN PERSONNEL
12   POST OFFICE
13   OFFICERS GARAGE
14   N.C.O. GARAGE
15   HOSPITAL GARAGE
14   ELEV. WATER TANK
15   POST ENGINEER
16   RESERVOIR
17   PUMP HOUSE
18   WAREHOUSE
19   SERVANTS QRS.
20   STABLES
21   SHED
22   LOADING RAMP
23   TEAMSTER QRS.
24   ORDNANCE SHOP
25   POST ENGINEER WHSE.
26   OIL GAS STATION
26   DISPATCHERS OFFICE
27   GREASE RACK
28   Q.M. GARAGE
29   WASH RACK
30   TRUCK GREASE
31   LATRINE
31A LATRINE & WAREHOUSE
32   POST ENGINEER WHSE.
33   STORAGE
34   MECH. BLACKSMITH SHOP
35   ELECTRICAL SHOP
36   PROVOST MARSHAL
37   POST FIRE HOUSE
38   POST BAKERY
39   AGENT FINANCE OFFICE
40   BAND STAND
41   FLAG STAFF
42   Q.M.C. OFFICE
43   Q.M.C. WAREHOUSE
44   SCALE HOUSE
45   ORDNANCE OFFICE
46   STORE HOUSE
47   PAINT WAREHOUSE
48   PRISON OFFICE
49   TOOL SHED

72   CADDY HOUSE
73   KINDERGARDEN
74   JASPER HALL (OFFICERS C.)
75   CHAPEL
76   RECREATION HALL
78   POST THEATER
79   POST LIBRARY
80   RED CROSS QTRS.
81   CIVILIAN QTRS.
82   COAST GUARD DOCK
83   RED CROSS BLDG.
100 RADIO STATION
102 PAINT SHOP
103 Q.M. WAREHOUSE (SHOE REPAIR)
104 Q.M. SEWING ROOM
105 POST LAUNDRY
106 Q.M. DOCK
107 SUPPLY OFFICE
108 REGIMENTAL HEADQUARTERS
109 FINANCE
110 FINANCE
111 REFRIGERATOR PLANT
112 CHEMICAL WAR BLDG.
113 COAL STORAGE
114 ORDNANCE MACH. SHOP
115 HOSPITAL SUPPLIES
116 HOSPITAL LINEN SUPPLIES
117 ARTILLERY ENGR. OFFICE
118 ENGINEER WAREHOUSE
700 N.C.C. CLUB.
701 RECREATION BLDG.

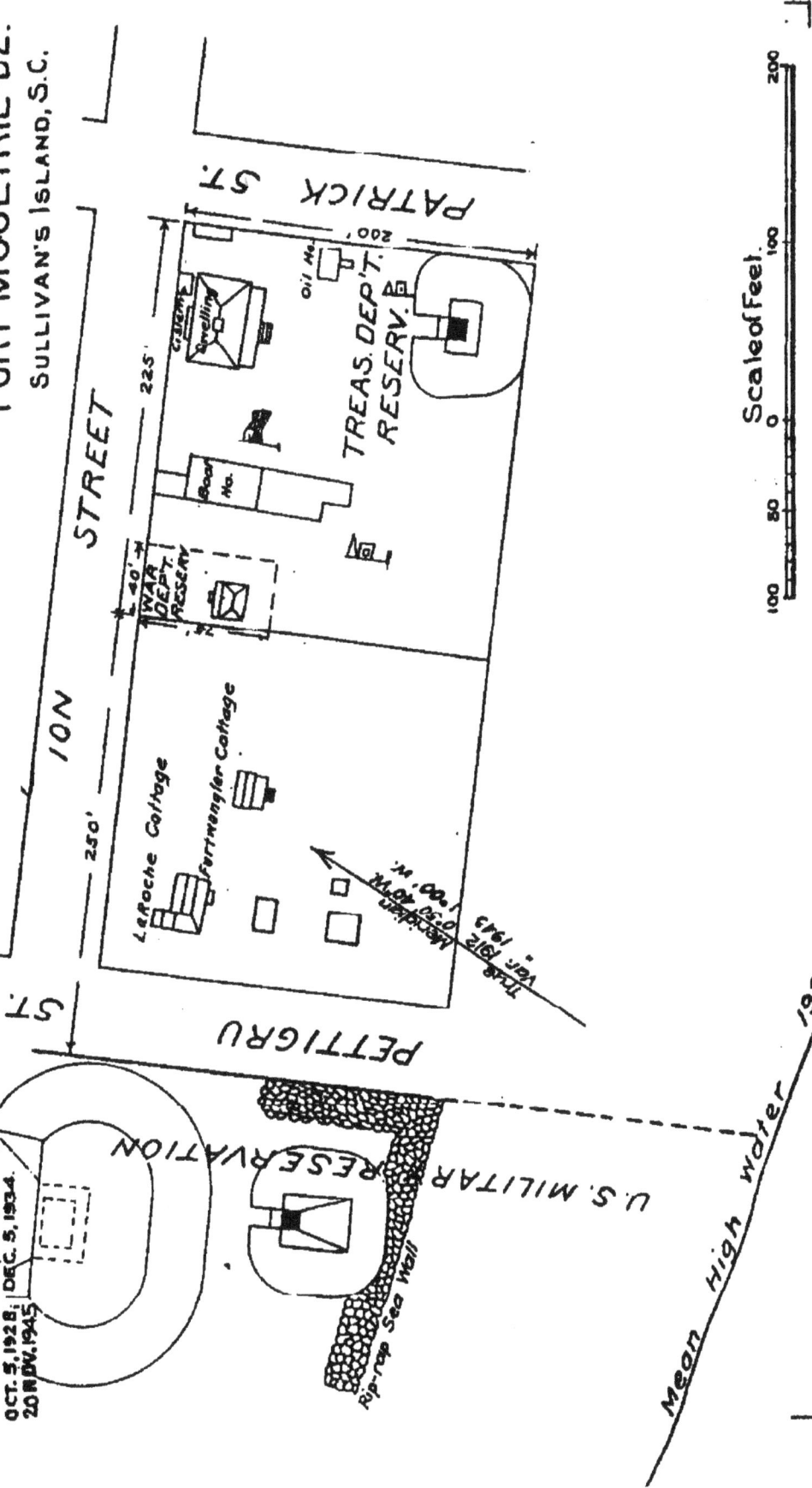

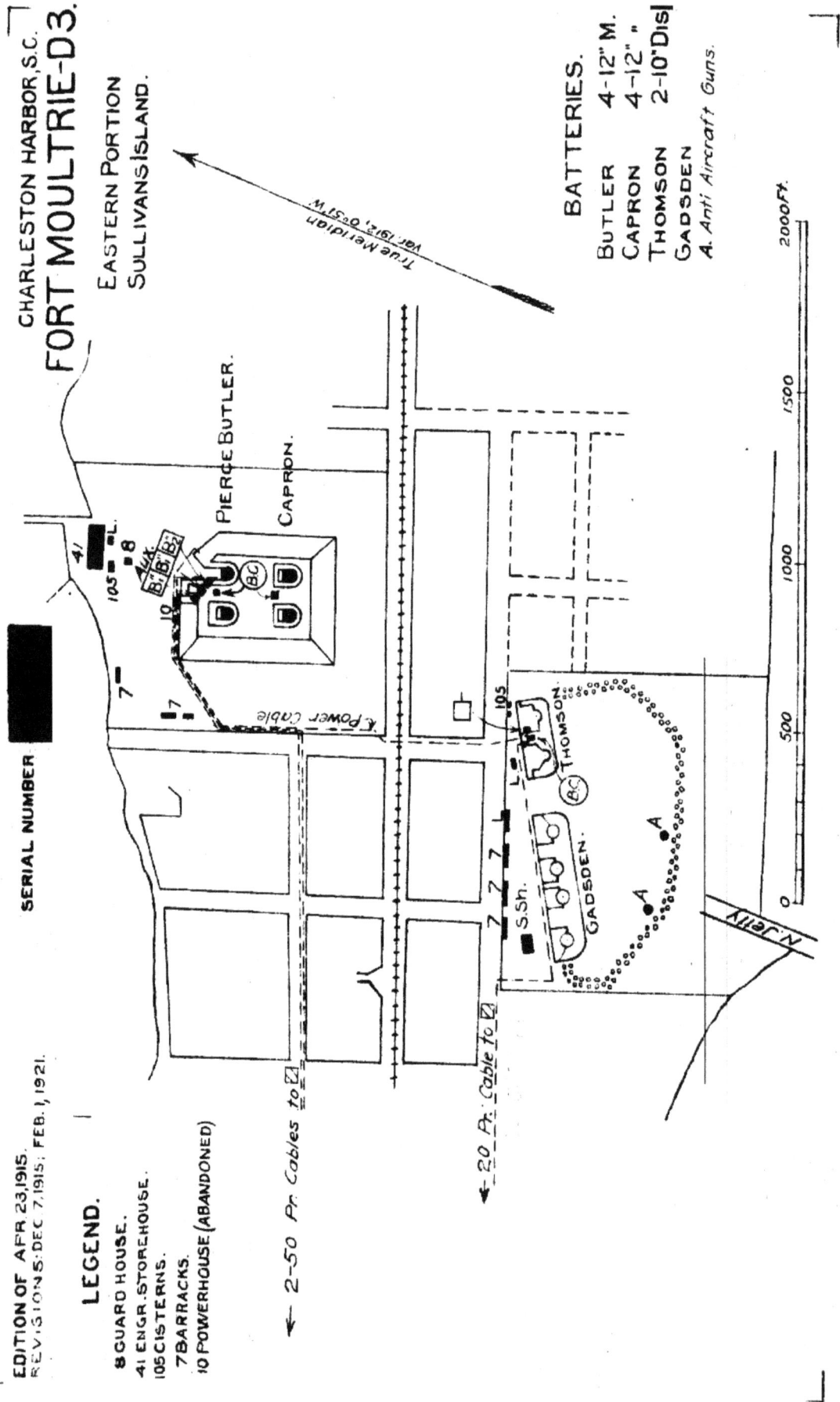

CHARLESTON HARBOR, S.C.
FORT MOULTRIE-D3.

EASTERN PORTION
SULLIVANS ISLAND.

EDITION OF APR 23,1915.
REVISIONS: DEC 7,1915; FEB. 1, 1921.

LEGEND.

8 GUARD HOUSE.
41 ENGR. STOREHOUSE.
105 CISTERNS.
7 BARRACKS.
10 POWERHOUSE (ABANDONED)

SERIAL NUMBER

True Meridian
Var. 1912, 0°-51' W.

BATTERIES.

BUTLER	4-12" M.
CAPRON	4-12"
THOMSON	2-10"Disp
GADSDEN	
A. Anti Aircraft Guns.	

PIERCE BUTLER.

CAPRON.

THOMSON.

GADSDEN.

2-50 Pr. Cables to

20 Pr. Cable to

Power Cable

N. Jetty

0   500   1000   1500   2000 Ft.

RE-DRAWN· 20 NOV.1945

SERIAL NUMBER

SCALE OF FEET

On Caretaking Status

CHARLESTON HARBOR, S.C.

FORT MOULTRIE D-4.

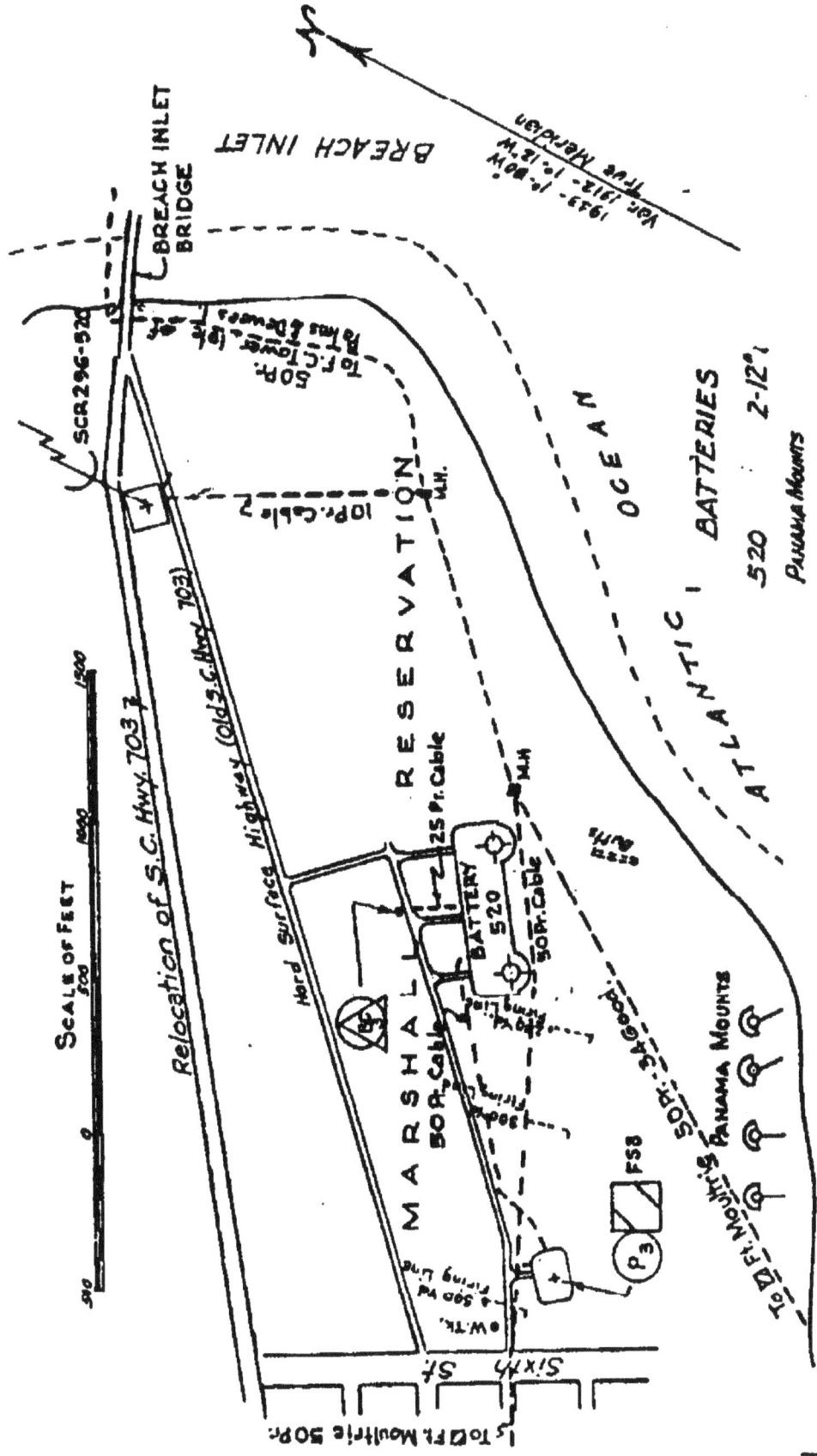

BREACH INLET BRIDGE

BREACH INLET

Var. 1943 - 1°·00'W
True Meridian

SCR 296-520

50R.
To F.C.Tower
To F.C.Tower
Palmetto Drawer
Hard Surface Highway (Old S.C.Hwy.703)
Relocation of S.C.Hwy.703

10 Pr. Cable

M.M.

RESERVATION

ATLANTIC OCEAN

BATTERIES

520   2-12"

Panama Mounts

M A R S H A L L
50 Pr. Cable

25 Pr. Cable

M.M.

BATTERY
520

50Pr.Cable

F.S.13

FSB

P 3

500 yd. Firing Line

SOR. 3A Cable

To Ft. Moultrie PANAMA MOUNTS

Sixth St.

To Ft. Moultrie 50Pr.

Is To Ft. Moultrie 50Pr.

THE HARBOR DEFENSES OF PORT ROYAL HARBOR – SOUTH CAROLINA

Port Royal Sound is the southernmost natural deep-water port on the Atlantic coastline. In 1890, U.S. Navy selected Parris Island for the construction of a graving drydock, at that time the largest naval drydock in the country, as part of the Port Royal Naval Shipyard. A small defensive work with two cement batteries was built on the opposite shore to protect a proposed mine field to guard the entrance to the harbor. The naval facility was moved to Charleston after 1901 (later became the U.S. Marine Corps Recruit Depot), and the harbor defenses were discontinued soon afterwards.

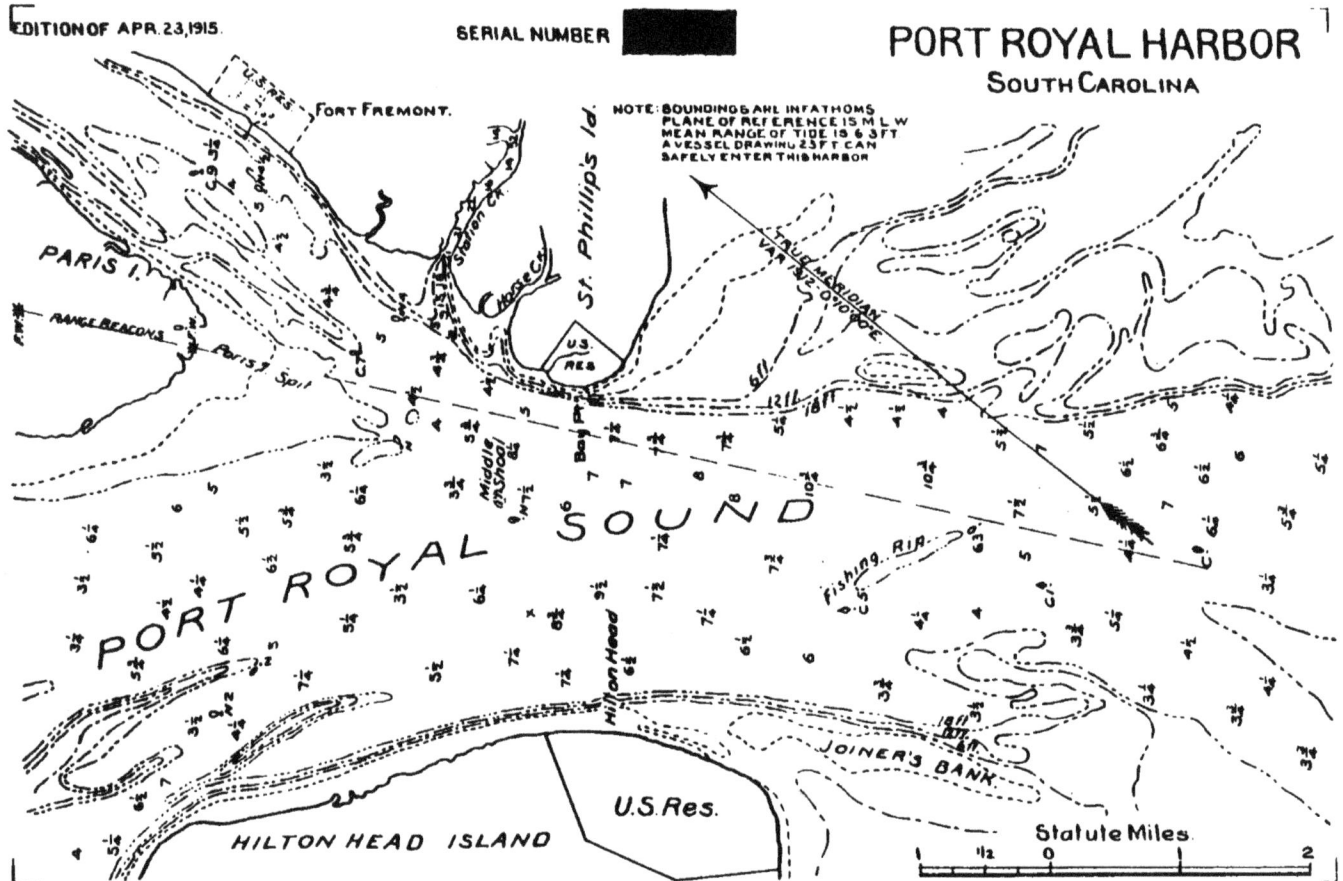

Fort Fremont (1899-1928) is located on southern end of St. Helena Island on the northern side of Port Royal Sound, about four miles southeast of Port Royal, South Carolina. This small Endicott Program fort consisted of a three gun 10-inch disappearing battery and a two gun 4.7-inch rapid-fire battery. It was named in General Orders 43 of 1900 for Maj. Gen. John C. Fremont, U.S. Army. The fort was declared surplus in 1928 and sold to private interests. The batteries remain, though the small battery is built-on. The two batteries are now part of the Fort Fremont Historical Preserve (owned by Beaufort County as a public park) which received a new visitor center in 2022. The post hospital remains as a private residence. Of special interest, was the construction in 1897 of a dynamite gun battery across the Port Royal Sound on Hilton Head Island. This short-termed battery mounted two 15-inch guns which used high pressured air to fire high explosive charges. The battery remains today but on private property.

FORT FREMONT

PORT ROYAL HARBOR, S.C.

St. Helena Island.

SERIAL NUMBER

EDITION OF Dec.7,1815;
REVISIONS: MAR. 8, 1919.

True Meridian
Var. 1912 - 0°15'00"E.

500' 0 500' 800 1000'

LEGEND.
3. OFFICERS QRS.
4. HOSPITAL.
10. OIL HOUSE.
11. SCALES.
12. LAVATORY.
13. STABLE.
14. SLEEPING QRS FOR STABLEMEN.

BATERIES
* JESUP
FORNANCE..2-4.7
*Dismantled

JESUP.

FORNANCE.

BEAUFORT

RIVER.

W.TK.

B'

S.Sh.

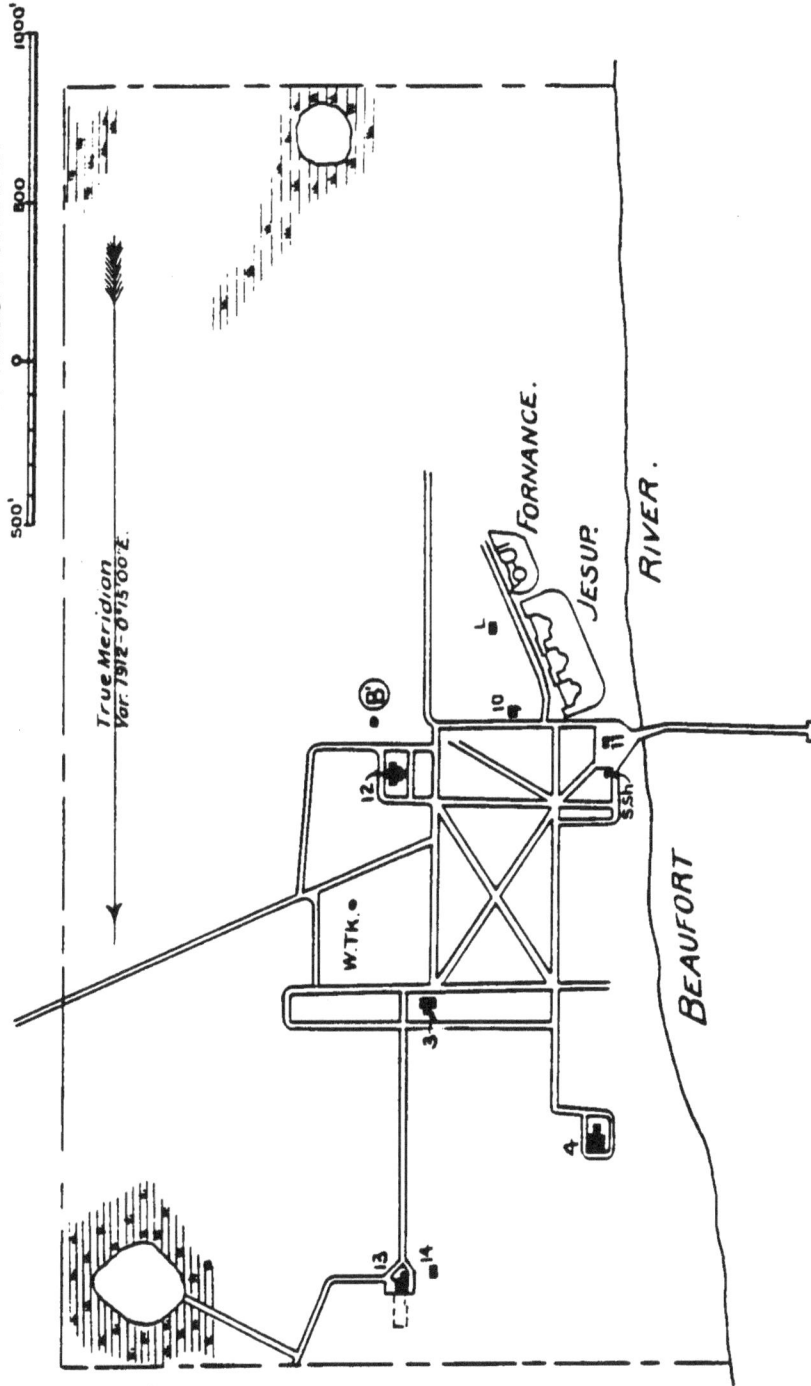

Fort Fremont Gun Batteries

- **JESUP:** A battery for three 10-inch disappearing guns was the primary battery for the defenses of Port Royal at the Fort Fremont reservation on St. Helena Island. The first two emplacements were authorized with National Defense Act funding on April 6, 1898, and the third unit shortly after on July 26, 1898. The plan was of standard type design. The two early emplacements were of internal design, the added third emplacement was of flank configuration—that carried a Model 1895 gun vs. Model 1888 in the other two. Work was done from July 8, 1898 to finishing on March 3, 1899. Transfer was made on September 2, 1899 for a cost of $124,293.85. The battery had its own power plant, along with a freshwater cistern for 30,000 gallons. It was named on General Orders No. 78 of May 25, 1903 for Brigadier General Thomas S. Jesup of War of 1812 and Indian War service. The battery was armed with two 10-inch Model M1888MII Bethlehem guns on disappearing carriages Model 1896 (#34/#32 and #15/#56) and one Model 1895 Watervliet gun on a Model 1896 disappearing carriage (#9/#52). The defenses of Port Royal were abandoned quite early. In fact, even the proposed 1900 modifications to the battery were put in abeyance because the permanency of the naval station at Port Royal was being debated by a commission. The Taft Board in 1907 recommended abandonment of the defenses, it appears that the battery was never permanently garrisoned. The armament was removed sometime between 1915 and 1918. The emplacement is now part of the Fort Fremont Historical Preserve (a Beaufort County Park) and open to the public.

- **FORNANCE:** A battery for two 4.7-inch rapid-fire guns emplaced on the left flank of Battery Jesup near the river, also firing with the same bearing to the west, southwest. Construction was authorized on April 6, 1898 using the emergency funding of the National Defense Act of the previous March. Concrete work was done rapidly from April 28 to June 30, 1898. For all practical purposes it was completed by August 20, 1898. Transfer was made on September 2, 1899 for a cost of $6,000. It was named on General Orders No. 78 of May 25, 1903 for Captain James Fornance who was killed in 1898 in the Spanish American War. The guns were mounted here on June 24, 1898. It carried two 4.7-inch Armstrong guns on pedestal mounts (#9423/#9081 and #9424/#11005). The gun blocks were augmented by a new BC station and improved magazine in 1905. The battery was abandoned with the defenses of Port Royal early, the armament being removed in 1913 and going into storage before eventual scrapping. The emplacement was later sold into private ownership, and a dwelling built upon it (which has subsequently been removed). The emplacement is now part of the Fort Fremont Historical Preserve (a Beaufort County park) and open to the public.

Hilton Head Military Reservation (1874-1927) is located on Hilton Head Island, South Carolina overlooking the entrance to Port Royal Sound. It can be reached from Fort Walker Drive, near Circle at Catesby Lane at Port Royal Plantation (Secure Gated Community-Restricted entrance). This reservation is about 200 yards north of the site of Fort Walker/Welles. The reservation contained the experimental Dynamite Battery (1901-1902) and an unnamed battery (1898-1899) (two M1888 8-inch BL guns on modified 15-inch Rodman carriages). Troops again garrisoned the post in World War I. The battery is in ruins, with no period guns or mounts in place. Two walls of the power plant are partially upright, only the foundations of the steam plant remain, and the gun emplacement is all broken up but recognizable.

Manuel, Dale A. "Fort Fremont and the Defenses of Port Royal Sound, South Carolina." *Coast Defense Journal* Vol. 16, No. 4, Nov. 2002, p. 29

PORT ROYAL HARBOR, S.C.

HILTON HEAD

HILTON HEAD ISLAND

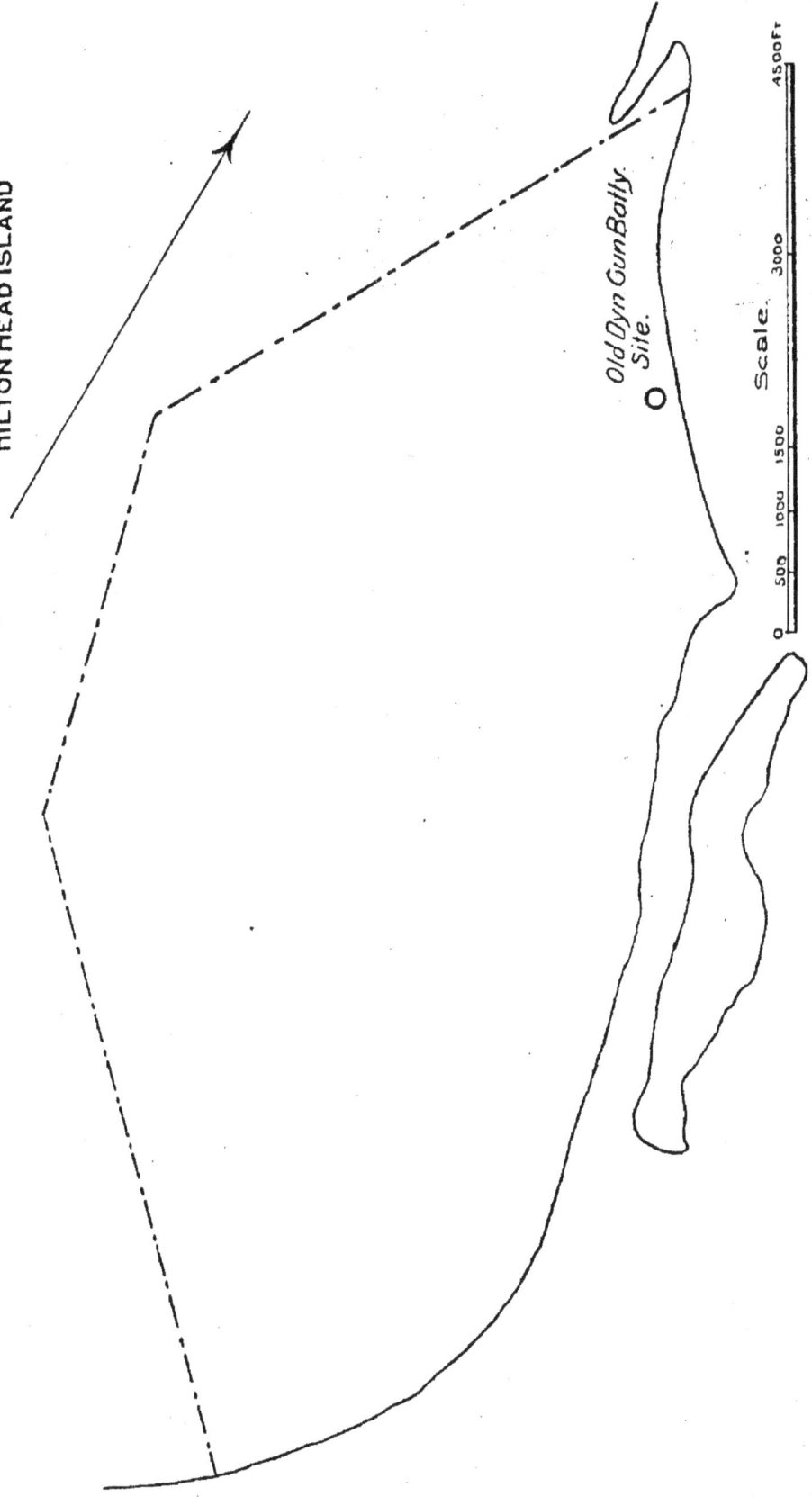

SERIAL NUMBER *124*

EDITION OF APR. 23, 1915.

Old Dyn Gun Batty. Site.

Scale.

0   500   1000   1500   3000   4500 Ft

Hilton Head M.R. Gun Batteries

- Hilton Head was selected as one of the sites to use the twenty-one modern 8-inch guns on reinforced 15-inch Rodman carriages emplaced as a temporary expediency during the Spanish American War. An emplacement for two such guns was selected on the eastern extreme of Hilton Head Island. Precise site location, close to the shore of the sound, was finalized on April 27, 1898. The site was still government land near the old Civil War fortification of Fort Walker (also known to Federal troops at Fort Welles). A nearby creek was used to bring in shallow boats with concrete and equipment from Fort Fremont. It consisted of two platforms holding the iron traversing rings behind breast-high concrete frontal walls. On the right flank was a single-room concrete magazine covered with sand. Work was done from June to July 1898 and reported complete on August 15, 1898. It was armed with two 8-inch Model 1888 guns (West Point Foundry #2 and Watervliet #42) shipped here on July 2, 1898. On August 31 of that year a gale flooded the magazine, one of many difficulties that made it hard to keep the battery in service. In April of 1899 they were reported as complete, but having never been put into service and with all ammunition removed from the magazine. It was soon disarmed. Somewhat surprisingly the two breast-high walls at the front of the pintles still exist, sticking up from the shore in the sand, on private property. The battery site is closed to the public.

- **Dynamite Battery:** Despite disappointing performance results from the two dynamite batteries built in the 1890s in New York and one in San Francisco, The Pneumatic Dynamite Gun Company convinced Congress to fund two new emplacements during the Spanish American War. Congress approved the expenditure and directed the Ordnance Department (on August 9, 1898) to work directly with the contractor to emplace two new, single 15-inch guns and carriages. One was to go to Hilton Head Island as part of the Port Royal defenses. Plans were submitted on August 23, 1900 for the battery, following the plan of the slightly earlier work at Fort H. G. Wright. It was sited near the shore just to the north of the temporary 8-inch battery, firing to the northeast. Concrete work began on the block and the adjacent large power steam plant and compressed air facilities. The Corps of Engineers were not directly involved in the planning or building—except to consult with the contractor. In fact, they were not enthusiastic at all about the type of weapon or work. It received 15-inch dynamite gun Model 1893 #4 on Model 1893 carriage #7. One of their biggest complaints was that the single gun, with no other light armament around it, was far too vulnerable. The gun was reported ready to turn over to the artillery on May 11, 1901. Just a year later the guns were declared obsolete and scrapped and the emplacement abandoned. It was never actually manned or put into service. For years the remains of the emplacement persisted, and there are in fact major concrete blocks from the steam plant still on the shore at the Hilton Head site. The battery site is open, but access to community is closed to the public.

The Hilton Head Dynamite Battery today (Terry McGovern)

THE HARBOR DEFENSES OF THE SAVANNAH RIVER – GEORGIA

The city of Savannah was settled in 1733, became part of the Royal Colony of Georgia in 1751 and by the outbreak of the Revolutionary war was the southernmost commercial port of the Thirteen Colonies. The British took the city in 1778 and retained control throughout the war. Defenses were built to guard the Savannah River during the Second and Third System of defense construction. During the Civil War the defenses came under Confederate control. Fort Pulaski was taken by siege by the Union forces in 1862, but the city did not surrender until Sherman's forces approached and placed the city under siege in 1864. The defenses at Fort Pulaski were repaired and renovated after the war and a new set of concrete batteries defenses were built further towards the sea on Tybee Island. These defenses were placed in caretaker status after World War I, ordered abandoned as coast artillery defenses in 1924. The fort's garrison facilities were transferred for use by the infantry and remained an active military post until 1947.

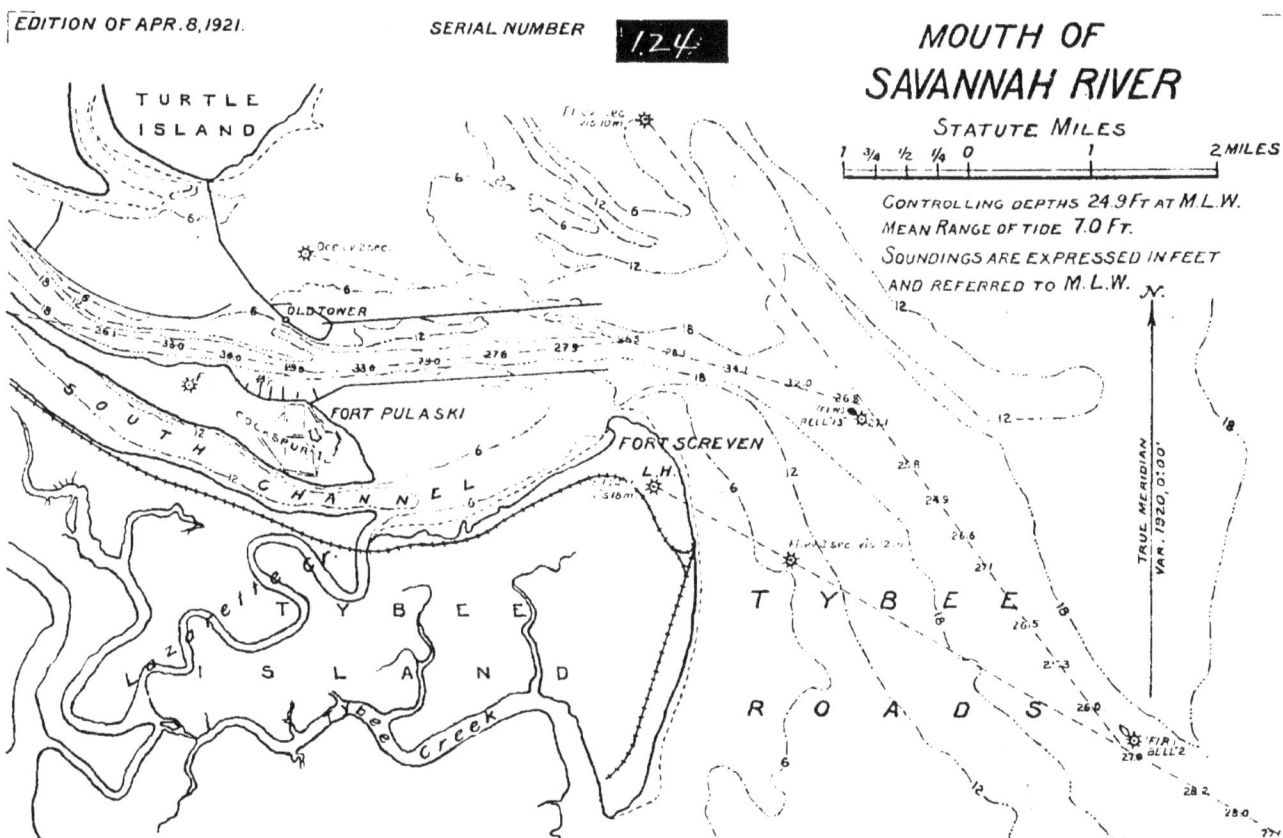

Fort Screven (1897-1924) is located on eastern end of Tybee Island on the southern side of the Savannah River, about eighteen miles southeast of Savannah, Georgia. Constructed as part of the Endicott Program, Fort Screven replaced Fort Pulaski, a Third System fort, as the primary defense of the entrance to the Savannah River. The fort armament consisted of six batteries with an additional battery located at Fort Pulaski. The fort also had a controlled submarine mine complex. It was named in General Orders 89 of 1899 for Colonel James Screven, Georgia Militia, killed in action at Medway Church in 1778. Fort Screven served as a coast artillery post from 1898 until declared surplus in 1924. During World War II, four 155mm GPF on Panama Mounts were installed at the fort. The fort continued in a military role until 1945 and sold to private interest. Several of the forts garrison buildings remain as private residents, and few of the concrete batteries have been built over for private homes and an American Legion Post. The Tybee Museum is located in Battery Garland and there is another museum associated with the lighthouse. The former reservation is open to the public but most remaining structures are private.

Fort Screven Gun Batteries

- **BRUMBY**: The main battery of four 8-inch disappearing guns erected at the Tybee Island reservation of Fort Screven. Plans were submitted on October 16, 1896, using funds from the Act of June 6, 1896. It was on the eastern side of the reservation, south of the wharf, firing to the east, northeast. The plan was for a conventional, straight emplacement for four disappearing guns. Magazines were on a lower level on the left flank of each platform, ammunition service by lifts. The initial contract was for just three emplacements, the fourth emplacment was added in 1898. the contract was awarded on November 30, 1896, with concrete work being finished on the first three emplacements by June 1898, and the fourth by March 1899. While efforts were pushed for mounting the armament, this was delayed when the four guns were lost at sea when the ship carrying them foundered in a hurricane. (The schooner *Agnes I. Grace*, carrying Watervliet 8-inch Model 1888 guns #21, #26, #46, and #50 was lost off Tybee Island on September 5, 1897). The emplacement was finished by July 1898, with only sodding work needed, though problems with serious cracks were already beginning to appear. Eventually the battery was armed with four 8-inch Model 1888 Watervliet guns on Model LF 1894 disappearing carriages (#33/#11, #47/#21, #49/#20, and #52/#13). The first of these was mounted in May of 1898, the last in March 1899. The battery was transferred on August 11, 1899 for a cost of $187,744. It was named in General Orders No. 43 of April 5, 1900 for Lt. Thomas M. Brumby, U.S. Navy of Spanish American War service. Like many other early batteries, it was soon modified with a battery commander station, improved loading platforms and new hoists in 1909. All four guns were dismounted on October 16, 1917 and shipped away on January 31, 1918. The emplacement was not subsequently used for armament. After 1945 the fort was sold to private ownership, and this emplacement was converted extensively to private residences. As such it still exists of private property. The battery is closed to the public.

- **GARLAND**: One of two single 12-inch barbette gun emplacements erected at Fort Screven. It was built on the southern, right flank of the 8-inch battery, and pointed and fired to the east. It was one of the emplacements built with National Defense Act funds to take advantage of the quickly ordere and quickly produced, new series of barbette carriages. Plans were submitted and approved on May 28, 1898. Plans were generally patterned on current disappearing types. It had a straightforward, "interior" shaped position, the loading platform was raised to the level of the gun platform to aid loading. The magazines were to the front, under the front parapet and shell handling was by hoists to the platform and then crane to the breech. Work began in mid-1898 and was complete by March 1899. The gun was mounted in February 1899 and was 12-inch Model 1888MII Watervliet gun on barbette carriage Model 1892 (#31/#12). Transfer was made on October 5, 1899 for a combined cost (with Battery Fenwick) for $120,768. It was named in General Orders No. 78 of May 25, 1903 for Colonel John Garland, 8th Infantry of Indian War and Mexican War service. After the post was discontinued as a Coast Artilery installation in 1925, the battery was not immediately disarmed. Apparently it was kept in a maintenance status, the gun and carriage were not actually removed and scrapped until 1942. The emplacement still exists, though somewhat modified to become the Tybee Island Museum. Access to the battery is open to the public when the museum is open.

- **FENWICK**: The second of the two 12-inch barbette emplacements. This was on the left flank of the 8-inch Battery Brumby, firing to the northeast. It was one of the emplacements built with National Defense Act funds to take advantage of the quickly ordered and quickly produced, new series of barbette carriages. Plans were submitted and approved on May 28, 1898. Plans were generally patterned on current disappearing types. It had a straightforward, "interior" shaped platform, the

SAVANNAH RIVER
FORT SCREVEN
TYBEE ISLAND, GA.
GENERAL MAP

SERIAL NUMBER 124

EDITION OF APR. 23, 1915.
REVISIONS: DEC. 7, 1915; JUNE 9, 1916.
MAR. 8, 1919; APR. 8, 1921.

For details of the area within this parallelogram, see the Map D-1.

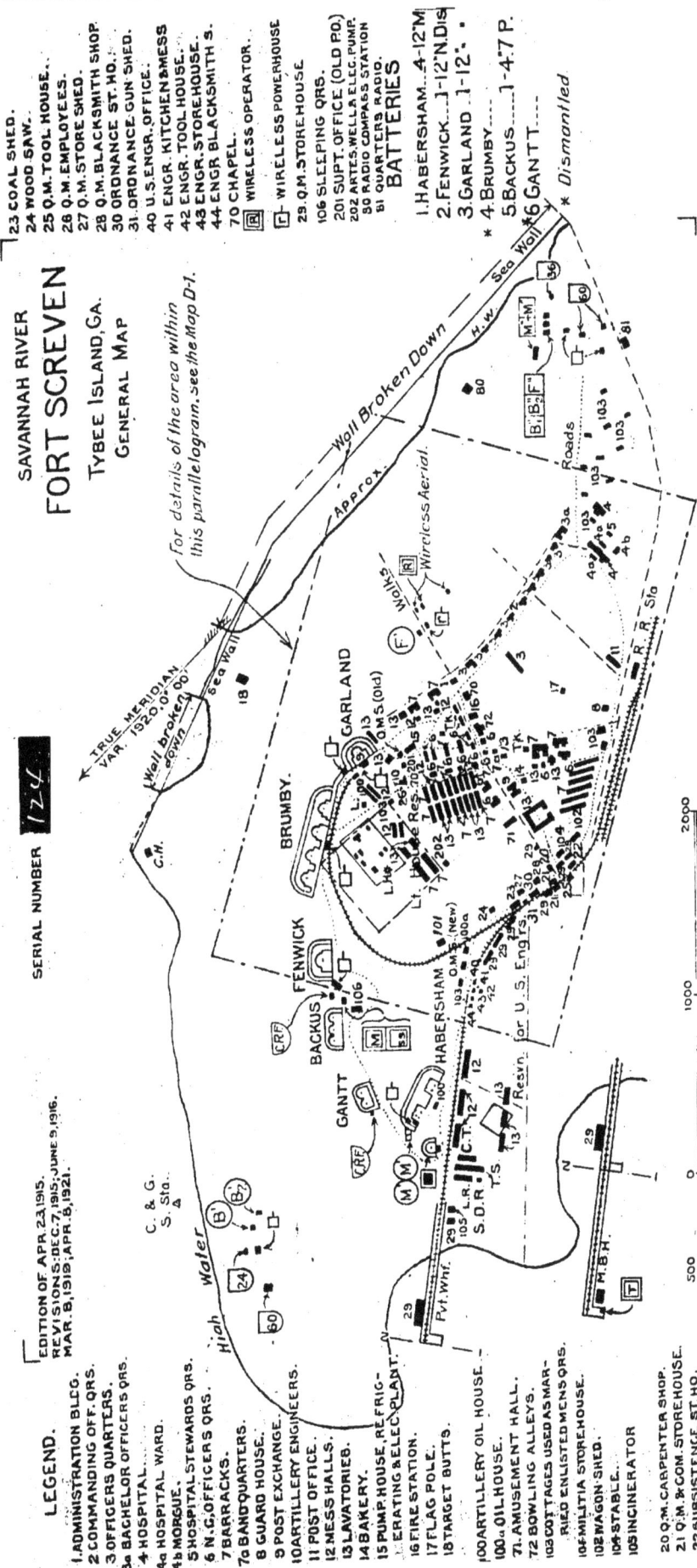

TRUE MERIDIAN
VAR. 1920 0°00'

Wall broken down

High Water

Sea Wall

Wall Broken Down

Approx.

Wireless Aerial

Walk 5

Roads

Resvn. for U. S. Engrs.

R. R. Sta.

GARLAND
BRUMBY
FENWICK
BACKUS
GANTT
HABERSHAM

Pvt. Wharf

M.B.H.

LEGEND.
1. ADMINISTRATION BLDG.
2. COMMANDING OFF. QRS.
3. OFFICERS QUARTERS.
3a. BACHELOR OFFICERS QRS.
4. HOSPITAL.
4a. HOSPITAL WARD.
4b. MORGUE.
5. HOSPITAL STEWARDS QRS.
6. N.C. OFFICERS QRS.
7. BARRACKS.
7a. BANDQUARTERS.
8. GUARD HOUSE.
9. POST EXCHANGE.
10. ARTILLERY ENGINEERS.
11. POST OFFICE.
12. MESS HALLS.
13. LAVATORIES.
14. BAKERY.
15. PUMP HOUSE, REFRIG-
     ERATING & ELEC. PLANT.
16. FIRE STATION.
17. FLAG POLE.
18. TARGET BUTTS.
100. ARTILLERY OIL HOUSE.
100a. OIL HOUSE.
71. AMUSEMENT HALL.
72. BOWLING ALLEYS.
103. COTTAGES USED AS MAR-
      RIED ENLISTED MENS QRS.
104. MILITIA STOREHOUSE.
102. WAGON SHED.
104. STABLE.
105. INCINERATOR.
20. Q.M. CARPENTER SHOP.
21. Q.M. & COM. STOREHOUSE.
22. SUBSISTENCE ST. HO.
23. COAL SHED.
24. WOOD SAW.
25. Q.M. TOOL HOUSE.
26. Q.M. EMPLOYEES.
27. Q.M. STORE SHED.
28. Q.M. BLACKSMITH SHOP.
30. ORDNANCE ST. HO.
31. ORDNANCE GUN SHED.
40. U.S. ENGR. OFFICE.
41. ENGR. KITCHEN & MESS
42. ENGR. TOOL HOUSE.
43. ENGR. STOREHOUSE.
44. ENGR. BLACKSMITH S.
70. CHAPEL.
[R] WIRELESS OPERATOR.
[R] WIRELESS POWERHOUSE.
29. Q.M. STOREHOUSE.
106. SLEEPING QRS.
201. SUPT. OFFICE (OLD P.O.)
202. ARTES. WELL & ELEC. PUMP.
50. RADIO COMPASS STATION
51. QUARTERS RADIO.

BATTERIES
1. HABERSHAM . . 4-12"M
2. FENWICK . . 1-12"N. Dis
3. GARLAND . 1-12" * *
* 4. BRUMBY . . . . . .
* 5. BACKUS . . 1-4.7" P.
* 6. GANTT . . . . . .
* Dismantled

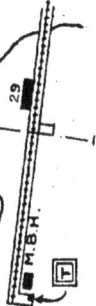

C. & G. S. Sta.

SAVANNAH RIVER
FORT SCREVEN D-1
TYBEE ISLAND, GA.

SERIAL NUMBER 124

EDITION OF APR. 8, 1921.
REVISIONS:

Wireless Aerial.

BATTERIES

FENWICK....I-12"N.
GARLAND I-12"
*BRUMBY

* Dismantled.

TRUE MERIDIAN
Y.P.R. 1920. 0:00'

Walks

GARLAND

BRUMBY

FENWICK

R. R. Sta.

Res'vn. for U.S. Engrs.

Lt. Ho. RES.

L.H.

O.M.S.
(NEW)

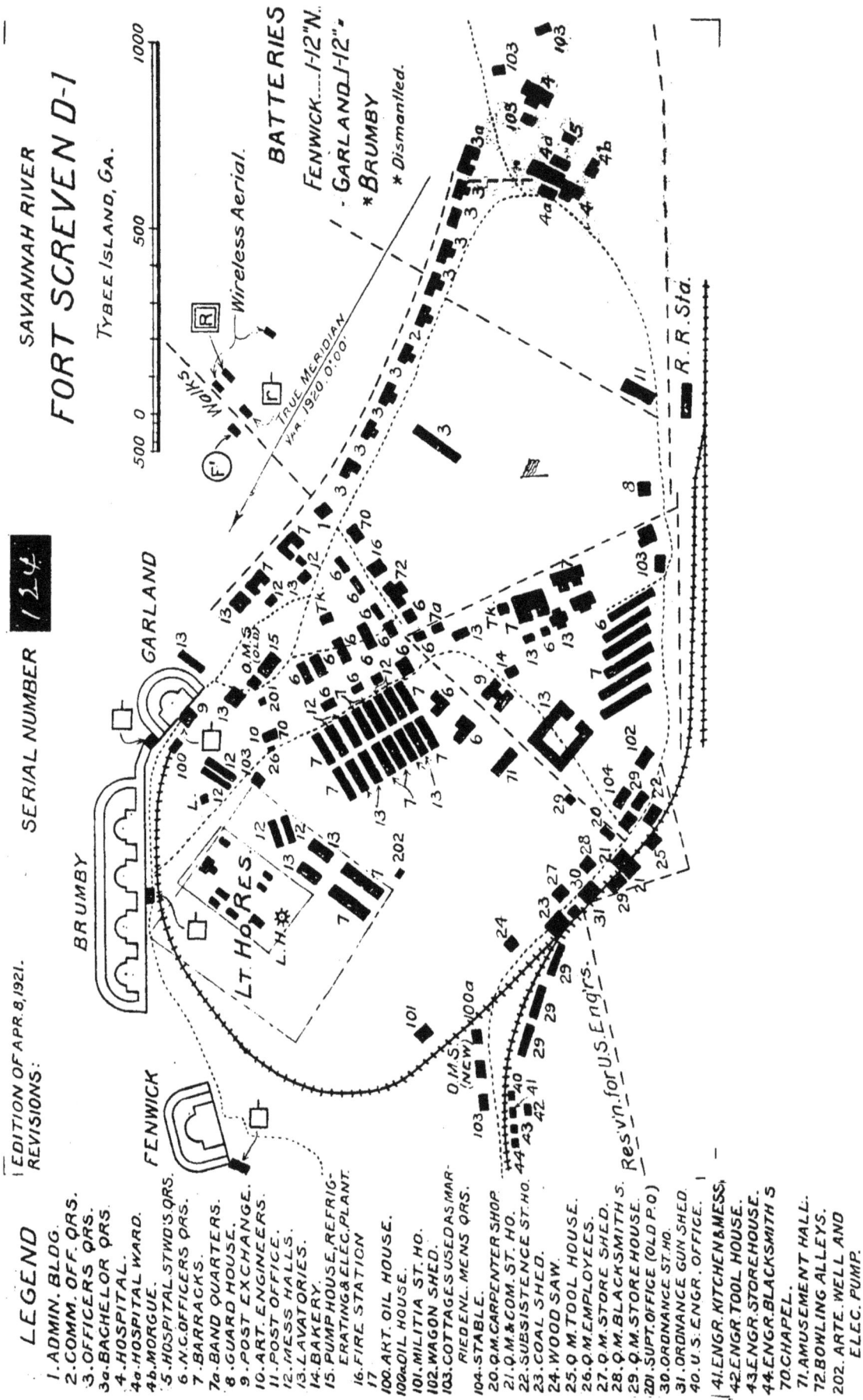

LEGEND

1. ADMIN. BLDG.
2. COMM. OFF. QRS.
3. OFFICERS QRS.
3a. BACHELOR QRS.
4. HOSPITAL.
4a. HOSPITAL WARD.
4b. MORGUE.
5. HOSPITAL STWD'S.QRS.
6. N.C.OFFICERS QRS.
7. BARRACKS.
7a. BAND QUARTERS.
8. GUARD QUARTERS.
9. POST EXCHANGE.
10. ART. ENGINEERS.
11. POST OFFICE.
12. MESS HALLS.
13. LAVATORIES.
14. BAKERY.
15. PUMP HOUSE, REFRIG-
    ERATING & ELEC.PLANT.
16. FIRE STATION
17
100. ART. OIL HOUSE.
100a OIL HOUSE.
101. MILITIA ST. HO.
102. WAGON SHED.
103. COTTAGES USED AS MAR-
     RIED ENL. MENS QRS.
104. STABLE.
20. Q.M.CARPENTER SHOP
21. Q.M.& COM. ST. HO.
22. SUBSISTENCE ST.HO.
23. COAL SHED.
24. WOOD SAW.
25. Q.M.TOOL HOUSE.
26. Q.M.EMPLOYEES.
27. Q.M. STORE SHED.
28. Q.M.BLACKSMITH S.
29. Q.M.STORE HOUSE.
201. SUPT.OFFICE (OLD P.O.)
30. ORDNANCE ST.HO.
31. ORDNANCE GUN SHED.
40. U.S. ENGR. OFFICE.
41. ENGR.KITCHEN & MESS.
42. ENGR.TOOL HOUSE.
43. ENGR.STOREHOUSE.
44. ENGR.BLACKSMITH S
70. CHAPEL.
71. AMUSEMENT HALL.
72. BOWLING ALLEYS.
202. ARTE. WELL AND
     ELEC. PUMP.

Fort Screven 1920s (NARA)

Fort Screven 1930s (NARA)

loading platform was raised to the level of the gun platform to aid loading. The magazines were to the front, under the front parapet and shell handling was by hoists to the platform, and then by crane. Work began in May 1898 and was completed by March 1899. It was armed with one 12-inch Model 1888MII Watervliet gun on Model 1892 barbette carriage (#24/#13). The battery was transferred along with Battery Garland on October 5, 1899 for a combined cost of $120,768. It was named in General Orders No. 78 of May 25, 1903 for Colonel John R. Fenwick the War of 1812 service. The battery was not disarmed after the defenses were discontinued in the early 1920s. Apparently kept in a maintenance status, the gun and carriage were not removed and scrapped until 1942. The emplacement property was sold for private use after World War II. The emplacement still exists with a private residence built upon it. The battery is closed to the public.

- **HABERSHAM:** The mortar battery constructed on Tybee Island at Fort Screven. It was the most northerly emplacement on the reservation, on the rail line near the wharf, pointing and firing to the east. Plans were submitted on January 26, 1899. The design followed the type recommended for later Endicott mortar batteries—wide pit dimensions opened to the rear, and magazines on the flank sides and in the central traverse between the pits. Plans initially recommended a complete parapet to the rear of the open pits, to include the location of the mining casemate; but these were disapproved in Washington and never built. Work was done from February 1899; concrete work being done by mid-April of that year. The mortars and carriages were mounted as they arrived between March and June 1900. Transfer was made on April 9, 1900 at a cost of $125,442. It was named on General Orders No. 43 of April 4, 1900 for Major Joseph Habersham, Continental Army and Postmaster General. The armament consisted of eight 12-inch Model 1890M1 mortars on Model 1896 carriages (Builders tubes #42/#197, #44/#220, Bethlehem tubes #3/#199, #41/#198, #46/#195, #54/#200, Niles tubes #11/#221, and Watervliet tubes #53/#222). In April 1908 mortar #54 was shipped back to Watervliet for repairs and replaced with Bethlehem mortar #52. All the mortars were temporarily dismounted in 1910 to allow the carriages to be modified to the M1896MII standard. It appears that four mortars (#3, #42, #41, #52) were removed in early 1918 during the battery reduction program, the other four mortars were then kept in reduced maintenance status until finally declared obsolete and authorized for disposal on May 1, 1942. The emplacement still exists, though partially demolished and without its earth embankment on Tybee Island city property. The battery site is open, but most of interior is closed to the public.

- **BACKUS:** A battery for intermediate British-type rapid-fire guns initially placed at Fort Screven during the Spanish American War. Orders were issued on April 30, 1898 for the building of a single emplacement for a purchased 6-inch Vickers gun. It was a simple emplacement for a single platform with a 36-foot interior crest, and magazine on lower left. Work was undertaken immediately and well underway by May 12th. It was placed on the left flank of Battery Fenwick, firing to the northeast. Work was transferred on August 11, 1899. It was armed with one Vickers 6-inch gun and pedestal mount (#12138/#11164). On March 2, 1899 plans were submitted for the erection of an emplacement for two 4.7-inch as an addition on the immediate left flank of the 6-inch platform. The guns were to be moved here from the temporary battery at Wassaw Island. The new emplacements followed type plans for 5-inch batteries, but with thicker magazine protection. They were built and ready for transfer on February 2, 1901. Transfer made on February 23, 1901 for a cost of $14,356. These were armed with two 4.7-inch Armstrong guns and pedestals (#11857/#10835 and #11858/#10846). Naming came in General Orders No. 78 of May 25, 1903 for Lieutenant Colonel Electus Backus, who was mortally wounded in 1813 at Sackett's Harbor during the War of 1812. The single 6-inch gun was removed in March 1905 for transfer to Battery Bankhead at

Fort Adams. That platform was then modified for another 4.7-inch gun, transferred from Battery Madison at Fort Caswell. This gun and pedestal (#10854/#10980) were emplaced in late 1908. In 1910 there was a rearrangement of some of the guns on pedestals at the battery, emerging with gun/carriage arrangement of #11959/#10980, #11857/#10835, and #11854/#10846). Apparently, the guns were approved for removal and scrapping on July 22, 1919, and at least two were soon removed. One gun may have remained in place as late as 1942, though undoubtedly not in active service. The emplacements still exist, though at one point a residence was built upon it and subsequently burned down in 1980 but has been replaced. The battery is closed to the public.

- **GANTT**: A battery for two 3-inch, 15-pounder rapid-fire guns emplaced at Fort Screven on the north flank of Battery Backus firing north, northeast to cover the mine field. Plans were submitted on July 11, 1898 using funds from the Fortification Act of July 7, 1898. Both this battery and the similar one at Fort Pulaski (Battery Hambright) were constructed at the same time. The plans followed the standard mimeograph type plans. Work was begun on December 9, 1899 and completed in February 1900. Two guns (3-inch Model 1898 Driggs-Seabury guns and masking parapet mounts #116/#116 and #117/#117) were mounted in September 1903. Transfer was made on January 6, 1904 for a cost of $10,220. It was named on General Orders No. 78 of May 25, 1903 for Lieutenant Levi Gantt, 7th US Infantry who was killed in 1847 at Chapultepec. The pillars and emplacement were modified to the M1898M1 pedestal status around 1916. The armament was removed in common with other Model 1898s in the summer of 1920. The emplacement still exists on private property at Tybee Island, with a home built on top of the emplacement. The battery is closed to the public.

Fort Pulaski (1829-1880, 1898-1903) is located on Cockspur Island between Savannah and Tybee Island, Georgia. Construction of Fort Pulaski was begun in 1829, was named in General Order 32 of April 18, 1833, and was not completely finished by 1861. Confederates occupied the fort between 1861 and 1862. The Union recapture in April 1862 proved the obsolescence of static forts. The outer walls are pockmarked with many battle scars, testifying to the ferocity of the battle. It was then used by the Union as a POW camp, and later as a federal prison. The demilune was rebuilt in 1873-1875 with nine new gun platforms and four magazines. The fort's war damaged walls were repaired, but no other modifications were done. The demilune was rearmed with one M1888 8-inch gun on M1892 barbette carriage, and a mining casemate was added (replacing two platforms), during the Spanish-American War. Became a National Monument in 1924, restored by the CCC in 1933. Battery Hambright (1899-1903) is just outside the fort. The entire island became a U.S. Navy and Coast Guard base during World War II and was reopened to the public in 1947 as the Fort Pulaski National Monument under the NPS control.

Fort Pulaski Gun Batteries

- A Spanish American War emergency battery for an 8-inch barbette gun emplaced in the outer triangular demilune of old Fort Pulaski. With National Defense Act funding one of the existing 15-inch Rodman platforms (from the 1870s modernization of the fort) was modified to hold a standard Model 1892 barbette carriage. The emplacement used the old, adjacent magazine for ammunition. Work was done in the summer of 1898. The carriage (Watertown #5) was shipped here on April 27, 1898 while the gun tube (Model 1888 Bethlehem #11) arrived on April 25th. It was apparently mounted for a short time, but by July 1, 1899 was reported as dismounted by ordnance troops. Both the tube and carriage were directed to go to Fort Baker in the San Francisco defenses for possible use in a new battery there (in fact never built). The emplacement was not formally transferred or

SAVANNAH RIVER

FORT PULASKI

COCKSPUR ISLAND.

SERIAL NUMBER 124

EDITION OF DEC. 7, 1915.
REVISIONS: JUNE 9, 1916.

BATTERIES.
HAMBRIGHT.—2-3" B.P.

LOW WATER LINE

HORACE HAMBRIGHT
(No armament)

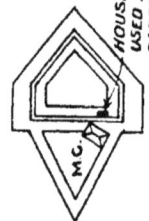

HOUSE ON TERREPLEIN
USED AS RESIDENCE OF
CARE TAKER.

M.C.

COCKSPUR ISLAND

BOAT LANDING

{ No armament; emplacements
would require extensive repairs
to render them serviceable. }

ever named. The site is now sodded over, though the bolt circle and block are presumably present just under the surface at the Fort Pulaski National Monument.

• **HAMBRIGHT**: A battery for two 3-inch masking parapet guns emplaced 800-yards to the northeast of old Port Pulaski along the river front near what was at the time the north wharf. Plans were submitted on April 1, 1899. It followed type plans, the right platform was interior, and left was a flank type. Each platform had its own magazine on the lower left flank. Work was done from 1899 to 1900. It was never transferred. The emplacement was named on General Orders No. 194 of December 27, 1904 for 2nd Lieutenant Horace G. Hambright, who died in North Dakota in April of 1896. The proposed armament of two 3-inch, 15-pounder Model 1898 guns and masking parapet mounts was never received. In 1910 it was reported that the emplacement had settled badly and would need extensive repairs, and as the primary mine field would no longer be located near Fort Pulaski, it was of little utility and would probably never be used. The battery was abandoned unarmed and unrepaired. The emplacement still exists at the Fort Pulaski National Monument. The battery is open to the public.

Wassaw Island: A temporary emplacement for two purchased 4.7-inch Armstrong guns built on Wassaw Island to cover the southern water entry into Savannah. Work was done in the late spring of 1898, the emplacement being reported ready for guns on April 30th, and the guns mounted on May 24th. It was a fairly substantial emplacement, with two flank platforms and a single large magazine on a lower level in the traverse between the guns. Otherwise, it generally followed the plan for 5-inch emplacements in Mimeograph No. 16. It was constructed of Portland cement. It carried two 4.7-inch guns and pedestals (#11857/#10835 and #11858/#10846). These served only a short time, being transferred to a more permanent emplacement of Battery Backus at Fort Screven in 1899. Considerable remains of the original concrete work, though in damaged and eroded condition, survive on private property at Wassaw Island. The battery is closed to the public.

Community Center on Tybee Island (Fort Screven Guardhouse) (Mark Berhow)

Tybee Island (Glen Williford)

Battery Habersham Tybee Island (Glen Williford)

Adams, Jack M. *A History of Fort Screven*. JMA2 Publications. Tybee Island, GA, 1996.

Tybee Island Historical Society. *Fort Screven 1897-1945*. Tybee Island, GA, 1988.

Gaines, William C. "Fort Screven: the Modern System of Defense at Savannah 1886-1946." *CDSG Journal* Vol. 11, No. 3, Aug. 1997, p. 30.

The main gun line of Fort Screven, Tybee Island (Glen Williford)

Battery Backus Tybee Island (Glen Williford)

THE HARBOR DEFENSES OF THE ST. JOHN RIVER AND JACKSONVILLE FLORIDA

Fort Clinch (1847-1900, 1942-1944) is located on a peninsula near the northernmost point of Amelia Island in Nassau County, Florida. The fort lies to the northeast of Fernandina Beach at the entrance to the Cumberland Sound, in the northeast part of the state. Never fully completed, construction was finally halted in 1867. In 1898 an unnamed battery of one M1888 8-inch BL gun on a modified 15-inch Rodman carriage was built on the fort's parapet. The old fort became a state park in 1936. U.S. Army platoon-sized detachments from the Camp Brunswick, GA coastal defense shore patrol base camp was posted here at the state park during World War II. Open to the public on a day use basis.

Fort Clinch Gun Battery

- At the start of the Spanish American War in April 1898 the Savannah Engineer office was authorized to emplace a single, modern 8-inch gun on a modified 15-inch Rodman carriage at Fernandina Beach. This was one of the 21 such guns hastily emplaced using an excess of model 1888 guns on older, but still serviceable carriages. The site selected was inside old Fort Clinch, at the northeast bastion. A simple concrete platform with rails for traversing the Rodman carriage and adjacent magazine with crane hoist was constructed in the summer of 1898 with NDA funds. Slow delivery of materials delayed completion, the emplacement not being ready for armament until August 15, 1898. The carriage was received and mounted in September and the gun tube (Bethlehem Model 1888 #6) was received on December 18, 1898. The battery was not formally named or turned over to troops; but apparently cost all of the appropriated funding of $7,325. It served only a short while, the armament removed and shipped out on November 20, 1900 the tube was later reused at Fort McKinley, Portland. The emplacement still exists at the Fort Clinch State Park. The battery is open when the fort is open to the public.

The emergency battery at Fort Clinch (Terry McGovern)

EDITION OF JAN'Y. 2,1914.
REVISIONS MARCH. 4,1914.
NOV. 8,1916.

SERIAL NUMBER 24

ST. JOHNS RIVER
FLORIDA.

SCALE OF FEET

RANGE OF TIDE AT ST. JOHNS BLUFF 3.4
" " " END OF JETTIES 5.22
MAXIMUM DRAFT TO ST. JOHNS BLUFF AT M.L.W 27.5

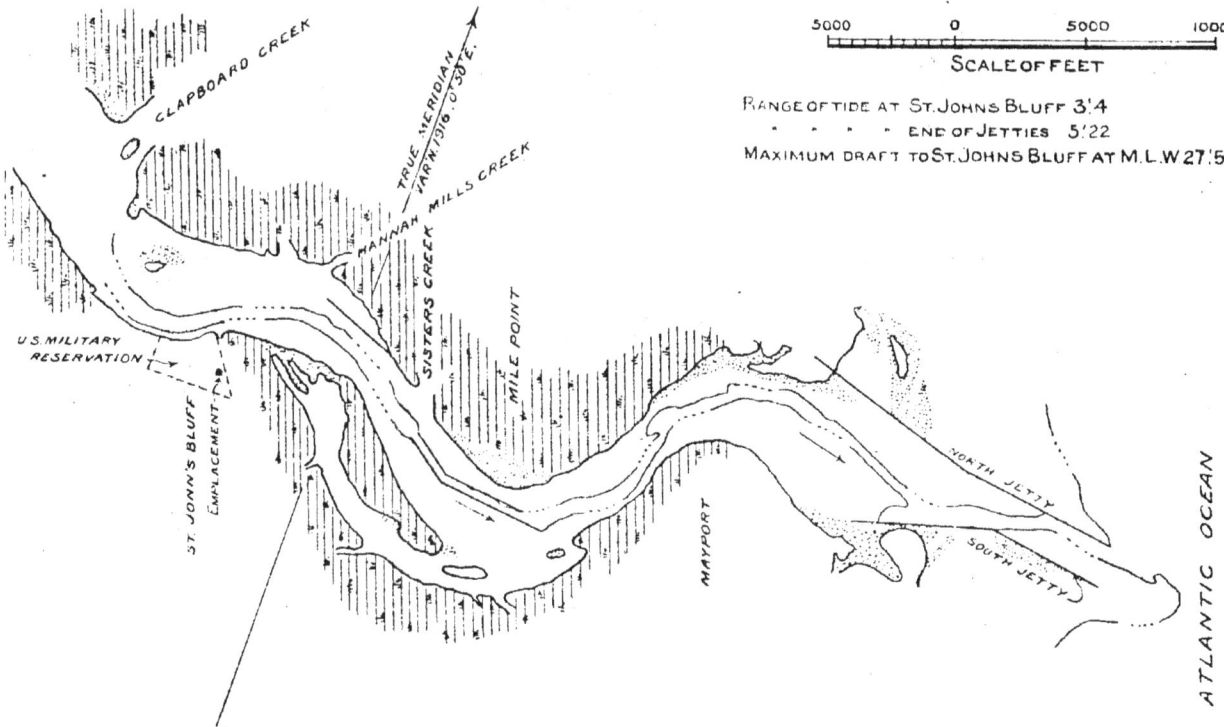

CLAPBOARD CREEK

TRUE MERIDIAN VAR'N 1916-0°50'E.

HANNAH MILLS CREEK

SISTERS CREEK

U.S. MILITARY RESERVATION

ST. JOHN'S BLUFF EMPLACEMENT

MILE POINT

MAYPORT

NORTH JETTY

SOUTH JETTY

ATLANTIC OCEAN

EDITION OF JAN'Y. 2,1914.
REVISIONS MARCH. 4,1914.
NOV. 8,1916.

SERIAL NUMBER 124

ST. JOHNS RIVER, FLA.
ST. JOHNS BLUFF.

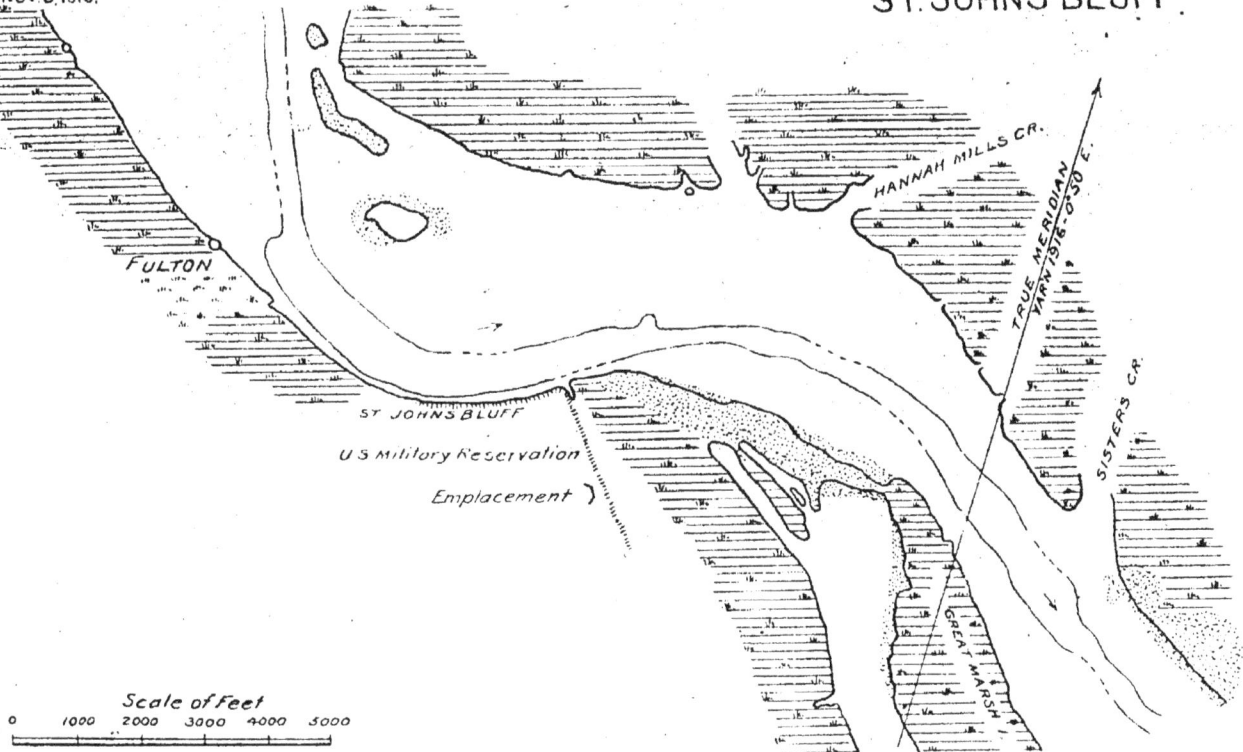

FULTON

HANNAH MILLS CR.

TRUE MERIDIAN VAR'N 1916-0°50'E.

SISTERS CR.

ST JOHNS BLUFF

U.S. Military Reservation

Emplacement

GREAT MARSH

Scale of Feet
0   1000   2000   3000   4000   5000

St. Johns Bluff Reservation (1898-1899/1925) is located bluff on the St. John's River about 15 miles downriver from Jacksonville, Floridia, near the site of Fort Caroline. An unnamed concrete two-gun battery (M1888 8-inch BL guns on modified 15-inch Rodman carriages) was built in 1898. The land was formally purchased and reserved by the Federal government in 1901, sold in 1925. The concrete battery, now located on National Park Service property, is one-third mile southeast of the Ribaut Memorial. It is in excellent condition.

St. Johns Bluff Gun Battery

- At the start of the Spanish American War in April 1898 the St. Augustine Engineer office was authorized to emplace a pair of modern 8-inch guns on a strengthened 15-inch Rodman carriages at a new location on St. John's Bluff. These were two of the 21 such guns hastily emplaced using an excess of model 1888 guns on older, but still serviceable carriages. Colonel Benyaurd began to fortify the new site. Plans were submitted on June 14, 1898 and were approved on June 17th. The site also received an emergency earthen battery for siege guns and a small mining station. The emplacement design was a relatively simple concrete platform with traversing rails and a breast-height wall for two adjacent 15-inch Rodman carriages. A concrete magazine was constructed on the flank of the gun platform. Brush clearing began on April 11, 1898 and concrete work done from June to August of that year. The emplacement was never officially transferred but was basically complete by November for an estimated cost of construction of $32,000. The two carriages arrived on May 7 and the tubes on July 16, 1898 (8-inch Model 1888 Bethlehem #10 and Watervliet #41). Some delay occurred in mounting, the armament was not reported as ready for service until January 1899 (and even then, one mount was reported as still missing some small parts). As the concrete magazine roof was never poured to its full extent, the battery was also missing its earthen cover. At the conclusion of the conflict the armament was declared surplus, and the tubes were removed and sent to Fort McRee in Pensacola for use in Battery Slemmer on June 29, 1899. The site was abandoned in the early 1900s and eventually sold to private owners. Today the emplacement still exists on National Park Service property, but is open to the public

St. John's Bluff Battery (Terry McGovern)

THE HARBOR DEFENSES OF KEY WEST – FLORIDA

The harbor area to the west of Key West Island is one of the few relatively deep-water harbor areas along the southern Florida coast. A Naval Station was established at Key West in 1823 as part of an effort to deal with Caribbean pirates. Along with the anchorage out at the Dry Tortugas, it received new fortifications during the late Third System construction period. Naval operations were expanded during the Mexican War, the Civil War and the Spanish American War. An U.S. Army barracks had been established in 1831 to house the soldiers based on the island. The Key West Defenses were modernized during the Endicott Program, and again during World War II. As the Army phased out its use of Key West after 1947, its properties were turned over to the U.S. Navy.

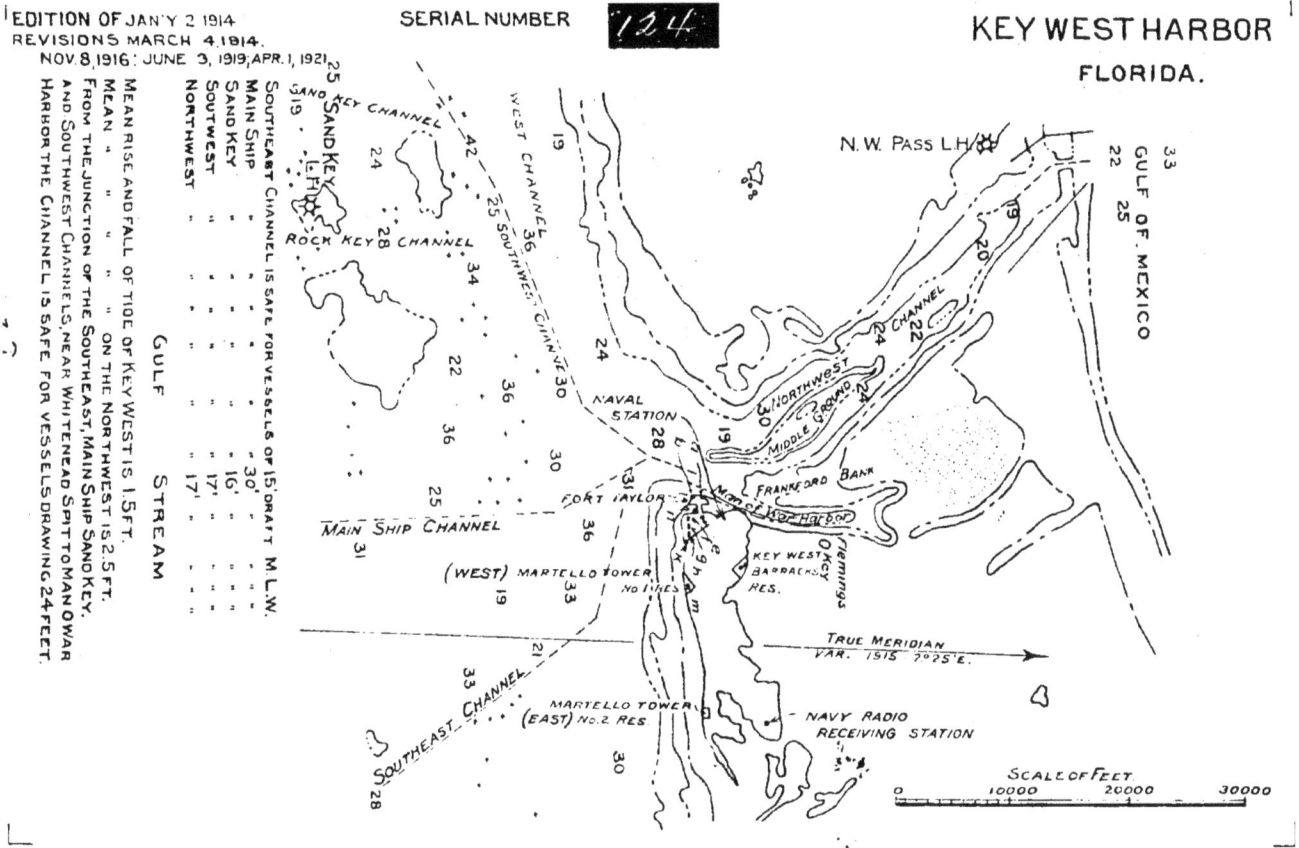

Fort Taylor (1846-1947) is located on the far western end of Key West Island at the end of Florida Keys. Construction of this five-sided, three-tiered, Third System fort on a sand shoal about a quarter of mile from shore was begun in 1846 and completed in 1861. It was named in General Order 38 of 1850 for Major General Zachary Taylor, the twelfth President of the United States. The fort was held in Union hands throughout the Civil War as Key West played an important role in supporting the Union Navy's blockade of Confederate ports. Key West also played a significant role in the Spanish-American War as a supply base. The Endicott Program saw the construction of ten batteries from 1898 to 1906. The largest being Battery Osceola, two 12-inch barbette guns, which was built into the walls of the cut-down Third System fort. The other primary battery was Battery Seminole, eight 12-inch mortars, which was built on Key West itself. The importance of Key West's naval station resulted in the construction two #200 Series batteries as part of the 1940 Program. Battery #231 was built on the Fort Taylor military reservation, while Battery #232 was built on eastern end of the island, near the airport. Also, four 155mm GPF guns were installed on Panama mounts on top of the Endicott batteries at Fort Taylor. Two more Panama mounts were built at the East Martello Tower. Two AMTB batteries were also installed. After World War II, the U.S. Army turned its reservation over to the U.S. Navy to be part of the naval base. The value of Key West grew again during Cuban Missile Crisis causing the U.S. Navy to destroy most of the Endicott batteries to make more room for naval activities. Today, the Third System fort has become the Fort Zachary Taylor Historic State Park with two Endicott batteries inside the fort, and the mortar battery remains part of the Truman Annex of Naval Air Station Key West. The park is open during daylight hours, entrance fee required. The fort is known for its large collection of Civil War cannons that were buried into the fort when the Endicott Program batteries were constructed. Two other defenses of note are the East and West "Martello" Towers, built during the Third System to protect the southern shoreline.

Fort Taylor/Key West Gun Batteries

- **OSCEOLA**: A battery for two 12-inch barbette guns emplaced within old Fort Taylor with funding from the National Defense Act at the beginning of the Spanish American War. This was one of the batteries built to use an excess of 12-inch tubes and quickly produced barbette carriages in order to accelerate the pace of adding new armament. Plans were submitted on May 14, 1898. It was placed parallel to the southwest face of the old work, with a parapet height of 27-feet. That required destruction of some of the rest of the old fort's third tier. Broken brick from that work was used as aggregate in the concrete, and old pieces of armament (including intact Rodman cannons) were also dropped into the concrete to add to its strength. The space available for a two-gun heavy battery was limited, so the plan adopted a compact design with all magazines sharing the space under the central traverse (and being advanced forward for better protection). Ammunition service was by hoists. There was only 147-feet between gun centers. The battery fired to the south. Work was started on June 30, 1898 and essentially completed (except for overhead trolley rails, doors and ladders) by July 1899. The battery was transferred on February 3, 1900 for a cost of $112,951.48. It was named on General Orders No. 43 of April 4, 1900 for the famous Seminole Indian chieftain Osceola. The guns had been mounted in late 1899 and consisted of two 12-inch Model 1888 Watervliet guns on standard Model 1892 barbette carriages (#39/#10 and #23/#11). The battery later received modifications to its power plant, hoists, commander's station and telephone system. It had a long service life with its original armament, the guns not being removed until October 16, 1944. The emplacement still exists at the Fort Zachary Taylor State Park. The battery site is open, but the interior is closed to the public.

- **SEMINOLE**: The mortar battery for Key West, emplaced directly east of the old fort on the mainland. Plans were submitted on October 25, 1896. Originally envisioned as a four-pit battery for sixteen mortars, it was authorized for just eight mortars and never increased. It was typical of early mortar designs—with two adjacent, small pits and the magazines forward under the front parapet. Overhead trolleys were installed in the shot and shell magazines to bring shells to carts. Rooms were later built into the right flank for the post telephone switchboard rooms. It was built by contract labor by the Venable Construction Co. under a contract dated March 4, 1897. Concrete was placed between April and October 1897, with details completed as late as January 1898. Final completion (fill, sodding, building the concrete retaining walls) was delayed by the Yellow Fever outbreak of September 2, 1898. The carriages arrived in mid-1898 and the tubes in January 1899 for mounting by March 1899. Transfer was made on June 30, 1904 for $112,001.65. It was named on General Orders No. 32 of April 4, 1900 for the Seminole Indian tribe. It was armed with eight 12-inch Model 1890M1 mortars on Model 1896 mortar carriages (Watervliet tubes #41/#100, #42/#101, #44/#102, #40/#130, #43/#92, Builders tubes #25/#93, #26/#132, and #24/#131). The flank observation stations were increased in height in 1902, and telautograph booths added in 1905.

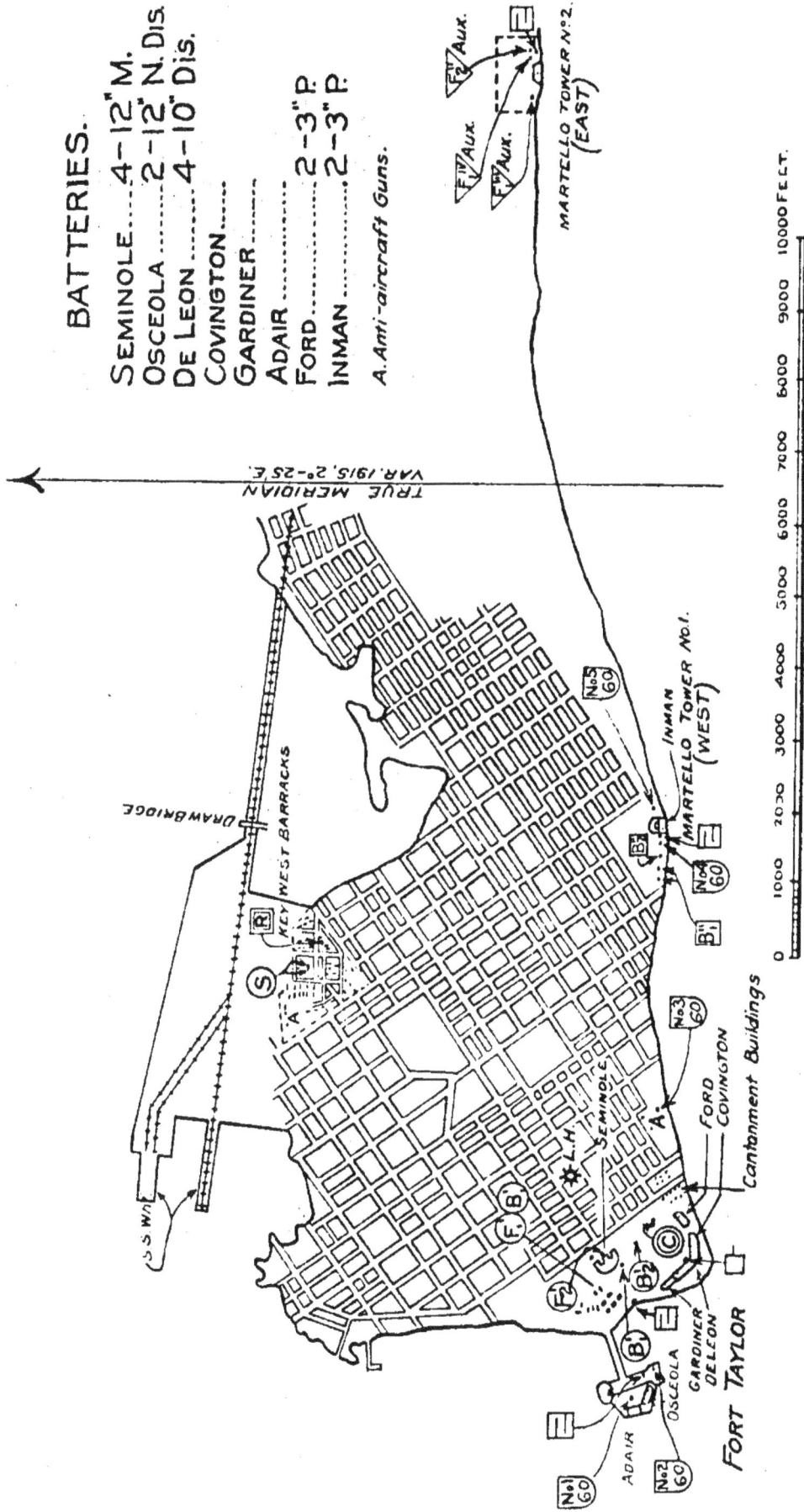

KEY WEST ISLAND

FLORIDA.

BATTERIES.

SEMINOLE....4–12" M.
OSCEOLA....2–12" N. Dis.
DE LEON....4–10" Dis.
COVINGTON....
GARDINER....
ADAIR........
FORD........2–3" P.
INMAN.......2–3" P.

A. Anti-aircraft Guns.

SERIAL NUMBER

1244

EDITION OF JAN'Y 2. 1914.
REVISIONS MARCH. 4. 1914.
NOV. 8. 1916; JUNE 3, 1919; APRIL 1, 1921.

TRUE MERIDIAN
VAR. 1915, 2°–25' E.

MARTELLO TOWER Nº 2.
(EAST)

F'' Aux.
F'' Aux.
F'' Aux.

0   1000 2000 3000 4000 5000 6000 7000 8000 9000 10000 FEET.

DRAWBRIDGE

KEY WEST BARRACKS

(S.S. WM.)

Nº 5 GO
Nº 4 GO
B''
B'
INMAN
MARTELLO TOWER Nº 1.
(WEST)

L.H.
SEMINOLE
Nº 3 GO
FORD
COVINGTON
A.
Cantonment Buildings

F. B.
F.²
B.²
B.'
GARDINER
DELEON
OSCEOLA
FORT TAYLOR
ADAIR
Nº 1 GO
Nº 2 GO

FIRE CONTROL STATIONS

- F₁ Taylor
- B₁ Seminole
- B₂ De Leon
- M Mine Command
- M" "

60 36 30 Searchlights

BATTERIES

Seminole	8 – 12"	M
Osceola	2 – 12"	N.D.
De Leon	4 – 10"	D.
Covington	2 – 8"	D.
Gardiner(Engr'r)	2 – 4.7"	P.)
		Transferred.
Adair	4 – 3"	B.P.
Dilworth	2 – 3"	B.P.
John DeKalb	2 – 6"	P.
Mahlon Ford	2 – 3"	P.

FORT TAYLOR
KEY WEST ISLAND, FLORIDA

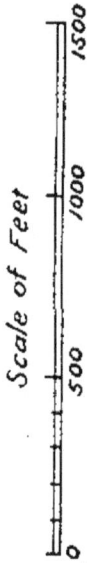

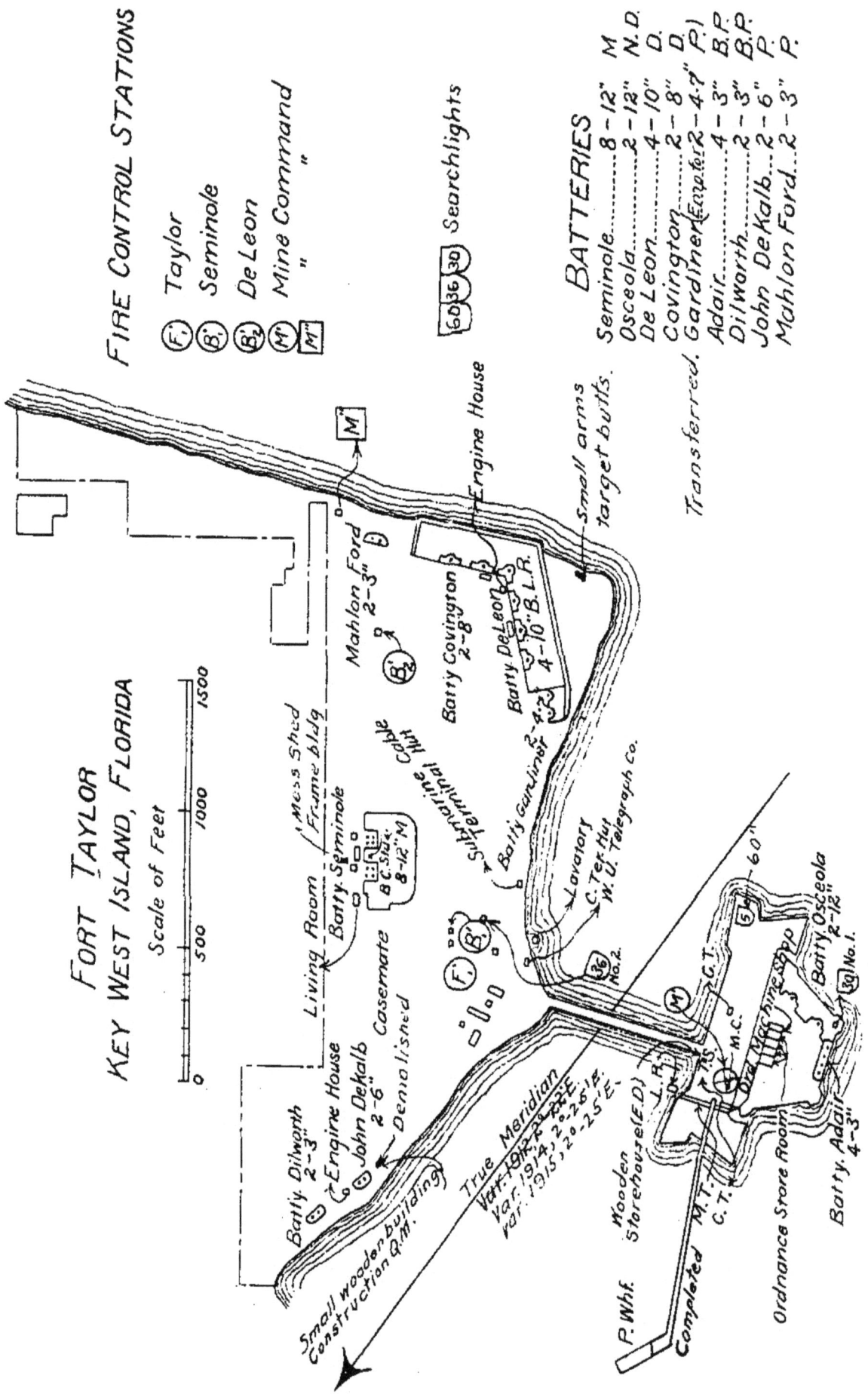

Scale of Feet

0   500   1000   1500

M"

Engine House

Small arms target butts.

Mahlon Ford 2 – 3"
B₂'

Batty Covington 2 – 8"

Batty DeLeon 4 – 10" B.L.R. 2 – 4.7

Batty Gardiner 2 – 4.7

Submarine Cable Hut

Mess Shed Frame bldg

Living Room

Batty Seminole
B.C.Stn.
8 – 12" M.

Casemate

Lavatory
C. Tel. Hut
W.U. Telegraph Co.

F₁

B₁'

36 No. 2

Batty. Dilworth 2 – 3"
Engine House
John DeKalb 2 – 6"
Demolished

Small wooden building
Construction Q.M.

True Meridian
Var. 1913, 2°22'E.
Var. 1914, 2°26'E.
Var. 1915, 2°25'E.

P.Whf.
Wooden Storehouse(E.D.)

Completed M.T.T.
C.T. M.
C.T.N.

6"

5" C.T.

M
M.C.
De Machine Shop

Batty Osceola 2 – 12"

30 No. 1

Ordnance Store Room

Batty. Adair 4 – 3"

6"
6.0"

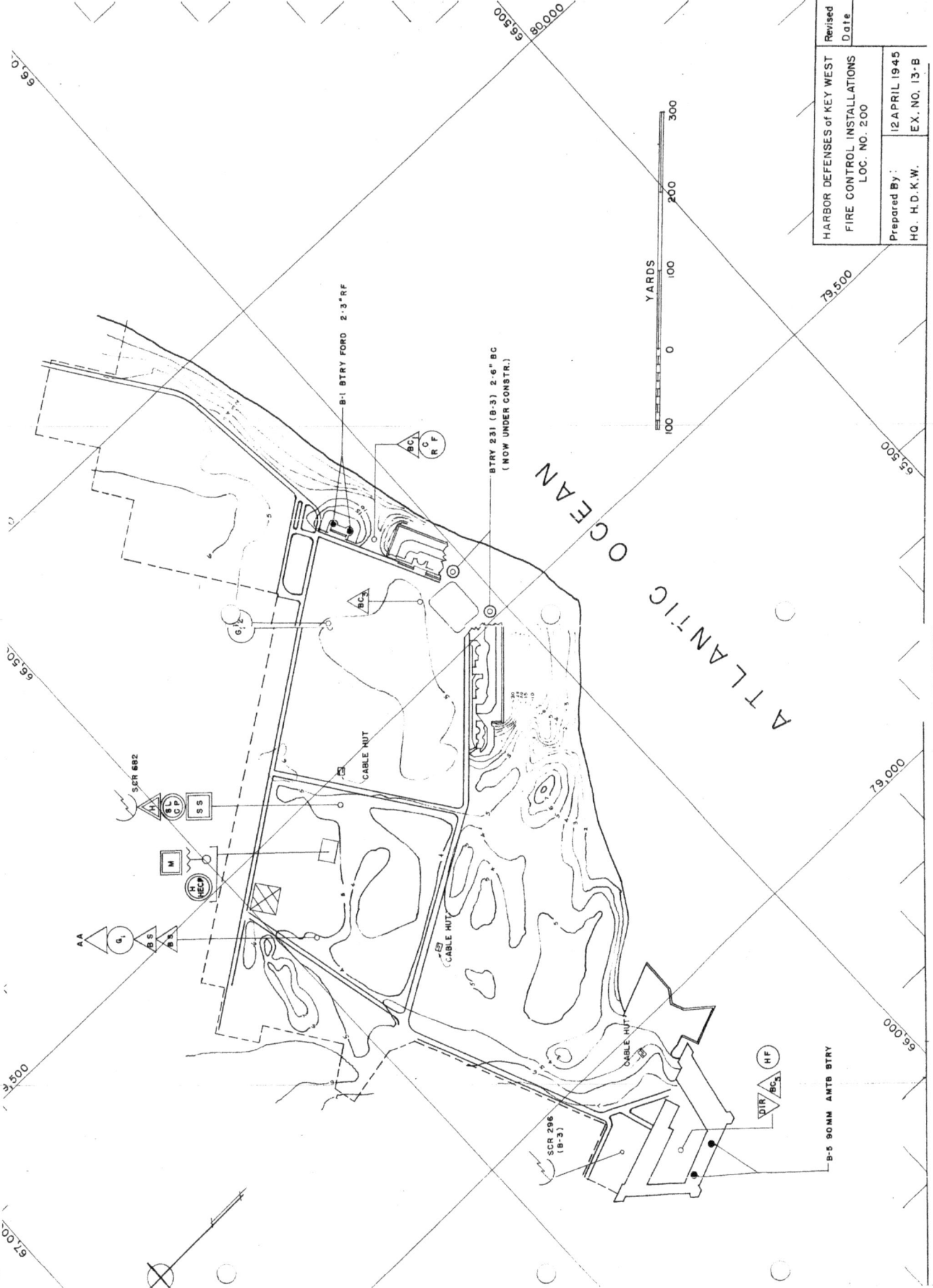

B-I BTRY FORD 2·3"RF

BTRY 231 (B-3) 2·6" BC
(NOW UNDER CONSTR.)

ATLANTIC OCEAN

YARDS

300 200 100 0 100

CABLE HUT

CABLE HUT

CABLE HUT

SCR 682

SCR 296
(B-3)

CABLE HUT

B-5 90MM AMTB BTRY

HARBOR DEFENSES of KEY WEST
FIRE CONTROL INSTALLATIONS
LOC. NO. 200

Prepared By:
HQ. H.D.K.W.

12 APRIL 1945

EX. NO. 13-B

Revised

Date

Four mortars (the forward ones in each pit) were removed in 1917. The remaining four served until authority for deletion given on November 11, 1942. The emplacement still exists at the Truman Annex of the Naval Air Station Key West property, though missing much of its original earthen embankment. The battery is closed to the public and access to the base is restricted.

- **DeLEON**: A battery for six disappearing guns emplaced on the mainland to the southeast of old fort Taylor near the 1870s South Battery. Site plans were submitted on August 8, 1896 and detailed cost estimates on October 25, 1896 using funds from the Act of June 6, 1896. As planned it consisted of a straight line for four 10-inch disappearing guns and on the left flank, bent more to the east a contiguous set of two 8-inch disappearing guns. The battery was named in General Orders No. 43 of April 4, 1900 for Spanish explorer Ponce de Leon, however in 1903 the two 8-inch guns were tactically split off to become Battery Covington. Plans for the series of 10-inch emplacements generally followed design mimeographs. The guns had a spacing of 128-feet, the right flank, No. 1 platform was a flank emplacement, and the No. 4 was at the "bend" of the series and had a field more to the south and a constricted loading platform. Magazines were on the lower right flank, ammunition service by lifts. It fired to the southwest. Work was begun in April 1897 along with Battery Seminole as the first of the Key West modern batteries. The basic concrete work was complete by the summer of 1898, for mounting armament in September. It was armed with four 10-inch Model 1888 Watervliet guns on Model LF M1896 disappearing carriages (#46/#55, #65/#42, #64/#39, and #34/#40). Troops manned the emplacement as early as March 1900, but official transfer did not come until June 30, 1904 at a cost of $198,446.40. The battery received the usual modifications to its hoists and widened platforms prior to 1915. Two guns were temporarily dismounted between 1917 and 1919. The battery was declared excess in the 1932 Review, and the armament removed in November 1941. The parapet was used for two 155mm Panama mounts of a battery erected there in late-1941. Most of emplacements No. 3 and 4 were destroyed to make room for the building of Battery #231 in 1943. The rest of the emplacement was destroyed about 1962 to allow for construction of a naval training facility at the post. No remains exist.

- **COVINGTON**: A battery for two 8-inch disappearing guns originally part of the Battery DeLeon sequence on the mainland to the southeast of old Fort Taylor. Detailed plans were submitted on October 25, 1896 with funds coming from the Act of June 6, 1896. Plans were based on type designs. It had one internal and one flank emplacement. There was a 124-foot gun spacing, the pair was oriented more to the south than the 10-inch sequence and fired to the south. Magazines were on the lower right and ammunition service was initially with lifts, replaced by new hoists in 1907. Changes were also made to the loading platforms and communication routes. Work was done in 1897. Carriages and guns were received in May and June 1898, the mounting being completed by July. Transfer did not come until June 30, 1904 for a cost of $99,276.45. At first it was part of Battery DeLeon but separated and named Battery Covington in General Orders No. 78 of May 25, 1903 for Brigadier General Leonard Covington mortally wounded during the of War of 1812. It carried two 8-inch Model 1888M1 Watervliet guns on disappearing carriages LF Model 1894 (#40/#25 and #39/#24). Troops manned the guns since July 1898, during the Spanish American War. The battery served until disarmed in 1917. In 1941 two Panama mounts for 155mm guns were emplaced on the battery's parapet between the old gun platforms and used the old magazines and rooms of the 8-inch emplacement. Portions were also destroyed for the erection of Battery #231 nearby. The entire emplacement was destroyed in 1962 along with Battery DeLeon to allow construction at the site. No remains exist.

- **DeKALB:** An emplacement for two 6-inch pedestal guns sited to the north of old Fort Taylor near the 1870s North Battery. Plans were submitted on October 12, 1903. The emplacement design conformed to the recommendation of mimeograph No. 40. The two gun platforms were spaced at 94-feet. They shared a magazine housed in the protected traverse between platforms. There were no hoists, ammunition service was by hand. The site had previously been used for a battery of Rodman guns in the 1870s and then utilized for three 8-inch converted rifles at the start of the Spanish American War. This emplacement was destroyed for the construction of the new 6-inch battery. It fired to the west, northwest. Work was done in 1903-1904. Guns were shipped here on November 9, 1905 and it was reported ready for transfer on February 5, 1906. Actual transfer was made on April 26, 1906 for a cost of $26,700. It was named in General Orders No. 194 of December 27, 1904 for Major General John B. DeKalb of the Continental Army, mortally wounded in 1780. The construction prompted the destruction of the rest of Fort Taylor's old barracks on the gorge side in order to clear the field of fire. The battery mounted two 6-inch Model 1900 guns on Model 1900 pedestals (#21/#40 and #31/#41). These were temporarily removed on February 24, 1905 and sent to Watervliet Arsenal for alteration but returned and remounted shortly thereafter. The battery was disarmed in 1917 the guns eventually being re-installed in Battery Chamberlin at San Francisco. The emplacement itself was soon surrounded by fill associated with the building of the new naval yard and was abandoned and eventually destroyed in the early 1930s. No remains exist.

- **GARDINER:** A battery for two 4.7-inch rapid-fire guns emplaced at Fort Taylor just to the northwest flank (right flank) of 10-inch Battery DeLeon. Plans were submitted as soon as March 24, 1898 with instructions from Washington to prepare an emplacement for these guns being purchased in England. As it was perceived that there was a deficiency of coverage of the southern approaches, the location in the old Southern Battery was selected. The plans reflected a simplified 5-inch battery design, with a 45-foot spacing of guns, a single magazine room on the lower left flank of each platform, and no battery commander station. It had increased concrete thickness due to its proximity to the heavy 10-inch Battery DeLeon. It fired to the southwest. It was authorized on March 19, 1898 with an appropriation of $6,000. Due to the emergency of the war, the guns were emplaced when received on temporary wooden grillage platforms not far from where they were ultimately to go. The guns themselves (Armstrong 4.7-inch guns and pedestals #11002/#9082 and #11003/#11004) with 400 rounds of ammunition had arrived on April 2, 1898. The temporary battery was transferred on May 18, 1898. Work on the permanent emplacement continued and was completed by the end of 1898. Guns were relocated by March of 1899 and this permanent site was transferred on December 24, 1898 for a cost of $18,000. The battery was named in General Orders No. 78 of May 25, 1903 for Captain George W. Gardiner killed in action with the Seminole Indians in 1835. The armament was in turn removed in 1913 for shipment to a new battery being built in Hawaii. The emplacement was later used to house the plotting room for Battery DeLeon. The structure was destroyed in the early 1960s to make room for local construction. No remains exist.

- **ADAIR:** A battery for four 3-inch masking parapet guns emplaced on the western curtain wall of old Fort Taylor. On December 13, 1898 plans were submitted for a battery of two 3-inch model 1898 rapid-dire guns for the curtain wall on the second tier of old Fort Taylor. In 1900 a third emplacement was added, and then a final fourth one on April 13, 1901. Though built over several years, they were all eventually treated as a single, 4-gun battery emplacement. The plans called for four emplacements (three internal, No. 1 being for flank fire), spaced 29-feet apart. Each had a separate, one-room magazine on the lower left, ammunition service was by hand. These were on the old floor of the original second tier, with the curtain wall above that level being removed. It fired to the west, southwest. Work was started in June 1899 and the first two emplacement finished by that

September. After three guns and carriages had arrived, they were mounted in October 1900 (three 3-inch, 15-pounder Model 1898 guns and M1898 Masking Parapet Mounts #42/#42, #43/#43, and #44/#44). Transfer came on April 23, 1901 for a total cost of $18,375. The fourth gun was delayed, not arriving until May 1903 for mounting on August 18, 1903 and transfer in 1905 for $9900, with carriage M1898 (#120/#120). The battery was named in General Orders No. 78 of May 25, 1903 for 1st Lieutenant Lewis D. Adair, mortally wounded in action against the Sioux Indians in 1872. The emplacements and carriages were modified to M1898M1 pedestals around 1916. A new CRF station was built adjacent to the emplacement and transferred in 1919. The four guns were removed in June 1920, and the carriages thrown into the moat. During World War II one pit was modified for a 90mm fixed mount of AMTB #5. The emplacement still exists at the Fort Zachary Taylor State Park. The four carriages were recovered from the fort site in the 1980s and have been reconditioned, the only remaining examples of this carriage type. They are currently on display disassembled in a room at the fort. The battery is closed to the public.

- **DILWORTH**: A battery for two 3-inch guns on masking parapet mounts emplaced on the northern flank of the fort covering Man O'War Bay. It was sited on the right flank of 6-inch Battery DeKalb to cover the mine field blocking the northwest passage. Plans were submitted on December 13, 1898. It was of standard masking parapet battery type, with two interior platforms with barrel niches, and guns spaced with 29-foot centers. Each platform had a single magazine room on its lower right, ammunition service being by hand. Work was done from January 1899 but suspended in August pending first the delayed arrival of the carriages and then the yellow fever outbreak and season. Work resumed in December 1900 and was completed by February 1901. Carriages and guns were received on October 8, 1900 and installed by February 17, 1901. It was armed with two 3-inch Model 1898 Driggs-Seabury guns and M1898 Masking Parapet Mounts (#39/#39 and #41/#41). The battery was finally transferred on April 13, 1901 for $13,375. It was named in General Orders No. 78 of May 25, 1903 for 2nd Lieutenant Rankin Dilworth who was killed in action at the Battle of Monterey in 1847. The carriages and emplacement were modified for the M1898M1 pedestal configuration around 1916. The battery served until June of 1920 when the guns were declared obsolete and ordered removed. The emplacement itself was destroyed in the 1920s for the new naval station, no traces remain today.

- **FORD**: An emplacement for two 3-inch pedestal guns erected on the southern side of Fort Taylor, to the east of the main gun line, firing to the south. Plans were submitted on November 13, 1903. The site was east of Battery Covington. The plans followed standard mimeograph recommendations. The two platforms were separated by 62-feet. Two magazines and a storeroom were under the central traverse. Excess broken brick from old Fort Taylor was used in the concrete aggregate. Work was done from 1903-1904. Transfer was made on April 26, 1905 for a cost of $14,100. It was named on General Orders No. 194 of December 27, 1904 for Major Mahlon Ford, of the 1st Artillery who served during the Revolutionary War. It was armed in 1907 or 1908, receiving two 3-inch Model 1903 guns and M1903 pedestal mounts (#72/#45 and #74/#46). The battery had a long service life, not being declared obsolete until right after World War II. The emplacement itself was destroyed in the 1960s during construction. Insignificant remains are left at the site.

- **Battery #231**: A 1940 Program standard dual 6-inch barbette battery emplaced at Fort Taylor. It was built at the site of old Battery Leon, requiring the destruction of emplacements No. 3 and 4 and major portions of the rest of that battery. Initially it was given a relatively low priority and not funded until the FY-1943 Budget. Construction was started on January 23, 1943. It had a standard design for the 200-series batteries, with guns spaced at the required 210-feet. Concrete work was finished in early 1944. The battery was never really completed, work being suspended before armament was mounted. It was to be armed with two 6-inch guns M1(T2) on Model M4 barbette carriages. The guns were never received but the carriages were delivered to the site (carriages #59 and #60 allocated). Further work was abandoned postwar. The emplacement still exists at the Truman Annex of the Naval Air Station Key West. The battery is closed to the public and access to the base is restricted.

- **AMTB #5**: A 1943 AMTB Program battery for two 90mm fixed and two 90mm mobile guns emplaced atop the western parapet of old Fort Taylor. The two fixed 90mm guns were given gun blocks, one to the north of old Battery Adair on the second floor of the old fort and the other was a conversion of Adair's emplacement No. 4. The work was done in 1943, the battery serving until disarmed about 1946. It served during World War II as local Tactical Battery No. 5. The two gun blocks and ready ammunition shed still exist at the Fort Zachary Taylor State Park. The battery is open to the public.

- **INMAN**: A battery for two 3-inch pedestal guns erected at the old West Martello Tower. It occupied the space between the inner and outer wall of the tower, facing and firing to the south. Site selection was decided on September 18, 1903, and plans submitted on November 13, 1903. Despite the limitations of the site, it followed standard plans. The guns were separated by 62-feet, and there were three rooms (two magazines and a storeroom) under the protected central traverse. Guns had only a 17.8-foot trunnion height. The outer brick wall of the tower was retained. Work was done from 1903-1904, being reported ready for transfer on February 3, 1906. That transfer was made on April 26, 1906 for a building cost of $14,300. It was named on General Orders No. 20 of January 25, 1906 for Captain Shadrach Inman, Georgia Militia who was killed in 1780 during the Revolutionary War. The original platforms had been constructed for Model 1902 guns, but they had to be changed for the Model 1903 type causing a delay in mounting armament. Not until 1917 was the armament finally erected, consisting of two Model 1903 guns and pedestal mounts (#63/#74 and #64/#75). The battery remained in service between the wars and was not finally disarmed and deleted until late 1945. The emplacement still exists at the Key West Garden Center. The battery is open to the public when the garden center is open.

- **AMTB #6**: A 1943 Program AMTB battery of two 90mm fixed and two 90mm mobile guns. It was emplaced just west of the West Martello Tower. The concrete gun blocks were built in late 1943. Guns were emplaced and then probably dismounted in late 1945. It served during World War II as local Tactical Battery No. 6. It was armed with gun M1 on carriages M3 (#5571/#215 and #7153/#231). The gun blocks still exist in the park (concrete mounts now built over by the Higgs Beach pavilions) to the west of the West Martello Tower. The battery site is open to the public.

- **Battery #232**: A standard 1940 Program dual 6-inch barbette battery. It was built on the site of a small island in the salt marshes (also called the Salt Lake) located on the eastern side of the island, near the commercial airfield. It was issued a relatively low initial priority and not funded until the FY-1943 Budget. The battery was of standard 200-series design. Actual construction was done from September 28, 1942 until October 31, 1943. Most of the equipment was installed by May 1944, the battery then being ready for armament. It received two 6-inch Model M1(T2) guns on Model M3 barbette carriages (#9/#10 and #99/#11). The guns fired for calibration on October 5, 1944. The battery was damaged by a hurricane on October 28, 1944. The gun pits were flooded with salt water, new wiring was required and replaced in November. The battery served only for a short while. It was never named, just known as Battery Construction No. 232. It was disarmed soon after the end of the war, probably in 1947. The emplacement still exists on Key West Airport property. The battery is closed to the public.

Fort Taylor 1930s (NARA)

Fort Taylor State Park (Glen Williford)

SERIAL NUMBER **124**

MARTELLO TOWER Nº1.
(WEST TOWER)

KEY WEST HARBOR, FLA.

EDITION OF JAN'Y. 2, 1914.
REVISIONS MARCH, 4, 1914; DEC. 7, 1915;
NOV. 6, 1916; APRIL 1, 1921.

BATTERIES
INMAN............2-3"PED.

TRUE MERIDIAN
VAR. 1915, 2°-25' E.

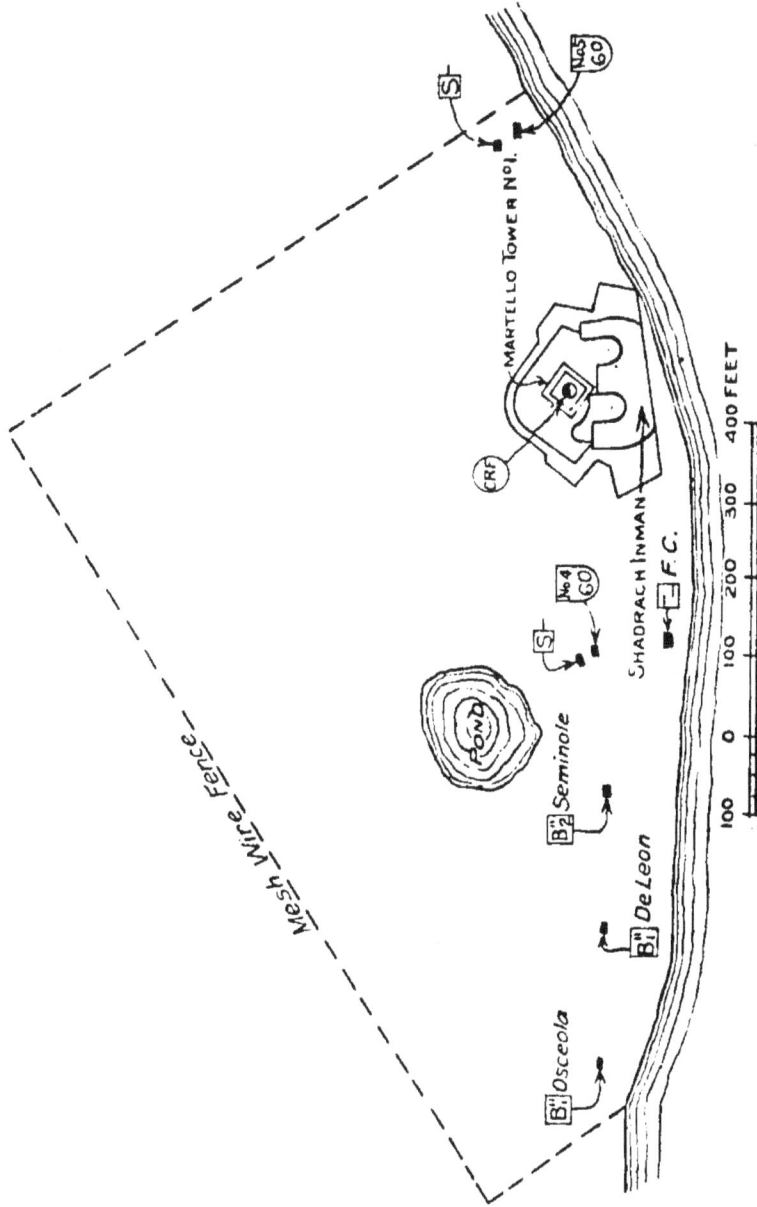

Mesh Wire Fence

POND

Bᴺ Osceola

Bᴺ De Leon

Bᴺ Seminole

S

No.4
GO

S

CRF

MARTELLO TOWER Nº1.

SHADRACH INMAN

F.C.

Nº5
GO

100   0   100   200   300   400 FEET

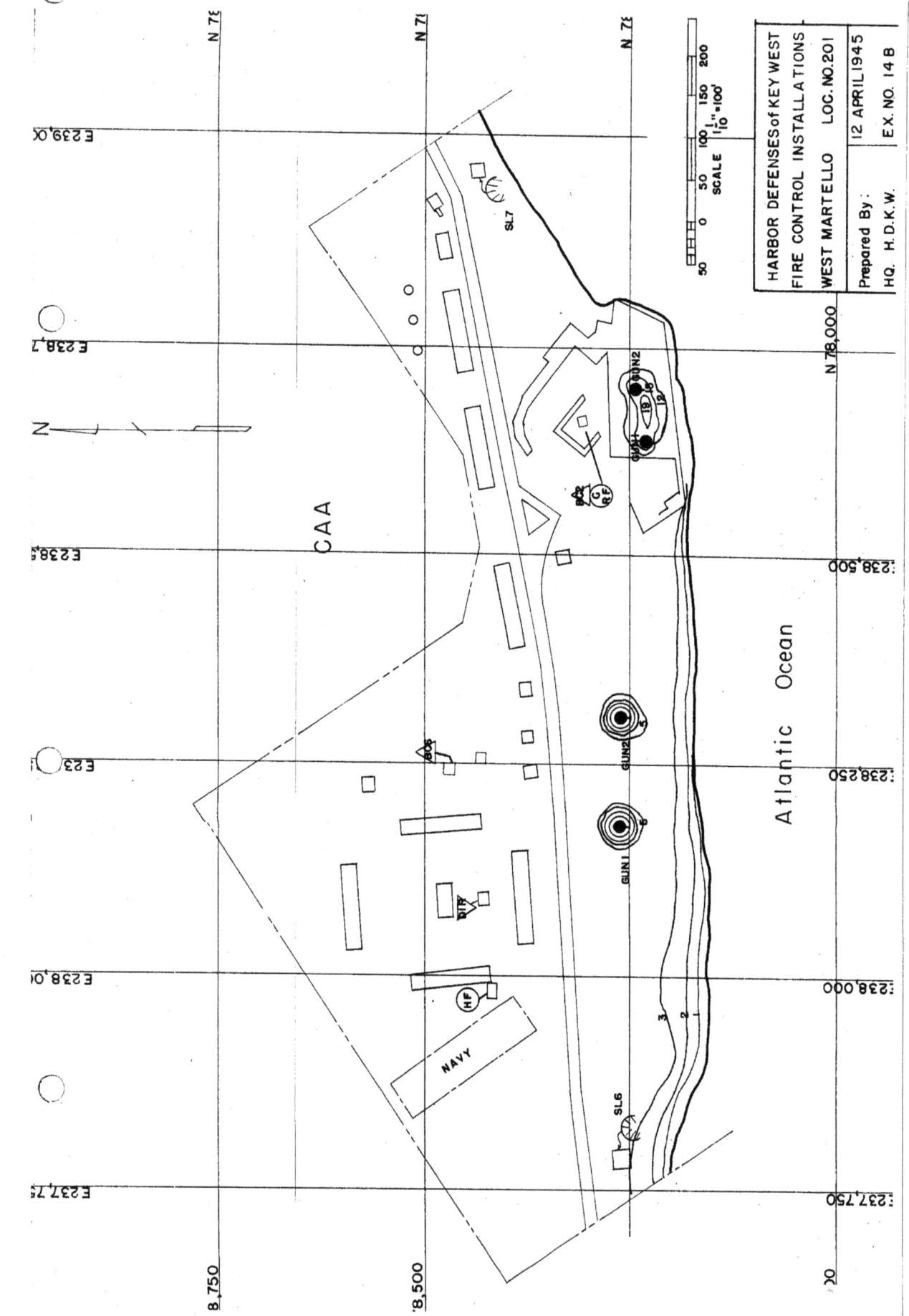

HARBOR DEFENSES of KEY WEST
FIRE CONTROL INSTALLATIONS
WEST MARTELLO   LOC. NO. 201

| Prepared By : | 12 APRIL 1945 |
| HQ. H.D.K.W. | EX. NO. 14 B |

Atlantic Ocean

NAVY

CAA

KEY WEST HARBOR, FLA.

MARTELLO TOWER No 2.

(EAST TOWER)

SERIAL NUMBER

Sec tiles

C.E. 602.1-1142. Aug. 20, 1931

C.E. 660 K (Key West) 43, July 6.

Marker in place but platform removed

On Caretaking Status.

Area declared surplus

10 A. ±

C of E 660.4 (Key West) East Martello Tower -14 by Pan American Airways Inc -

G.III Aux.

G.IV Aux.

MARTELLO TOWER N.2

Navy Anc. compass 16

F.C.

100   0   100   200   300   400 FEET.

TRUE MERIDIAN. VAR. 1915.2° 25E.

EDITION OF JAN'Y. 2, 1914.
REVISIONS MARCH, 4, 1914.
NOV. 8, 1916; APRIL 1, 1921; FEB. 28, 1825,
NOV. 27, 1928, DEC. 5, 1934.

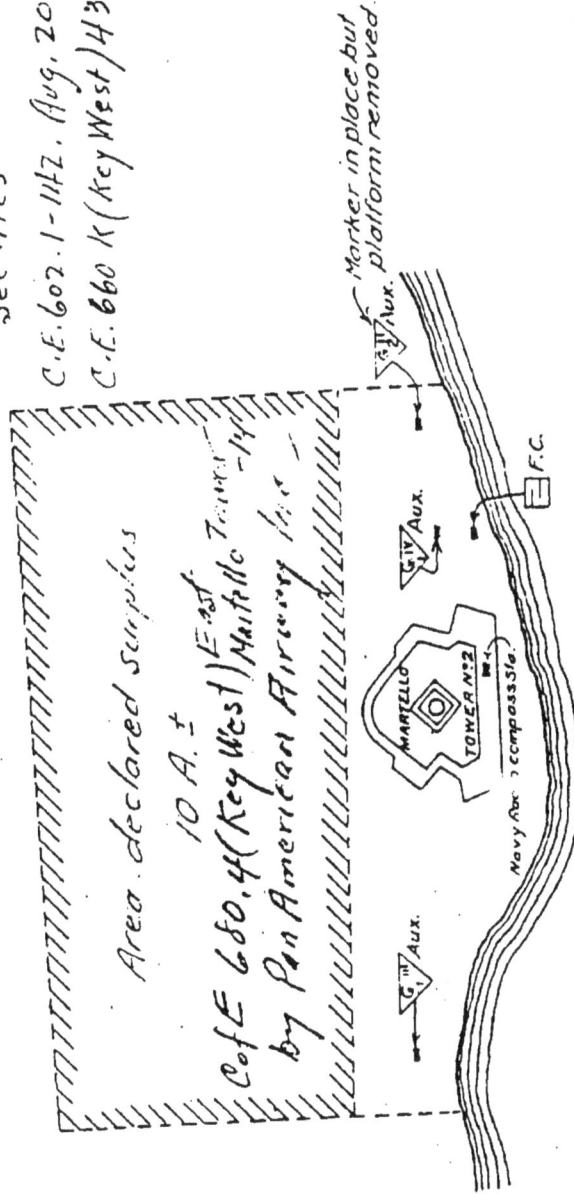

Martello Tower No. 2 (East Tower) Withdrawn from sale by order of the A.G. April 23, 1928.

N

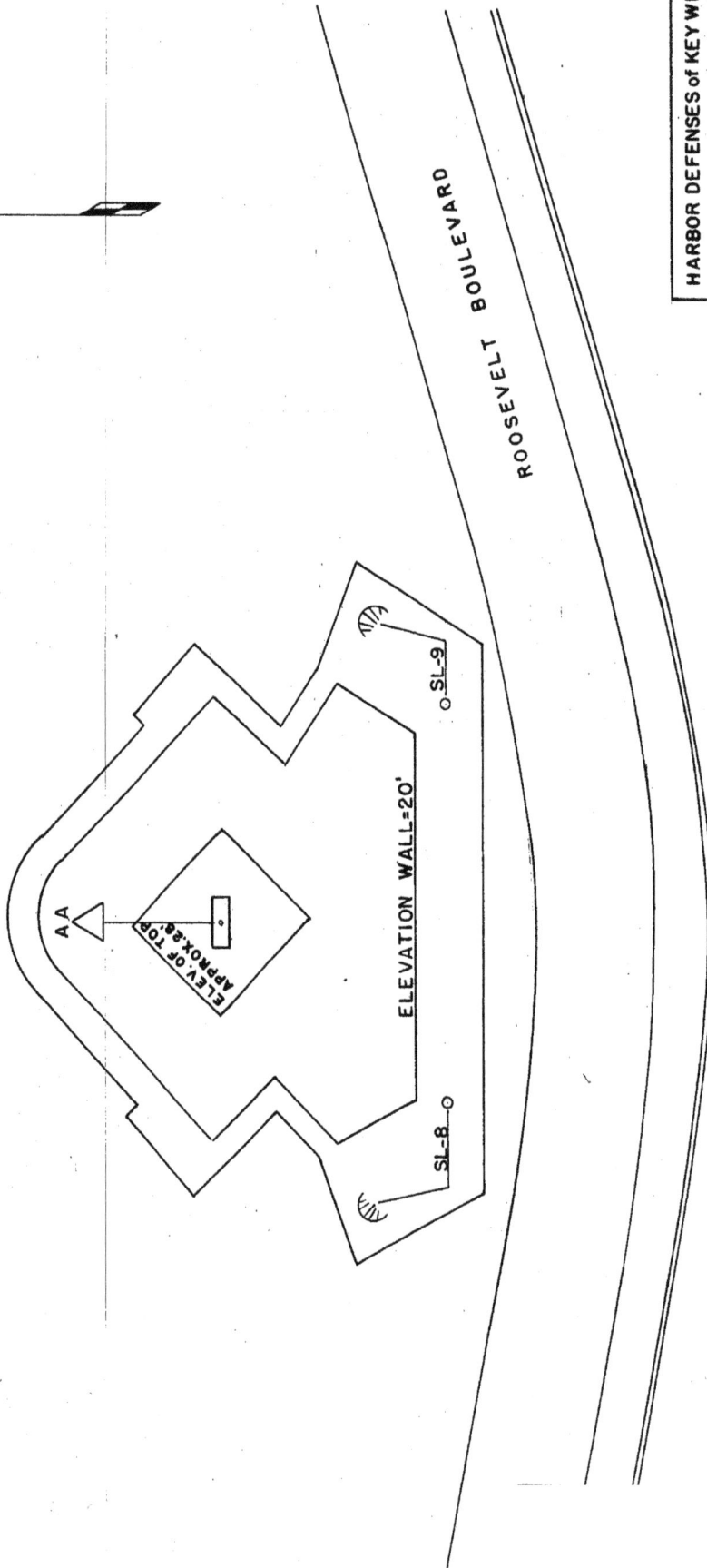

HARBOR DEFENSES of KEY WEST
EAST MARTELLO LOC: NO.
202 FIRE CONTROL INSTALLATION

Prepared By:
HQ. H.D.K.W.

10 April 1945

EX. NO. 15-B

ROOSEVELT BOULEVARD

SL-9

ELEVATION WALL=20'

AA

ELEV. OF TOP
APPROX. 86'

SL-8

YARDS

10   0   10   20   30   40   50

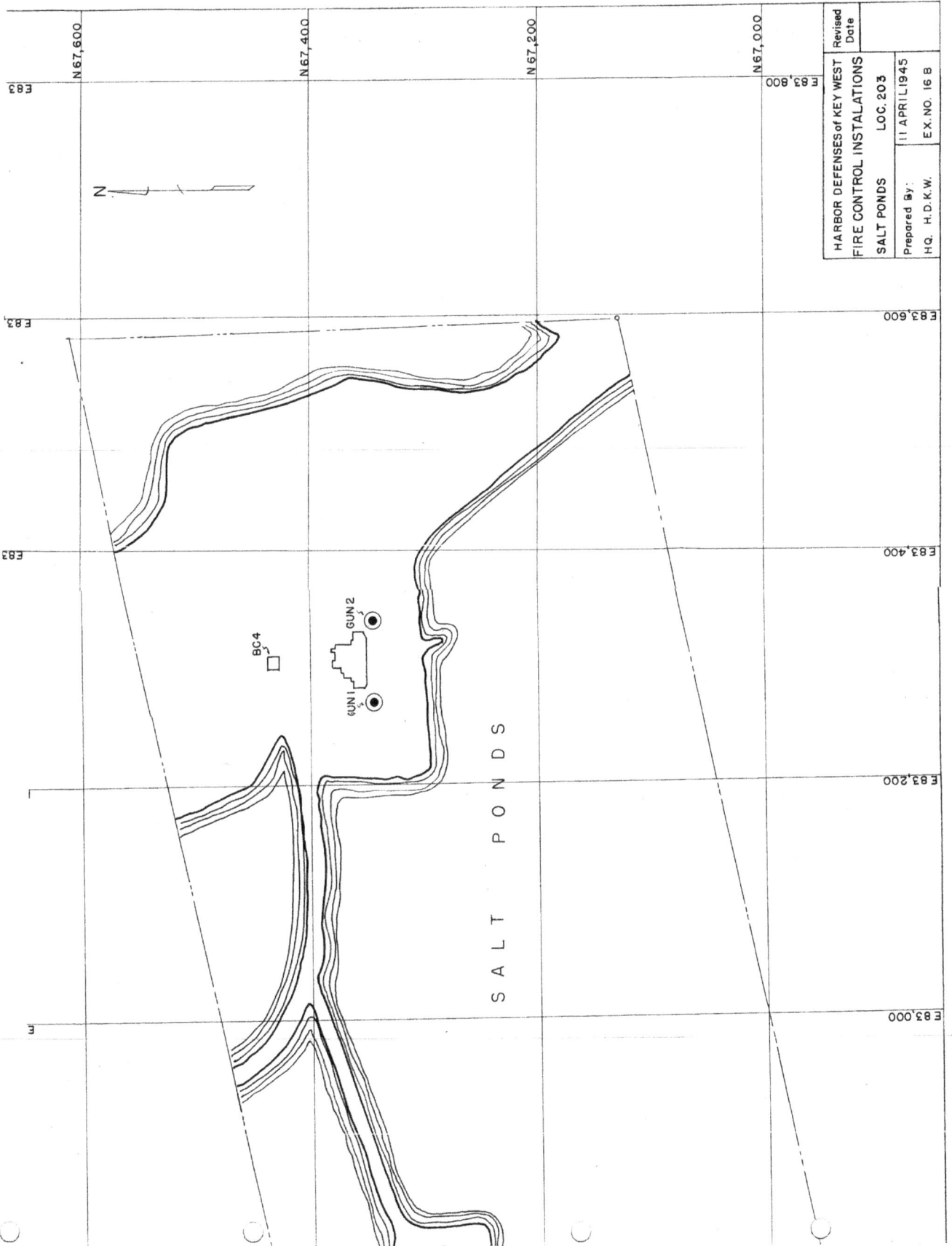

HARBOR DEFENSES of KEY WEST
FIRE CONTROL INSTALATIONS
SALT PONDS   LOC. 203

| Prepared By: | 11 APRIL 1945 |
| HQ. H.D.K.W. | EX. NO. 16 B |

N 67,600
N 67,400
N 67,200
N 67,000

E 83,800
E 83,600
E 83,400
E 83,200
E 83,000

BC4
GUN 2
GUN 1

SALT PONDS

KEY WEST HARBOR FLA.

KEY WEST BARRACKS.

SERIAL NUMBER

N

True Meridian
Var'n 2°-45′ E. in 1935

Concrete Fence

Gate

Corral

R

500   400   300   200   100   0   500 FT.

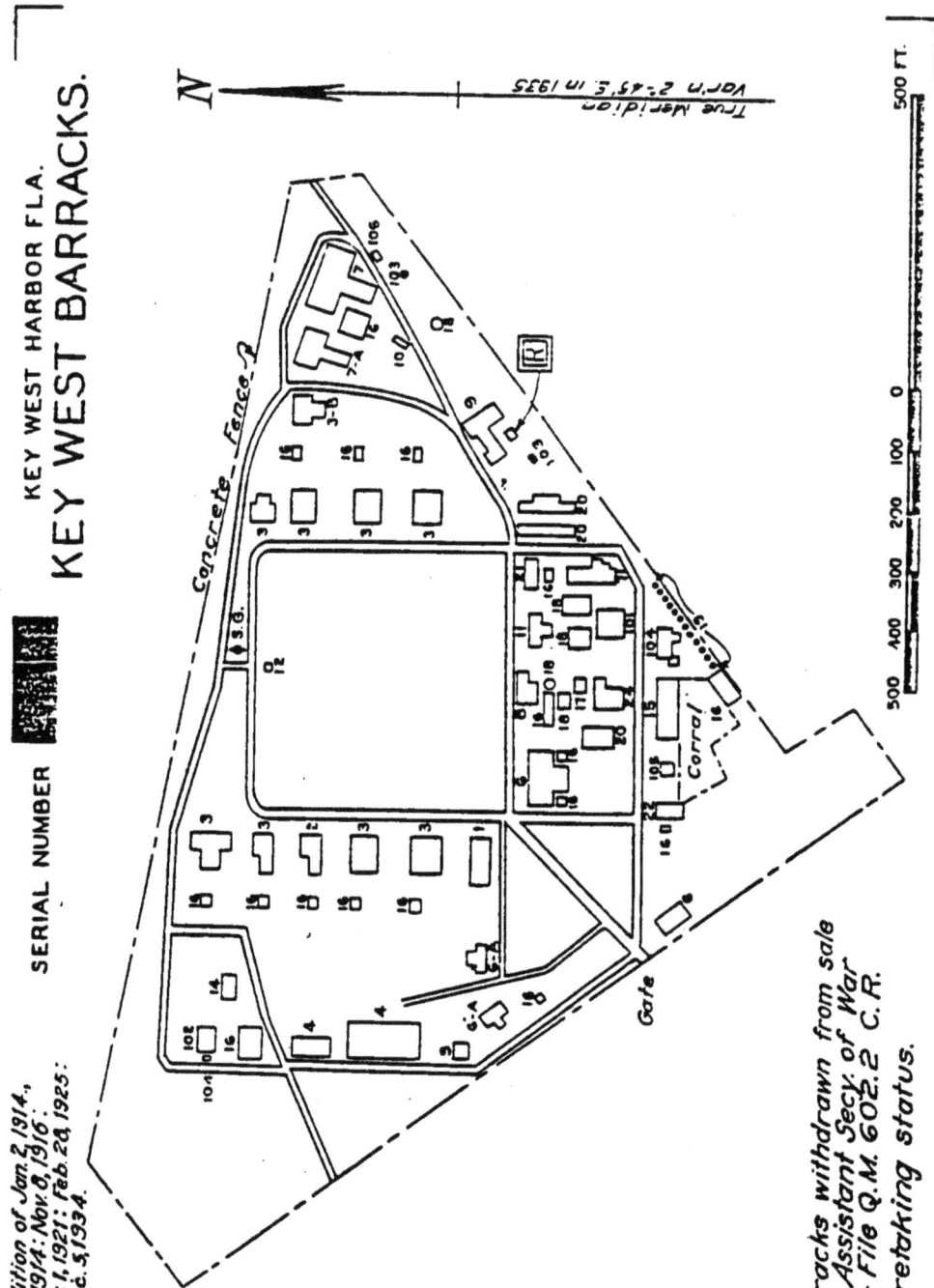

Redrawn from Edition of Jan.2,1914.,
Revised Mar.4,1914: Nov.8,1916:
June 7,1919: Apr.1,1921: Feb.20,1925:
Nov.27,1926: Dec.5,1934.

1   ADMINISTRATION BLDG.
2   COMMANDING OFFICERS QRS.
3   OFFICERS QUARTERS.
3-B BACHELOR OFFICER QRS.
4   HOSPITAL & ATTACHMENT QRS.
5   HOSPITAL STEWARDS QRS.
6   N.C.O. QUARTERS.
6-A N.C.S. QUARTERS.
7   BARRACKS.
7-A COMPANY BARRACKS.
8   GUARD HOUSE.
9   POST EXCHANGE.
10   LAVATORIES.
11   BAKERY.
12   FLAGPOLE.
13   BLACKSMITH & CARPENTER SHOP
14   OIL HOUSE.
15   SHED.
16   CISTERN.
17   PUMP HOUSE.
18   WATER TANKS.
19   12 WELLS.
20   Q.M. STOREHOUSE.
21   COMMISSARY ST. ROOM.
22   Q.M. STABLE.
2A   FUEL SHED.
71   SERVICE CLUB.
101 GARAGE & REPAIR SHOP.
102 GARAGE.
103 RADIO TOWERS.
104 FILLING STATION.
105 FIRE HOUSE.
106 VACANT.

Key West Barracks withdrawn from sale
by order of Assistant Secy. of War
Dec.23, 1927 - File Q.M. 602.2 C.R.

On caretaking status.

Key West Barracks (1831-1947) was the main garrison reservation on the island for the troops posted to Fort Taylor.

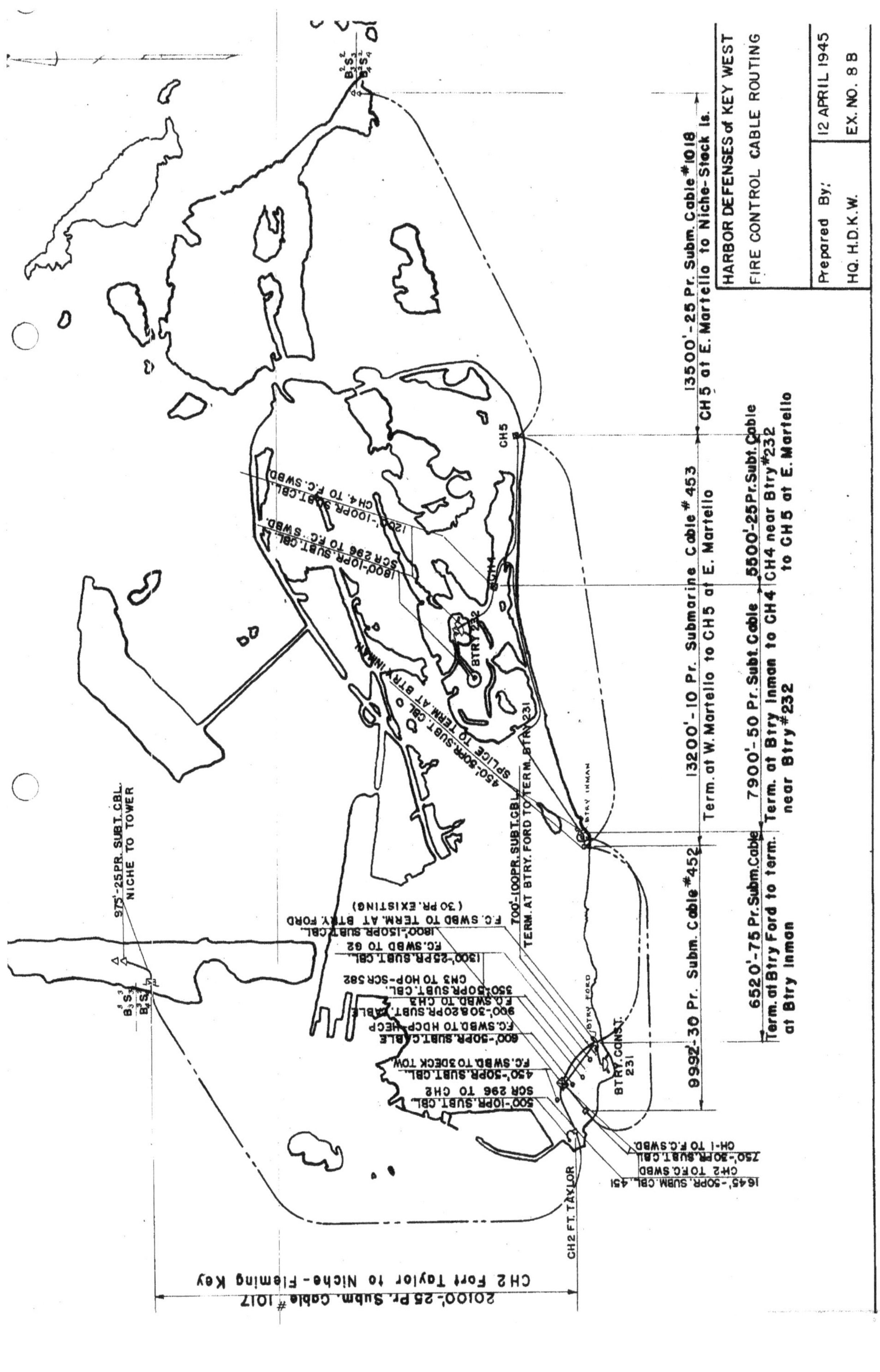

HARBOR DEFENSES of KEY WEST

FIRE CONTROL CABLE ROUTING

Prepared By:	12 APRIL 1945
HQ. H.D.K.W.	EX. NO. 8 B

13500'-25 Pr. Subm. Cable #1018
CH 5 at E. Martello to Niche-Stock Is.

5500'-25 Pr. Subt. Cable
CH4 near Btry #232
to CH5 at E. Martello

7900'-50 Pr. Subt. Cable
Term. at Btry Inman to GH4
near Btry #232

13200'-10 Pr. Submarine Cable #453
Term. at W. Martello to CH5 at E. Martello

6520'-75 Pr. Subm. Cable
Term. at Btry Ford to term.
at Btry Inman

9992'-30 Pr. Subm. Cable #452

975'-25 PR. SUBT.CBL.
NICHE TO TOWER

450'.50PR. SUBT. CBL.
SPLICE TO TERM. AT BTR. INMAN

1200'-100PR SUBT.CBL.
CH 4, TO F.C. SWBD.

1800'-10PR SUBT. CBL.
SCR 296 TO F.C. SWBD.

1800'-10PR. SUBT.CBL.
TERM. AT BTRY. FORD TO TERM. BTRY. 231

700'-100PR. SUBT.CBL.
TERM. AT BTRY. FORD
F.C. SWBD TO TERM. AT BTRY. FORD

(30 PR. EXISTING)

1800'-150PR SUBT.CBL.
FG. SWBD. TO 82

1300'-25PR. SUBT. CBL.
CH3 TO HDP-SCR 582

350'.50PR. SUBT.CBL.
F.G. SWBD TO CH4

900'-30&20PR. SUBT. CABLE
FG. SWBD. TO HDCP-HECP

600'.50PR. SUBT. CABLE
F.C. SWBD. TO DECK TOW

430'-30PR. SUBT. CBL.
SCR 296 TO CH2

500'-10PR. SUBT. CBL.
CH2 TO F.C. SWBD.

OH-1 TO F.C. SWBD.

750'-20PR. SUBT. CBL.
CH2 TO F.C. SWBD.

1645'-30PR. SUBM. CBL. 451

BTRY. CONST. 231

BTRY FORD

BTRY INMAN

BTRY 232

CH5

CH2 FT. TAYLOR

Key West World War II-era Site Locations. Stations housed in a single structure are connected by dashes (-)

location	Loc#	Purpose
Fort Taylor	200	Batt. Tact. #1 Ford, Batt. Tact. #3 BCN 231, Batt. Tact. #5 AMTB, G1, BC, BC/Ford, BC/231, HECP-HDCP, Met, SCR296-231
West Martello	201	Batt. Tact. #2 Inman, Batt. Tact. #6 AMTB, BC/Inman, SL 6,7
East Martello	202	SL 8,9
Salt Ponds	203	Batt. Tact. #4 BCN 232, BC/232
Stock Island	204	BS2/231, BS2/232
Key West Barracks	205	Military Reservation
Fleming Key	206	BS3/231, BS2/232

Key West and Fort Taylor in 1954 (NARA)

Gaines, William C. "Defending the Reef: A History of the Coastal nad Harbor Defenses of Key West, Florida 1895-1946." *Coast Defense Journal* Vol. 39, No. 1, 2025, p. 4; Vol. 39, No. 2, p. 35; Vol. 30 No. 3, p. 35, Vol. 39, No. 4, p. 47

THE HARBOR DEFENSES OF TAMPA BAY – FLORIDA

Tampa Bay is the largest, though originally shallow, inlet along the western coast of Florida. By 1890 it had been developed into a significant commercial port. The Endicott Program planned defenses for this harbor on two islands, and the harbor played a key role as an embarkation port during the Spanish American War as a staging area for the invasion of Cuba. The defenses were put into caretaker status after World War I. By 1923 the area had been dropped as defensive priority and defenses abandoned as harbor defenses in 1928. Of special note, is a one-of-a-kind 8-inch memorial gun on a barbette carriage is located at Plant Park in downtown Tampa.

Fort Dade (1898-1928) is located on Egmont Key at the entrance to Tampa Bay, about 35 miles southwest of the city of Tampa, Florida. It was fortified by a temporary battery in 1898 as part of the Spanish-American War. During the Endicott Program the post received five new gun batteries, a controlled mine complex and a large garrison complex between 1898 to 1906. It was named in General Order 43 of 1900 for Maj. Francis L. Dade, U.S. Army, killed with his command by the Seminoles on Dec. 28, 1835. Fort Dade was abandoned as a coast artillery post in 1923. Fires in 1925 and 1927 destroyed most of fort's buildings. During World War II, the island returned to the military as a harbor patrol station. This site, designated a National Wildlife Preserve in 1974, is jointly managed as such by the Florida Park Service and the U.S. Fish and Wildlife Service. All of the garrison buildings, except the guardhouse, have been removed. Erosion has nearly destroyed the two southern batteries and is encroaching on the three northern batteries. The park is accessible by seasonal ferry and private boats. The fort continues to suffer from hurricanes and shifting sands.

Fort Dade Gun Batteries

- A temporary, emergency battery for two 8-inch guns on 15-inch Rodman carriages emplaced at the start of the Spanish American War at the northern end of Egmont Key. This was one of the batteries authorized for armament with excess 8-inch Model 1888 guns and specially reinforced old 15-inch Rodman carriages hastily prepared at the start of the war. Construction funding was covered with the National Defense Act of March 9, 1898. This was the first battery for this new reservation and emplaced at the northwest end of the island to cover the main shipping channel into Tampa. On April 28, 1898 $6000 was allocated to Assistant Engineer J. H. Neff to begin work on the battery. Actual construction with concrete was begun in May and completed in September. While none of these batteries were "standard," the design embodied two side-by-side blocks for the guns with foundation traversing circles and an adjacent protected magazine, The battery received two 8-inch Model 1888M1 guns (Bethlehem tubes #1 and #2) shipped to Egmont on December 1, 1898. Guns and carriages were soon mounted. New Battery McIntosh was barely completed and the tubes from this temporary battery were dismounted to be transferred to this permanent battery at the post. This was done in mid-1900. The actual emplacement was then converted and used for 3-inch guns to become Battery Mellon in 1902-1903. As heavily modified it still exists at the Egmont Key State Park.

- **McINTOSH:** A battery for two 8-inch disappearing guns emplaced to the south of the temporary 8-inch battery on Egmont Key. Designed to cover the northwest channel, it fired more to the west than north. Plans were submitted on September 29, 1898 using funds from the Deficiency Act of July 7, 1898. The plan followed type recommendations. Both platforms were shaped as flank emplacements. Guns were placed on 116-foot centers. Trunnion height was just 25-feet. Magazines were on the lower left, ammunition service by hoists. Improvements to the wharf and rail track to the site had to be made before work could begin. Actual concrete work began in October 1898 and was finished by late December. Transfer was made on May 8, 1900 for a final cost of $118,949. It was armed with two 8-inch Model 1888MI guns moved here from the temporary battery just to the north on Model LF 1896 disappearing carriages (Bethlehem #1/#26 and #2/#31). It was named in General Orders No. 78 of May 25, 1903 for Lieutenant Colonel James S. McIntosh who was killed in action in 1847 during the Mexican War. It served until the defenses were abandoned. The guns and carriages were turned over to the state in 1923 for ornamental purposes and removed in 1927 to be shipped to other state locations. The emplacement still exists at the Egmont Key State Park. The battery is open to the public.

- **HOWARD:** A battery for two 6-inch disappearing guns also built at the northern end of Egmont Key, in a general gun line to the north of Battery McIntosh. Plans were submitted on April 27, 1903. Originally board plans called for a four-gun emplacement, by September 1st this was reduced to just a two-gun emplacement, both to be flank positions. It followed type plans, with a 125-foot gun spacing and a single, shared magazine in the central traverse. Ammunition service was by hand. The low-lying island resulted in a trunnion height of just 30-feet. Work was done in 1903-1904, the last concrete being poured in May of 1904. It was transferred on April 28, 1906 for a cost of $63,250. It was named in General Orders No. 194 of December 27, 1904 for Major Guy Howard, Quartermaster in the U.S. Volunteers who was killed in action during the Philippine Insurrection in 1899. Guns were shipped to Fort Dade in February of 1905 and mounted soon thereafter. It carried two 6-inch Model 1903 Watervliet guns on Model 1903 disappearing carriages (#10/#67 and #56/#68). The battery was disarmed to release guns for conversion to field carriages in October of

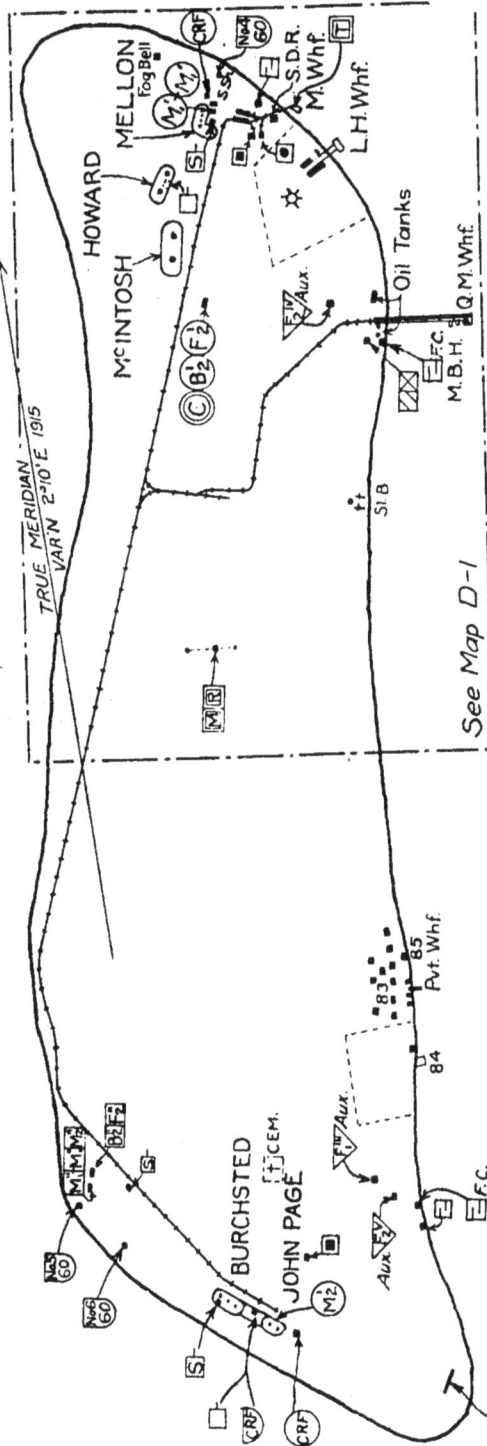

TAMPA BAY FLA

FORT DADE

EGMONT KEY

Scale of feet

1000   0   1000   2000

TRUE MERIDIAN
VAR'N 2°·10'E 1915

MELLON
Fog Bell

HOWARD

McINTOSH

Oil Tanks

M.B.H. Q.M.Whf.

S.D.R.
M.Whf.
L.H.Whf.

See Map D-1

St. B

Pvt. Whf.
85
84
83

BURCHSTED

JOHN PAGE

CEM.

F.C.

Aux.

Signboard for
Cable Crossing

SERIAL NUMBER 124

EDITION OF JUNE 3, 1919.
REVISIONS: APRIL 1, 1921.

LEGEND

83 PILOT STATION BLDGS.
84 PLANE HANGAR.
85 WEATHER BUREAU
   DISPLAY POLE.

BATTERIES

McINTOSH......2-8"DIS.
BURCHSTED...
HOWARD......
MELLON......
PAGE.........2-3"P

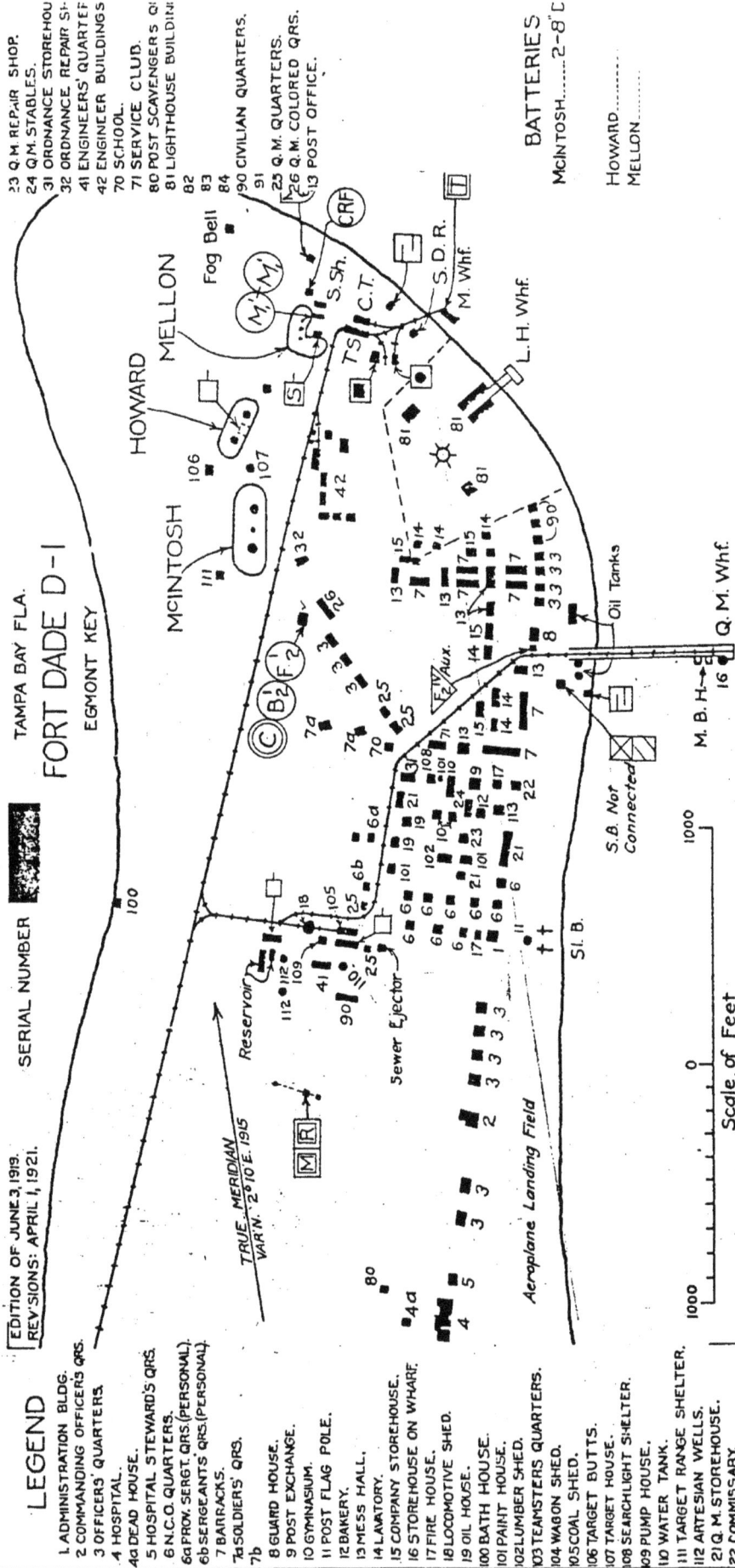

TAMPA BAY FLA.

FORT DADE D-1

EGMONT KEY

EDITION OF JUNE 3, 1919.
REVISIONS: APRIL 1, 1921.

SERIAL NUMBER ■■■

LEGEND

1 ADMINISTRATION BLDG.
2 COMMANDING OFFICERS QRS.
3 OFFICERS' QUARTERS.
4 HOSPITAL.
4a DEAD HOUSE.
5 HOSPITAL STEWARD'S QRS.
6 N.C.O. QUARTERS.
6a PROV. SERGT. QRS. (PERSONAL).
6b SERGEANTS' QRS. (PERSONAL).
7 BARRACKS.
7a SOLDIERS' QRS.
7b
8 GUARD HOUSE.
9 POST EXCHANGE.
10 GYMNASIUM.
11 POST FLAG POLE.
12 BAKERY.
13 MESS HALL.
14 LAVATORY.
15 COMPANY STOREHOUSE.
16 STOREHOUSE ON WHARF.
17 FIRE HOUSE.
18 LOCOMOTIVE SHED.
19 OIL HOUSE.
100 BATH HOUSE.
101 PAINT HOUSE.
102 LUMBER SHED.
103 TEAMSTERS QUARTERS.
104 WAGON SHED.
105 COAL SHED.
106 TARGET BUTTS.
107 TARGET HOUSE.
108 SEARCHLIGHT SHELTER.
109 PUMP HOUSE.
110 WATER TANK.
111 TARGET RANGE SHELTER.
112 ARTESIAN WELLS.
21 Q.M. STOREHOUSE.
22 COMMISSARY.

23 Q.M. REPAIR SHOP.
24 Q.M. STABLES.
31 ORDNANCE STOREHOU
32 ORDNANCE REPAIR S
41 ENGINEERS' QUARTER
42 ENGINEER BUILDINGS
70 SCHOOL.
71 SERVICE CLUB.
80 POST SCAVENGER S Q
81 LIGHTHOUSE BUILDIN
82
83
84
90 CIVILIAN QUARTERS.
91
25 Q.M. QUARTERS.
26 Q.M. COLORED QRS.
13 POST OFFICE.

BATTERIES

McINTOSH........ 2-8" C
HOWARD.........
MELLON.........

McINTOSH

HOWARD

MELLON

Fog Bell

CRF

Reservoir

Sewer Ejector

Aeroplane Landing Field

Scale of Feet

1000   0   1000

TRUE MERIDIAN
VAR'N. 2°10 E. 1915

Oil Tanks

S.B. Not Connected

S.B.

Q.M. Whf.

M.B.H.

L.H. Whf.

M. Whf.

S.D.R.

C.T.

T.S.

S.Sh.

Aux.

1917. The emplacement was used subsequently as a site for a new northern power plant, transferring for this purpose in June 1921 at a cost of $1,783.77. The emplacement still exists, though damaged by shore erosion, at the Egmont Key State Park. The battery is open to the public.

- **MELLON**: A battery for three 3-inch, 15-pounder masking parapet guns erected on the northern end of the Egmont Key Fort Dade reservation. Correspondence in May of 1900 suggested that advantage should be made by using the abandoned temporary battery (only two years old) as the foundation for a new emplacement of rapid-fire guns due to be placed on the northern end to cover the mine field here, firing to the northwest. Approval was granted and plans submitted on July 3, 1900. These were approved a month later and funds from the Act of May 25, 1900 allocated to build three emplacements. The peculiarities of the previous work required variance from the standard Mimeograph No. 30 recommendations. Gun centers were still spaced at 29-feet. Platforms were rounded, flank emplacements. Magazines were on the lower left flanks, though for the No. 3 emplacement it was withdrawn under the flank traverse. In the same general emplacement, a power plant and shelter for mobile searchlight were also built. Work was done in 1900-1901, being essentially completed by March of that year for the mounting of the balanced pillars. Transfer was made on June 25, 1904 for a cost of $15,333.94. It was named in General Orders No. 194 of December 27, 1904 for Captain Charles Mellon, 2nd U.S. Artillery who was killed in action with the Seminole Indians in 1837. The battery held three 3-inch, 15-pounder Model 1898 Driggs-Seabury guns and masking parapet pillar mounts (#110/#110, #111/#111, and #112/#112). The emplacement and carriages were modified in about 1916 to the M1898M1 fixed pedestal standard. A new CRF was built on the right flank of the battery in 1919. The guns and pillars were declared obsolete in June 1920 and ordered removed. The emplacement still exists, though at times overgrown, at the Egmont Key State Park. The battery is open to public.

- **BURCHSTED**: A battery of rapid-fire guns emplaced at the opposite, southern end of Egmont Key to protect the minefield covering the southern channel. Local engineers were asked to submit plans on April 25, 1898 for a pair of 6-inch, rapid-fire Vickers guns being purchased in England at the start of the Spanish American War. Plans were duly submitted on June 4th. Using a modified plan for a 5-inch battery, they featured two separate gun platforms with short parapets spaced by 58-feet. The low site meant a trunnion elevation of just 20-feet. An allocation of $8000 was supplied from the March 1898 National Defense Act. Actual work was done from May to November 1898. It was transferred on January 28, 1899 for a total cost of $33,955. It was armed with two 6-inch Vickers guns (sometimes mistakenly described as Armstrong guns) on pedestal mounts (#12139/#11162 and #12140/#11157). Although originally envisioned as only a temporary work, the battery was fully integrated into the defenses. Subsequent minor modifications were made. In 1902 a seawall 400-feet long costing $2400 was built to help prevent the serious erosion taking place on this corner of the island. In 1903 cracks in the structure had to be repaired, and in April 1904 ammunition hoists were added. In late 1900 it was decided to add a single 3-inch Model 1898 masking parapet emplacement on the battery's left flank. Plans were submitted on April 30, 1901. The single platform for the gun was just 37-feet from No. 2 of the 6-inch pair. A magazine was placed on the lower left flank, ammunition service was solely by hand, the entire battery was not electrified. Work was done on this part in mid-1901 and was finished by year's end. The gun was installed in 1903 and transfer of this unit made on November 10, 1904 for $9,968.32. It was armed with a 3-inch, 15-pounder Driggs-Seabury gun on a masking parapet mount (#60/#60). The three-gun battery was named in General Orders No. 78 of May 25, 1903 for 1st Lieutenant Henry A. Burchsted who was killed in action with the Creek Indians in 1813 in Alabama. The exposed 6-inch guns developed mechanical problems. In May 1918 they were reported as both working poorly, experiencing particular difficulty being with the traversing mechanism. New parts and cleaning of all moving parts ensued. In August 1919 it was decided to declare all the guns of this type obsolete, they were abandoned in-place soon thereafter. The 3-inch gun type was discontinued in mid-1920 and was removed. After the abandonment of the Tampa defenses in the late 1920s the two 6-inch guns rusted away while the emplacement itself was heavily eroded by the sea and left as a detached ruin offshore. During the 1980s the guns were finally dismounted and moved for restoration and display purposes at the nearby Fort DeSoto State Park. Only ruins of the emplacement still exist but are surrounded by waters of the Gulf at the Egmont Key State Park.

- **PAGE**: An Endicott emplacement for two 3-inch pedestal guns erected on the south side of Egmont Key to add to the coverage of the southern channel and mine field. Plans were submitted on May 25, 1903. It was located about one mile from the engineer wharf, a construction rail line was needed to move material to the site. Construction funding came from the Gun and Mortar Battery Act of March 3, 1903. The plan was according to recommended designs, with a 62-foot spacing between the guns and with two magazines and a storeroom in the protected traverse between the gun platforms. The battery was a little southeast of Burchsted, near the shore. It pointed and fired to the southwest. Work was done from late 1903 to April 1904, and transfer made on April 26, 1906 for a cost of $18,700. It was named in General Orders No. 194 of December 27, 1904 for Captain John Page, 4th U.S. Infantry who was a casualty of the Mexican War in 1846. The battery carried two 3-inch Model 1903 pedestal guns and Masking Parapet carriages (#85/#47 and #86/#48). There was a considerable delay in mounting the armament, even by 1910 the carriages were set up, but the tubes were still undelivered. The armament was removed about 1923 with the abandonment of the Tampa Bay defenses. One mount was reused in Puerto Rico during World War II. In recent years shore erosion has left the collapsed remains of the battery considerably offshore and surrounded by the waters of the Gulf.

Fort Dade Guard House, Egmont Key State Park (Mark Berhow)

TAMPA BAY FLA.
FORT DE SOTO.
MULLET KEY.

SERIAL NUMBER 124

Laidley

Bigelow

True Meridian
Var. 1915, 2° 18' E.

Scale of Feet
0 500 1000

Q. M. whf.

LEGEND.

EDITION OF JAN'Y 2, 1914.
REVISIONS MARCH, 4, 1914; DEC. 7, 1915;
NOV. 8, 1916; JUNE 3, 1919; APRIL 1, 1921.

1 ADMINISTRATION BLDG.
2 COMMANDING OFFICER'S QRS.
3 OFFICER'S QUARTERS.
4 HOSPITAL.
5 HOSPITAL STWD'S. QRS.
6 N.C.S. QUARTERS.
6A PROVOST SERGT. QRS.
7 BARRACKS.
8 GUARD HOUSE.
9
10 STOREHOUSE.
11 FREIGHT SHED.
12 BAKERY.
13 MESS HALL.
14 LAVATORY.
15 FIRE STATION.
16 TANK (60 000 GAL.)
17 SAWMILL.
18 BLACKSMITH AND
   CARPENTER SHOP.
19 OIL HOUSE.
100 WAGON SHED.
101 STABLE.
102 POST FLAGPOLE
103
104
105 TARGET RANGE SHL.
106
21 Q.M. STORE HOUSE.
22 Q.M. & COM. ST. HO.
31 ORDNANCE ST. HO.
41 ENGINEER ST. HO.
23 TOOLHOUSE.

BATTERIES.
LAIDLEY.......4-12" M.
BIGELOW.......

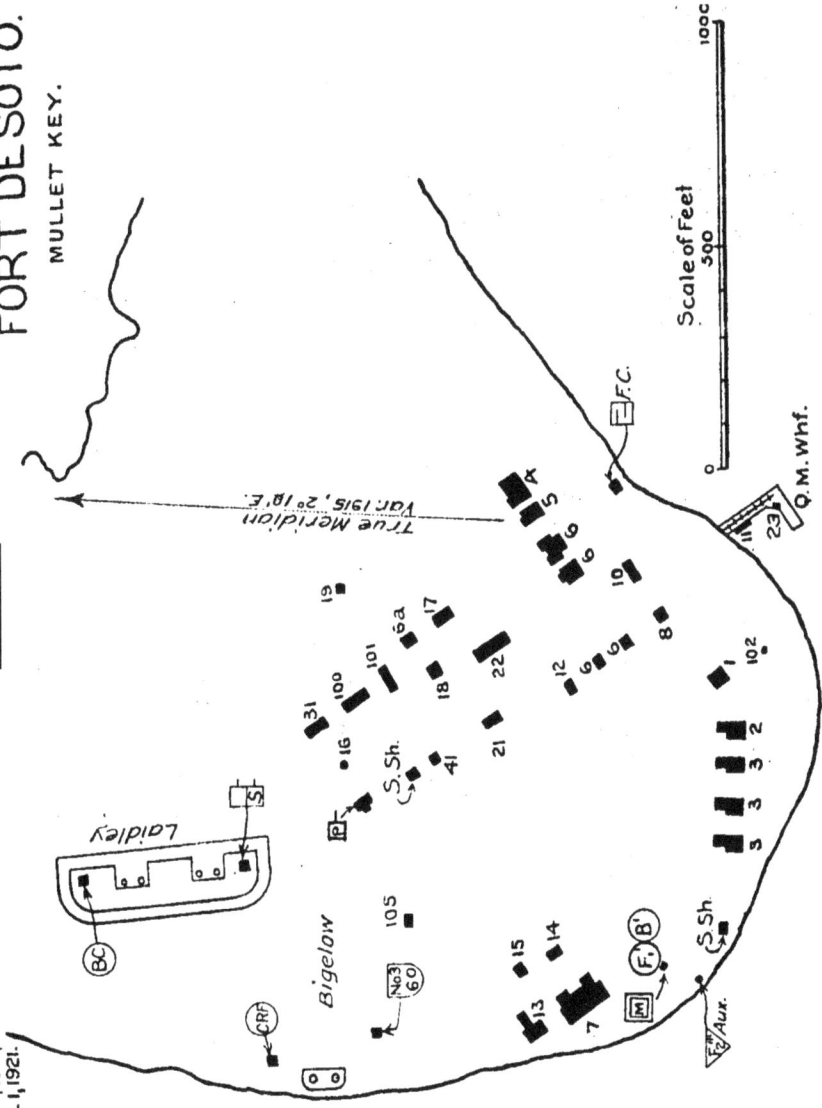

Fort De Soto (1898-1928) is located on Mullet Key at the entrance to Tampa Bay, south of the city of St. Petersburg, Florida. This small Endicott Program fort had two batteries built between 1900 to 1904. It was named in General Order 43 of 1900 for the early Spanish explorer Ferdinand De Soto. Fort De Soto served primarily as a sub-post of Fort Dade and was abandoned as a coast artillery post in 1928. Four 12-inch mortars were left in Battery Laidley when the military left the island. The 271-acre reservation was purchased by Pinellas County in 1938. During World War II, the U.S. Army repurchased the fort to be used a gunnery and bombing training center. Pinellas County again purchased the key in 1948 for use as a county park. In 1962, a toll road was complete to the mainland. Today the park has a large beach access area with a convenience store and has camping and boating facilities. Battery Laidley still retains the four 12-inch mortars. These are the only 12-inch mortars in the United States and the only modern seacoast artillery to remain in their original location in the United States. The two 6-inch Vickers guns on Armstrong carriages, originally located at Battery Burchstead at Fort Dade, are on display on concrete pads behind Battery Laidley. A replica of the Quartermaster's Store House was built in 1999 and now serves as the fort's museum. The park is open daily, toll bridge fees and parking fees required, with self-guided tours.

Fort DeSoto Gun Batteries

- **LAIDLEY:** The single mortar battery for the Tampa defenses was located at Mullet Key on the north side of the main shipping channel, across from Egmont Key. Plans were submitted on July 24, 1898 using funds from the Act of May 7, 1898. It was located on the southwest point of the island, pointing to a little west of due south. It followed the plans for the times, with the new, widened pits of four mortars each side-by-side, with magazines only on the flanks and under the central traverse between the pits. Plans were approved on July 30, and the site cleared by the end of December 1898. Concrete construction was done by the fall of 1899 and the battery essentially completed and eight carriages mounted by June 30, 1900. Transfer was made on May 8, 1900 for a total cost to date of $155,561. It was armed with eight 12-inch Model 1890M1 Watervliet mortars on Model 1896 mortar carriages (#22/#185, #132/#186, #107/#187, #85/#188, #76/#189, #86/#183, #127/#190, and #135/#184). The battery also included the post electric plant and has the usual attached observation station. It was named in General Orders No. 78 of May 25, 1903 for Colonel Theodore T. S. Laidley, Ordnance Department, of Mexican War and Civil War service. The four rear mortars were removed in April 1917 for use in a new battery being built in the defenses of San Diego. The remaining four mortars continued to serve until the early-1930s when the defenses of Tampa were abandoned. For some reason these four mortars and their carriages were never removed, and consequently, are still present and on display at the Fort DeSoto Pinellas County Park. The battery is open to the public.

- **BIGELOW:** A battery for two 3-inch rapid-fire guns on masking parapet mounts erected to defend the offshore mine fields from the Mullet Key reservation of Fort DeSoto. Plans were submitted on April 9, 1901 using funds from the Act of March 1, 1901. It followed recommended design form, with two platforms and guns separated by 29-feet and small magazines on the lower right flank of each platform. Ammunition service was by hand. The emplacement was placed in advance and on the left flank of the mortar battery, some 200-feet from the beach and 350-feet south of the barracks. Construction was accomplished from June to October 1901. Transfer was made on June 28, 1904 at a cost of some $13,980.18. It was named in General Orders No. 78 of May 25, 1903 for 1st Lieutenant Aaron Bigelow, who was killed in 1814 at the Battle of Lundy's Lane. It was armed with two 3-inch Model 1898 Driggs-Seabury guns on balanced pillar mounts (#65/#65 and #66/#66). These were two of the few such guns not to be modified to M1898M1 pedestal standard. The guns were dismounted in June 1920 when all such armament was declared obsolete. In subsequent years the fort became county property. Shore erosion has seriously damaged the battery, only partial, broken remains exist on the shoreline at the Fort Desoto Pinellas County Park.

Fort DeSoto 1945 (NARA)

McCall, Bruce E. "Coastal Defense of Tampa Bay." *CDSG Journal* Vol. 10, No. 3, Aug. 1996, p. 52.

Egmont Key State Park (Glen Williford)

Fort DeSoto County Park (Glen Williford)

Two of the 12-inch mortars in Battery Laidley, Fort DeSoto County Park (Mark Berhow)

6-inch Vickers gun and mount at Fort DeSoto County Park (Mark Berhow)

THE HARBOR DEFENSES OF PENSACOLA — FLORIDA

The largest deep-water harbor along what would become the American Gulf coast. An important harbor, Pensacola was initially settled by the Spanish in 1559, but was soon abandoned, then resettled in 1698 as a fortified town. Florida was seceded to the British at the end of the French and Indian War in 1763 but was retaken by the Spanish in 1781 during the Revolutionary War. Spain formally ceded the area to the United States in 1819. In 1825 the Pensacola Naval Yard was established, followed by the reconstruction of the Spanish battery at Fort Barrancas and the construction of three new forts to secure the harbor entrance and land defense of the naval yard. The defenses were modernized with concrete batteries after 1890, and a large garrison post was established next to old Fort Barrancas. The defenses were modernized in 1920 with a new long-range 12-inch gun battery and again with six new batteries during World War II. The defenses were deactivated and abandoned by the U.S. Army in 1947.

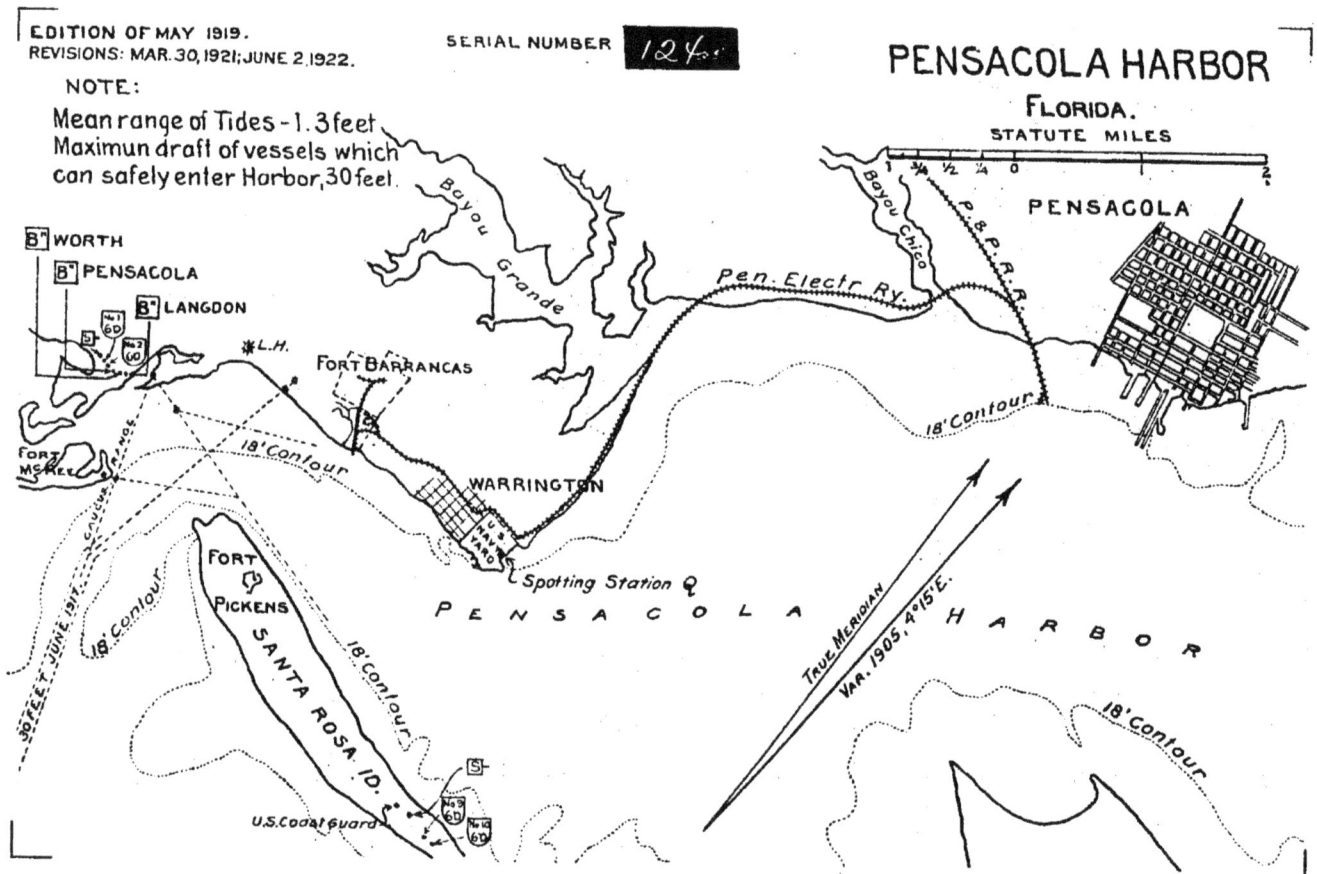

Fort Pickens (1845-1947) is located on the western end of Santa Rosa Island at the entrance to Pensacola Bay, about ten miles west of Pensacola Beach, Florida. The Third System fort was built from 1829 to 1834. It was named in General Order 32 of 1830 for Maj. Gen. Andrew Pickens, Continental Army. This masonry fort served as the primary defense of Pensacola Naval Station and the commercial harbor of Pensacola, along with its companion forts, Fort Barrancas and Fort McRee. While the naval station and other forts were occupied by Confederate forces during the Civil War, Fort Pickens was held by Union forces throughout the war. The fort served as a prison after the Civil War. Fort Pickens's reservation was used during the Endicott Program for eight concrete batteries and controlled submarine mine facilities. Installed inside the walls of the old fort was the fort's primary battery of two 12-inch guns on disappearing carriages. To the east of the old fort, a battery of eight 12-inch mortars was also installed in 1899. Fort Pickens served as a sub-post of Fort Barrancas, which was the headquarters of the Coast Artillery District. The fort's batteries were well used as training center for the National Guard and CMTC units. As part of modernization of the harbor defenses during the Interwar Period, a battery of two 12-inch guns on long-range barbette carriages was constructed to the east of the old fort in 1923. The importance of the naval yard and naval air station caused during the 1940 Program the addition of two #200 Series batteries with two 6-inch barbette guns each and the casemating of the 12-inch long-range battery. A 90mm AMTB battery and a 155mm battery on Panama mounts were also added to the defenses of Fort Pickens. After World War II and the demise of the Coast Artillery Corps, Fort Pickens was declared surplus and became a state park. Transferred to the National Park Service in 1972 as part of the Gulf Islands National Seashore, the fort offers an excellent view of the progression of American coast defenses. Of special interest, are surviving coast artillery weapons with a 6-inch gun on a disappearing carriage in Battery Cooper and two 6-inch guns on shielded barbette carriages at Battery #234, along with the large pre-Civil War fort with a 15-inch Rodman cannon, a visitor's center and bookstore. Nearby are the Third System Fort Barrancas and the Barrancas Advanced Redoubt. Outstanding location for seeing fortifications of all eras. Fort Pickens and Fort Barrancas are open daily, the Advanced Redoubt is open for scheduled tours. Fort Pickens is generally accessible by road from Gulf Shores for day use, though hurricanes occasionally damage it. The park is also accessible by ferry on the weekend.

Fort Pickens Gun Batteries

- **PENSACOLA**: A battery for two 12-inch disappearing guns erected within the confines of old Fort Pickens on Santa Rosa Island. The battery was authorized with the funding from the National Defense Act of March 23, 1898 as a two-gun emplacement on disappearing carriages Model 1897. It pointed and fired to the south, southwest. Plans were submitted on March 23, 1898. While generally of type plans, it did have certain modifications to fit the confines of the parade inside the old fort. The crest was rasied so it would extend to the level of the old parapet, trunnion height being 32-feet. It was designed with two flank emplacements, but the guns were spaced at just 147-feet apart. Magazines were on the lower left side, with shot galleries directly behind the gun blocks. Ammunition service was by hoists. Work was done by June 30, 1898 with the major concrete pours. Finishing of the battery took another year, transfer was made on June 30, 1899 for a cost of $128,707. It was armed with two 12-inch Model 1895 Watervliet guns on Model LF M1897 disappearing carriages (#6/#3 and #7/#4). In August 1898 another project authorized a new power plant and a tunnel on the left flank so a passage could access the old casemates used for storage otherwise cutoff from the remaining parade ground, at the cost of another $4000. It was named in General Orders No. 73 of May 29, 1900 for the city of Pensacola. Modifications were made in the battery's early years. Loading platforms were extended and a new BC station added in 1912. Hoists were replaced with ones from Fort Schuyler in 1918 and modified for long-point shells. A new powerhouse was added in 1920. The original armament appears to never have been changed, but the carriages were modified to allow higher elevation in about 1917. The battery was scheduled for deletion under the 1932 Review, and the breech blocks were removed on June 16, 1933. The tubes were shipped away on October 24, 1934 and the carriages scrapped in October 1942. The emplacement still exists at the Fort Pickens Area of the Gulf Islands National Seashore. The battery is open to the public.

- **WORTH**: The mortar battery assigned to Fort Pickens. It was located on a site to the east of the old fort, pointing and firing south, southwest. Plans were submitted on June 12, 1897 using funds from the Act of March 3, 1897. It generally followed type plan designs. It was one of the final small-pit dimension types and had the magazines under the frontal parapet. Within a year the concrete work was done and the mortar carriage base rings set. Transfer came on June 30, 1899

PENSACOLA HARBOR
FORT PICKENS
Santa Rosa Island.
General Map.
Scale of Feet.

U.S.C.G.

EDITION OF MAR. 3, 1919.
REVISIONS: MAR. 30, 1921.
JUNE 2, 1922.

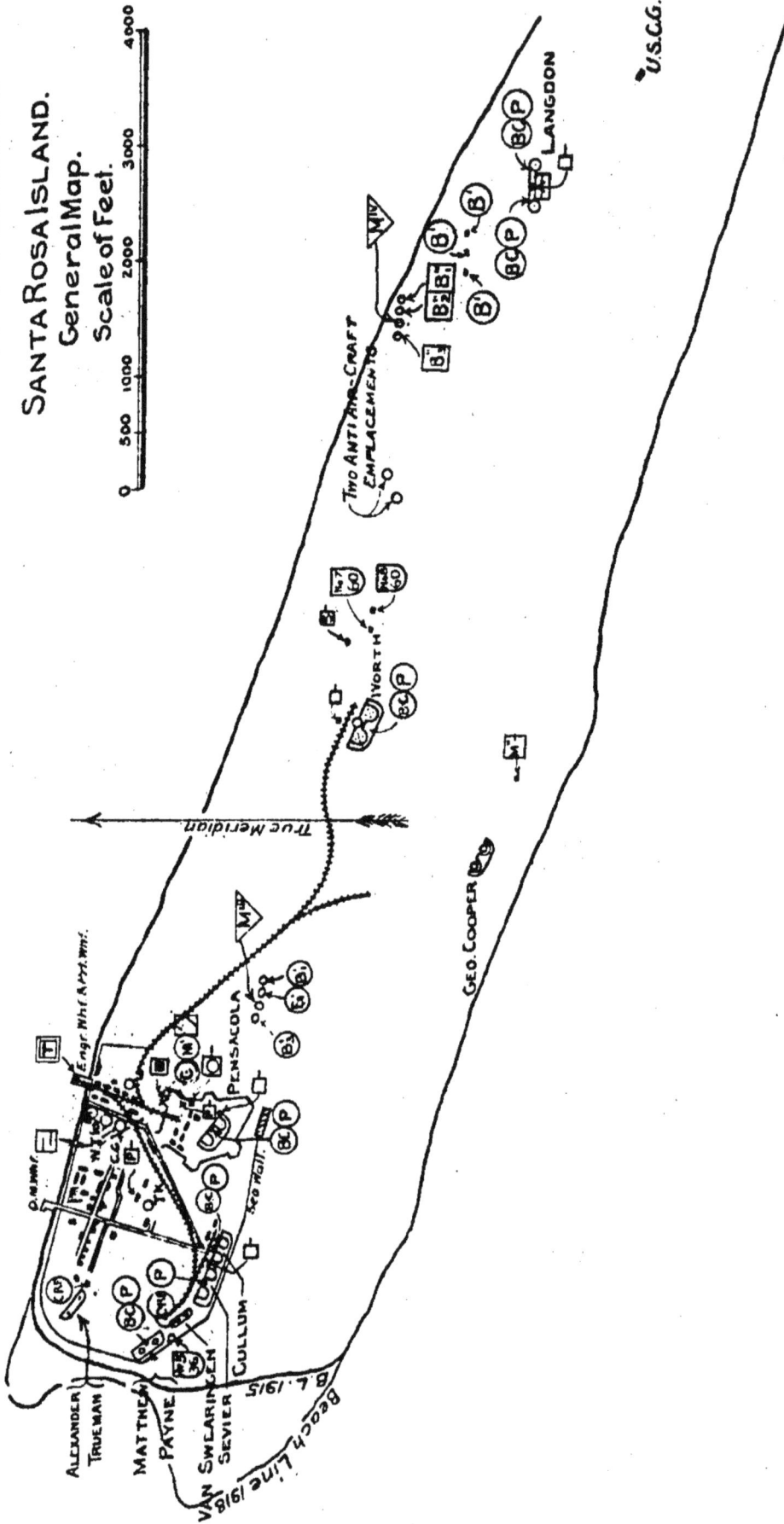

SERIAL NUMBER

124

PENSACOLA HARBOR
FORT PICKENS-DI.
SANTA ROSA ISLAND.
Scale of Feet.

SERIAL NUMBER

EDITION OF MAR. 3, 1919.
REVISIONS: JUNE 2, 1922.
REVISIONS: MAR. 30, 1921.

TRUE MERIDIAN

Beach Line 1918.

Beach Line 1815.

Sea Wall

WORTH

PENSACOLA

TRUEMAN

VAN SWEARINGEN

SEVIER

CULLUM

PAYNE

LEGEND

1. ADMINISTRATION BLDG.
2. COMMADG. OFF. QRS.
3. OFFICER'S QRS.
4. HOSPITAL STWD. QRS.
5. N.C. OFFICER'S QRS.
6. BARRACKS.
7. GUARD HOUSE.
8. POST EXCHANGE.
9. MINE STORE ROOM
10. Q.M. BOAT HOUSE.
11. ORDNANCE ST. HO.
12. Q.M. COAL SHED.
13. MESS HALLS.
14. TEMP. BARRACKS.
15. MESS HALLS.
16. U.S. ENGR. OFFICE.
17. U.S. ENGR. QRS.
18. U.S. ENGR. MESS HALL.
19. U.S. ENGINEER.
    DREDGE STORER.
20. U.S. ENGR. TOOL HO.
21. U.S. ENGR. ST. ROOM.
22. BOILER ROOM FOR
    CENTRAL POWER PLANT.

BATTERIES.

PENSACOLA	2-12" Dis.
* CULLUM	2-10" .
VAN SWEARINGEN	2-4.7 P.
PAYNE	2-3" P.
TRUEMAN	2-3" P.
SEVIER	2-10" Dis.
WORTH	4-12" M. .

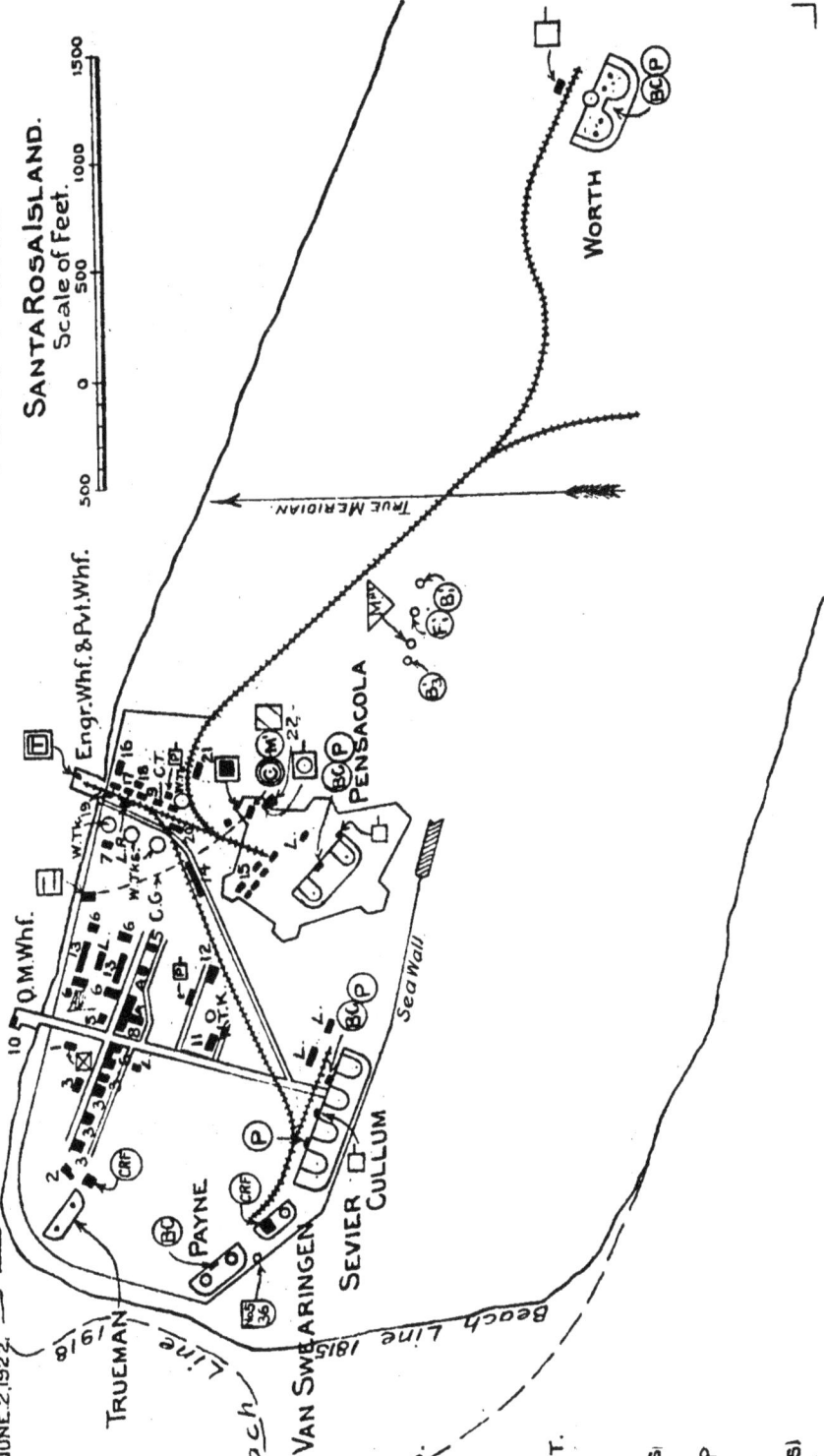

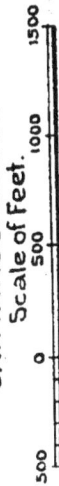

* *Guns dismounted*

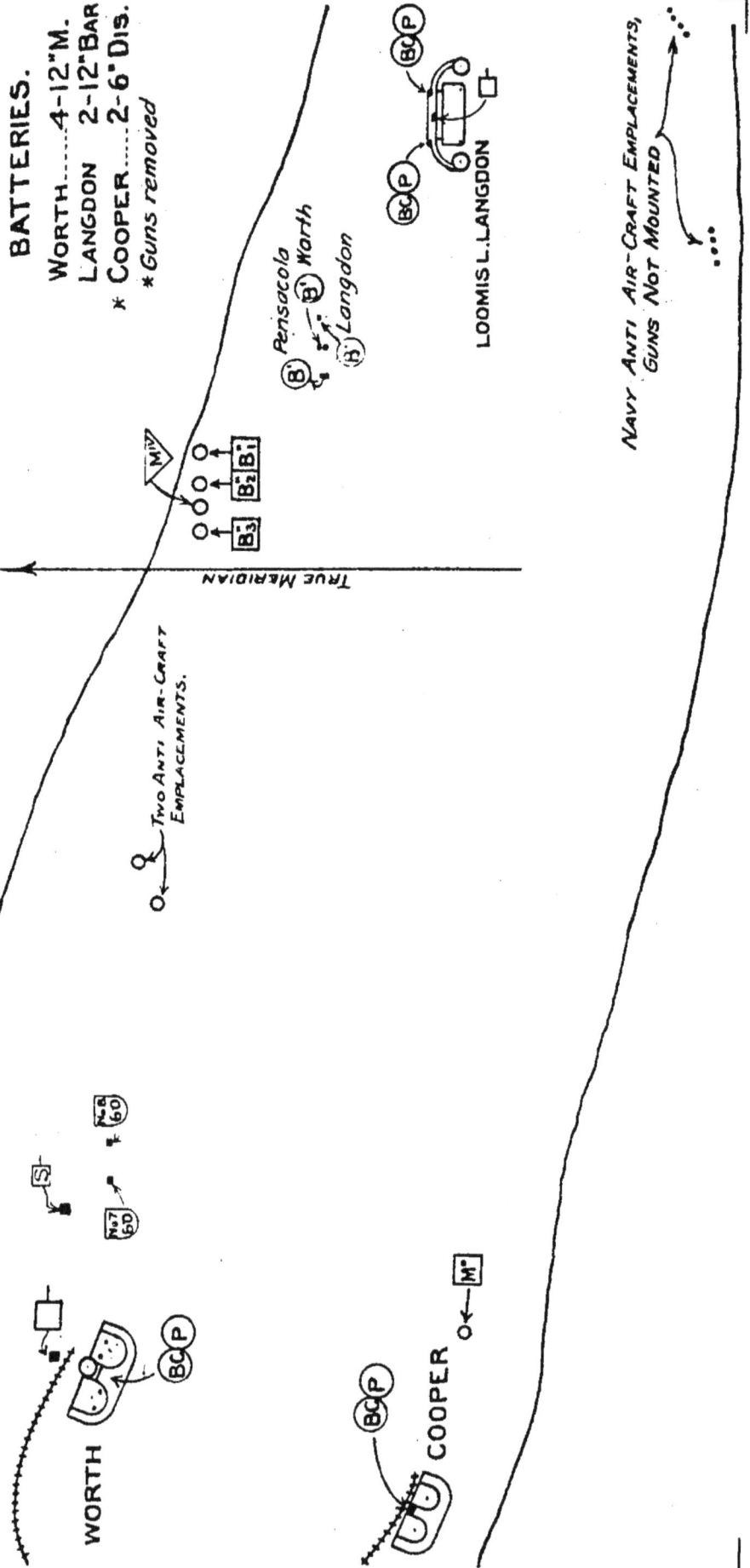

PENSACOLA HARBOR
FORT PICKENS-D2.

SANTA ROSA ISLAND.
Scale of Feet.

1500 1000 500 0 500

BATTERIES.

WORTH ...... 4-12"M.
LANGDON ... 2-12"BAR.
* COOPER .... 2-6" DIS.

* Guns removed

SERIAL NUMBER

EDITION OF MAY 1919.
REVISIONS: MAR. 30, 1921.
JUNE 2, 1922.

Fort Pickens East 1928 (NARA)

Fort McRee 1928 (NARA)

Battery Pensacola, Fort Pickens Unit Gulf Shores National Seashore (Terry McGoven)

Battery Cullum-Sevier Fort Pickens Unit Gulf Shores National Seashore (Terry McGoven)

6-inch gun on a disappearing carriage in Battery Cooper
Fort Pickens Unit Gulf Shores National Seashore (Terry McGoven)

Perido Key (Fort McRee) Gulf Shores National Seashore (Terry McGoven)

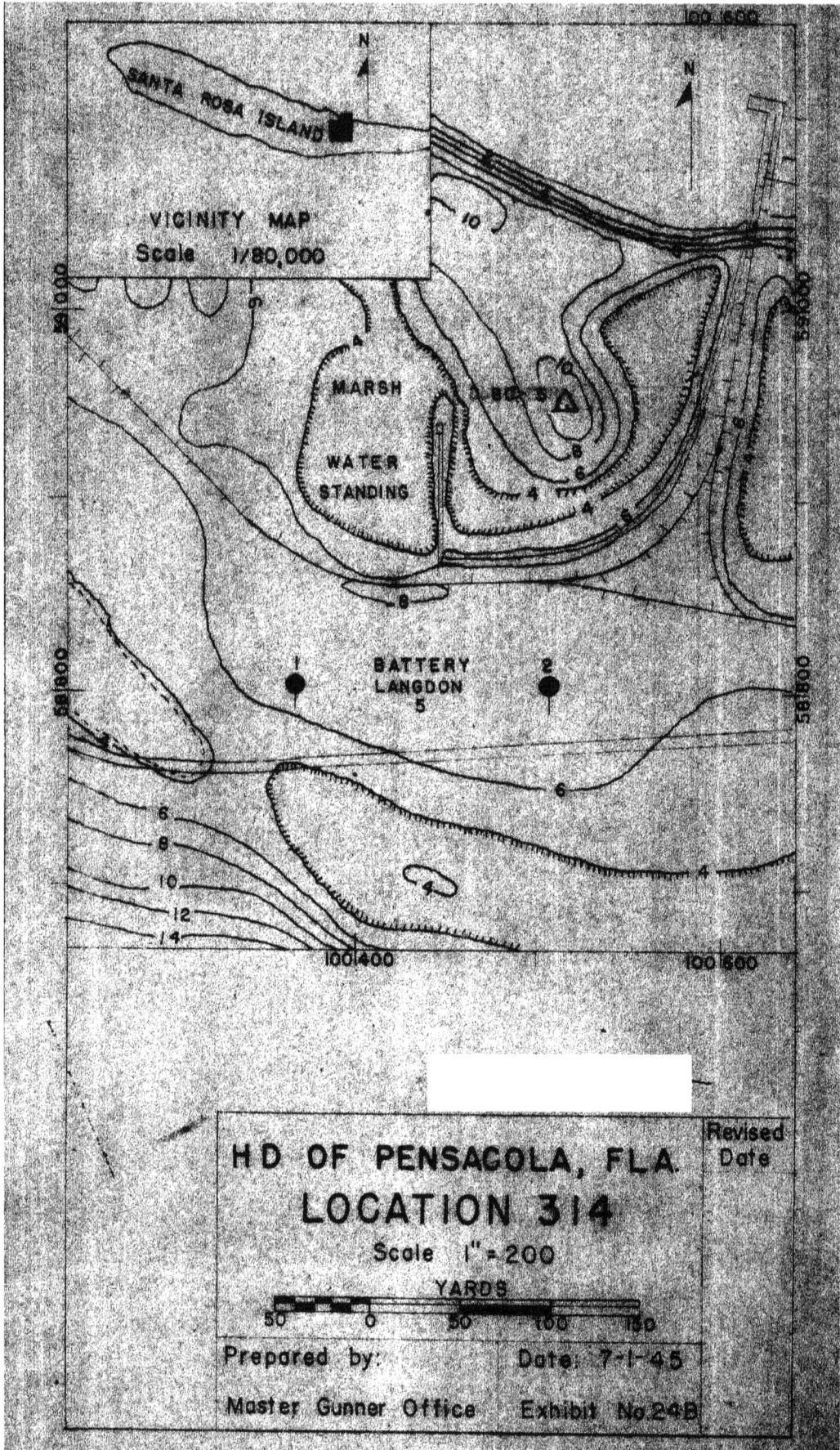

N

SANTA ROSA ISLAND

VICINITY MAP
Scale 1/80,000

MARSH

WATER
STANDING

BATTERY
LANGDON
5

1

2

6
8
10
12
14

100 400

100 600

HD OF PENSACOLA, FLA.
LOCATION 314
Scale 1"=200
YARDS
50 0 50 100 150

Prepared by:

Master Gunner Office

Revised
Date

Date: 7-1-45

Exhibit No.24B

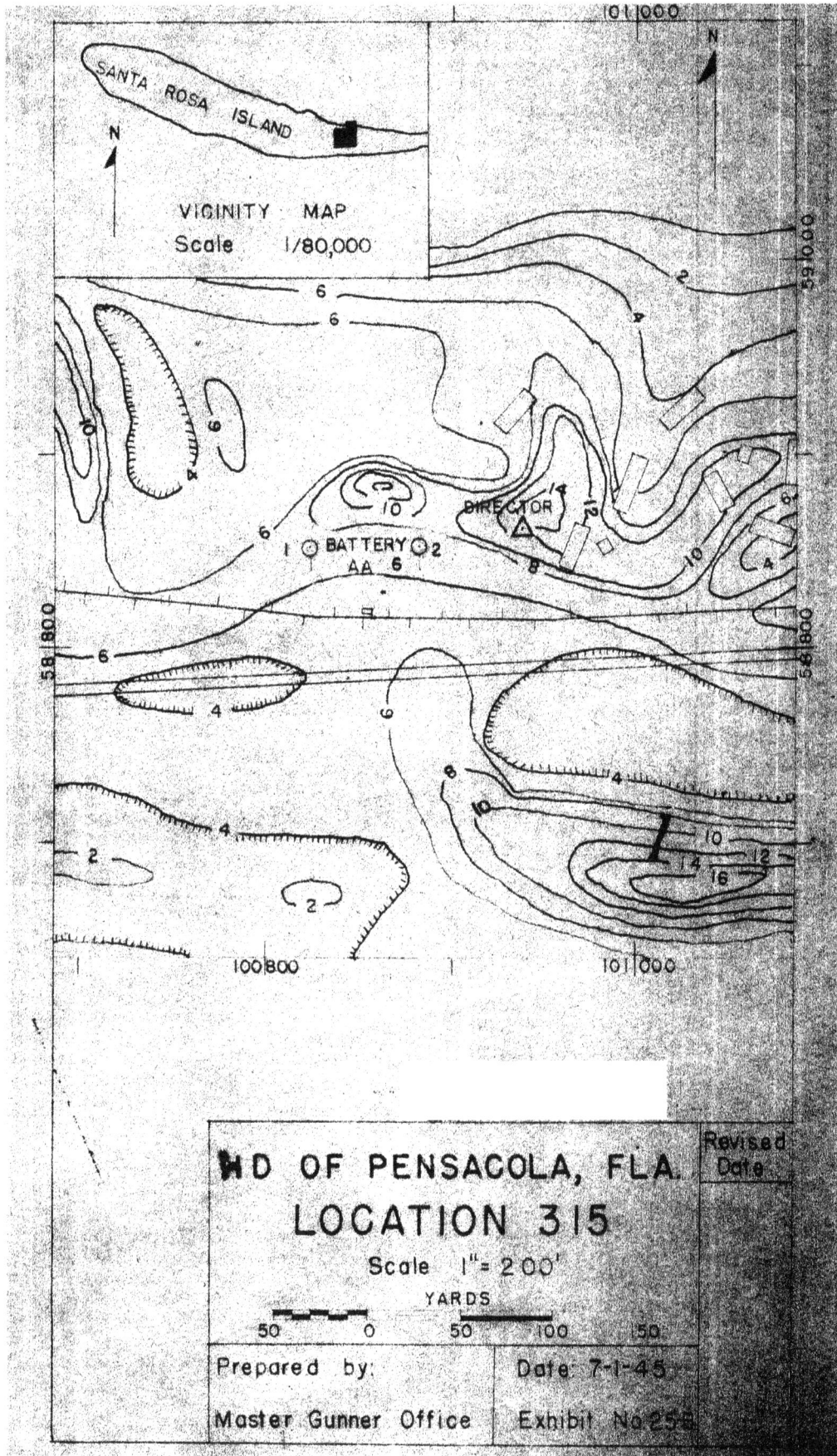

VICINITY MAP
Scale 1/80,000

SANTA ROSA ISLAND

N

101,000

N

59,000

59,800

DIRECTOR

1 BATTERY 2
AA 6

58,800

100,800

101,000

HD OF PENSACOLA, FLA.
LOCATION 315
Scale 1"= 200'

YARDS
50   0   50   100   150

Revised
Date

Prepared by: Date: 7-1-45

Master Gunner Office   Exhibit No. 25-6

for a cost of $123,093. It was named in General Orders No. 43 of April 4, 1900 for Brevet Major General William J. Worth of Mexican War service. It was armed with eight 12-inch mortars Model M1890M1 on Model 1896 carriages (Watervliet tubes #23/#109, #38/#97, #45/#98, Niles tubes #3/#99, #5/#129, #4/#111, and Builders tubes #21/#128, and #18/#110). After initial construction the usual telautograph booths were added and upgrades made to the fire control system. Four mortars were removed in April of 1918. The four remaining mortars served until removed under authority of May 1942—implemented in December of 1942. The emplacement itself was then modified to become the combined HECP/HDCP for Pensacola. As modified with that construction, the emplacement still exists at the Fort Pickens Area of the Gulf Islands National Seashore. The battery is open to the public.

- **LANGDON:** A 1915 Program dual 12-inch barbette battery selected to be emplaced at Fort Pickens. The site was further east than all of the previous works, being some two miles away from the old masonry fort. It was just to the north of the main east-west road (still 600-feet from the shoreline), and fired to the south, southwest. Local engineers were supplied instructions and typical standard plans on December 29, 1916. Plans were approved on May 24, 1917, started quickly and the structural work was finished by the end of 1917. It followed closely the standard plan and was of the open-back variety. Two totally open gun platforms had a heavily protected traverse magazine and support room build in between. Armament was erected by May 17, 1921. Transfer was made on March 3, 1923 for a construction cost of $308,786. It was named in General Orders No. 13 of March 27, 1922 for Brigadier General Loomis L. Langdon, who died in 1910. The armament was proofed in place on August 8, 1923 and consisted of two 12-inch guns Model 1895M1A4 on barbette carriages Model 1917 (Bethlehem tube #14/#18 and Watervliet tube #62/#25). It was retained as the most modern armament for Pensacola between the wars. The battery was given massive overhead protection and enclosed gun houses early in the war, being modernized from November 6, 1942 to July 31, 1943. The two-gun houses were each canted away from the main axis by 16.5-degrees in order to give a broad field of fire. The carriages got new gun shields installed in December 1943. The battery served until deleted in July 1947 and subsequently disarmed. The emplacement still exists, though generally closed to the public, at the Fort Pickens Area of the Gulf Islands National Seashore. The battery site is open, but the interior is closed to the public.

- **CULLUM – SEVIER:** The first Endicott battery for disappearing guns. It was located to the west of the old fort, near the seawall, firing to the southwest. Site plans were submitted for this location on a sandy ridge on January 4, 1896, to be erected on piles. Construction plans were submitted on April 11, 1896 for the first two emplacements for M1894 carriages. Two subsequent emplacements were added, though in this case one for a M1894 carriage and one for a M1896 disappearing carriages. The plans generally followed type plans. Guns were separated by 128-feet. Magazines were on the lower right, ammunition service initially by lifts, later modified to chain hoists. Preparations with clearing and driving piles were started for the first three emplacements in early 1896. All of these were for the M1894 carriage. Garrison troops took them over by August 31, 1898. The fourth and final emplacement (for a Model LF 1896 carriage) was added in 1897. All four emplacements were transferred on June 4, 1898 for a total cost of $188,920.24. All four were named in General Orders No. 43 of April 4, 1900 for General George Cullum, a Civil War engineering officer. The battery was armed with three 10-inch Model 1888 Watervliet guns on Model 1894 LF (#25/#6, #42/#7, #44/#8) and one 10-inch Model 1888 Watervliet gun on a Model 1896 LF disappearing carriages (#30/#38). In General Orders No. 15 of April 25, 1916, the battery was split to facilitate command and fire control. Emplacements No. 1 and No. 2 became Battery Sevier, named for General John Sevier, a North Carolina Revolutionary War officer. The other emplacements remained Battery Cullum. Numerous modifications were made to the battery before World War I. In 1900 owing to cramped conditions of the electric light plant and dampness in the dynamo room, two new rooms were added and the ventilation of lower level rooms improved. In 1904-1905 the loading platforms were enlarged. New concrete plotting rooms and battery commander's stations were also added. Battery Cullum was disarmed during World War I, the guns being removed in May of 1918 and sent to Watervliet for use on railway carriages on June 15, 1918. However, on March 18, 1919 it was decided to retain and rearm the battery, two replacement tubes (Model M1895 tubes Watervliet #48 and #49) were placed on the original carriages in May of 1921. Both batteries were recommended for disarmament under the 1932 Review, and the breech blocks were removed in 1934. Final salvage of the guns and carriages was authorized in November 1942. Sevier continued to serve at this location as a meteorological station, while Cullum's platforms were used for the relocated guns of Battery Trueman in 1942. With these modifications both emplacements still exist at the Fort Pickens Area of the Gulf Islands National Seashore. The battery site is closed to the public.

- **COOPER**: An emplacement for a pair of 6-inch disappearing guns, and the last Endicott battery built at Fort Pickens. Plans were submitted on May 25, 1903. It was practically identical to the type recommended in the current mimeographs. The two flank emplacements had guns spaced at 125-feet. A shared magazine was in the central traverse, ammunition delivery by hand. It was located in a hollow in the sand dunes in front (south) of the mortar battery. The low location (trunnion height of just 25-feet) allowed for a less conspicuous target. It fired to the south, southwest. Construction was done from December 1905 to August of 1906 for transfer on September 15, 1906 at a cost of $56,743.85. The armament was received in December 1905 and mounted during the following April. It was armed with two 6-inch Model 1903 guns and carriages (#42/#32 and #53/#31). It was named in General Orders No. 194 of December 27, 1904 for 2nd Lieutenant George A. Cooper killed in action during the Philippine Insurrection. The armament was removed in December 1917 and the carriages scrapped in 1920. The battery location was reused for a 155mm Panama mount battery using Panama mounts during World War II in front and to the side of the battery. The emplacement still exists at the Fort Pickens Area of the Gulf Islands National Seashore. One M1903 6-inch gun on a M1903 disappearing carriage was reemplaced here for display purposes in 1976. The battery is open to the public.

- Battery #234: A 1940 Program dual 6-inch barbette battery emplaced at Fort Pickens. It was located on the eastern reservation, just to the west of older 6-inch Battery Cooper. It was of standard 200-series design. A separate BC station on a steel lattice tower was built immediately to the rear right flank. Work was done from January 16, 1943 to completion on July 1, 1943. Transfer was made on June 6, 1944 for a cost of $212,482.42. It was never named, just being known as Battery Construction No. 234. No armament was ever mounted in this battery, but it was structurally brought to completion. It was assigned M4 carriages #51 and #62 but never received gun tubes. It was subsequently abandoned after the war and the carriages and other equipment sold for scrap. Today there are two 6-inch guns on shielded barbette carriages re-emplaced in the battery, brought here in the 1970s from the abandoned battery on Fisherman Island, defenses of the Chesapeake. It still exists as an interpreted display at the Fort Pickens Area of the Gulf Shores Islands National Seashore. The battery is open to the public.

- **VAN SWEARINGEN**: An emplacement for a pair of Armstrong 4.7-inch guns assigned to Pensacola under the emergency of the 1898 National Defense Act. Fort Pickens was ordered to build an emergency emplacement on March 19, 1898 and provided guns. Without submitting plans, the platforms were built and reported ready for armament on April 28, 1898. The design followed in general that for 5-inch guns. It had two separate flank platforms with a 45-foot spacing between guns. Each platform had a magazine on the lower right, ammunition service being by hand. The battery was built on the right flank of the 10-inch battery, contiguous to it and firing like it to the southwest. Trunnion height was 20.5-feet. It was transferred on June 29, 1898 for a cost of $7,497.94. It was armed with 4.7-inch Armstrong guns and carriages recently purchased in England (#9718/#10836 and #9719/#10841). The battery was named in General Orders No. 78 of May 25, 1903 for Captain Joseph Van Swearingen, who was killed in the Seminole War in 1837. The emplacement was disarmed in February 1918, though the guns and carriages were placed in storage at the post. In May of 1921 the guns and carriages were disposed of, being sent off as monument guns to southern communities. Gun and carriage #9718/#10841 went to Danielsville, Georgia where it still remains. In 1922-1923 a new CRF station as built in one of the gun pits for $2,000. As modified, the emplacement still exists at the Fort Pickens Area of the Gulf Islands National Seashore. The battery is open to the public.

- **PAYNE**: A battery for two 3-inch pedestal guns erected at Fort Pickens a short distance to the northwest of Battery Van Swearingen. Its field of fire included to the southwest to cover the mine fields protecting the entrance to the Pensacola harbor. Plans and estimates for both Payne and Trueman were submitted initially on October 9, 1902. Plans were rejected and resubmitted on January 2, 1903 using pedestal rather than masking parapet guns. It was a hybrid design, with only 48-feet between gun centers, magazine under the center traverse, and two flank-type platforms with a trunnion height of 20-feet. To protect it from fire from the rear (inside the bay) a parados was built entirely around the battery, with a tunnel entry into the lower level on the left flank. Also, the central traverse was given a high concrete cover to deflect possible incoming shells. The guns had the clearance to fire over the rear parados if necessary. Work was done in 1904 and finally transferred on January 7, 1908 at a cost of $28,302. Armament was two 3-inch Model 1902 Bethlehem guns and M1902 pedestal mounts (#16/#16 and #17/#17). It was named in General Orders No. 194 of December 27, 1904 for Colonel Mathew Payne of Mexican War service. In the spring of 1913 gun #16 was sent to Sandy Hook and replaced with gun #24 from Trueman on July 26, 1913. gun #16 was eventually returned to Pickens and remounted in Battery Trueman in 1916. On July 18, 1922 a severe accident occurred here. A detail of the Reserve Corps at summer camp, while firing the battery, several soldiers were severely injured when a recoil tore it loose and dismounted the tube. After repairs the battery was retained for a long period, finally being disarmed after the war on June 27, 1946. The emplacement still exists at the Fort Pickens Area of the Gulf Islands National Seashore. The battery is open to the public.

- **TRUEMAN**: A second emplacement for a pair of 3-inch guns erected at Fort Pickens at the same time as Battery Payne. It was located north of Payne to cover the inner harbor approaches, though still firing to the southwest. Initially planned as a battery for two masking parapet guns submitted on October 9, 1902, that was revised for an emplacement for two Model 1902 pedestal guns and resubmitted on January 2, 1903. It conformed with current gun spacing dimensions of 62-feet, and with two magazines and a storeroom under the protected center traverse, with ammunition service by hand. To prevent damage from possible fire from the rear, the battery was surrounded with an earthen parados, entry into the battery being through a tunnel in the berm on the left, rear flank. The guns at their 20-foot trunnion height could still turn and fire to the rear over the parados. Work was done in 1905, and transfer made on January 7, 1908 at a cost of $28,332.50. It was armed, after some delay in receiving the weapons, with two 3-inch Bethlehem Model 1902 guns and pedestal mounts (#24/#24 and #25/#25). It was named in General Orders No. 194 of December 27, 1904 for Major Alexander Trueman killed in action during St. Clair's Defeat in Ohio in 1792. In August of 1913 Trueman transferred gun tube #24 to Battery Payne and received in turn tube #16 from Sandy Hook in July 1916. It was retained throughout the interwar years, in 1923 receiving a new combined BC/CRF station just to the south costing $3,164. During World War II the armament was removed and re-emplaced in a new battery structure on top of old 10-inch Battery Cullum. The original emplacement still exists at the Fort Pickens Area of the Gulf Islands National Seashore. The battery is open to the public.

- **(NEW) TRUEMAN**: The two 3-inch Model 1902 guns and carriages from Battery Trueman were transferred to a new emplacement built into the east flank (on top of the old, unoccupied gun pits) of 10-inch Battery Cullum. It was a fairly elaborate work, using the shape and area of standard 3-inch designs for its platforms. New hoists were installed to allow it to use the old magazines of the previous disappearing battery. The work was done from March to December 1942. Transfer was made in May of 1944 for a cost of $20,330. It was armed with 3-inch Model 1902 guns and pedestals #25/#25 and #16/#16. This was retained until disarmed on June 27, 1946. The emplacement still exists atop Battery Cullum at the Fort Pickens Area of the Gulf Islands National Seashore. The battery is closed to the public.

- **AMTB #3A**: A 1943 Program AMTB battery of two 90mm fixed and two 90mm mobile guns. It was emplaced just to the southeast of Battery Cullum/Sevier, beyond the seawall. Work was done from April 7 to June 16, 1943, for transfer on July 7, 1943 at a cost of $16,810.85. Two fixed gun blocks were paced at 120-feet part and intervening and behind their line was wooden/earth magazine, plot and power room. It was then abandoned after the war and the armament removed about 1946. The two fixed blocks still exist at the Fort Pickens Area of the Gulf Islands National Seashore. The battery site is open to the public.

Fort McRee (1842-1947) is located on eastern end of Perdido Key at the entrance to Pensacola Bay, about five miles southeast of Fort Barrancas, Florida. A Third System fort, called Fort McRee, was built from 1834 to 1842. The masonry fort served as part of the defenses of Pensacola Naval Station, along with its companion forts, Fort Barrancas and Fort Pickens. Artillery attacks by Union forces in 1861 heavily damaged the fort and it was further damaged when Confederates forces burn the fort in 1862 when they abandoned Pensacola. The 3rd System fort was never repaired, and it became a source of bricks for Pensacola's other forts. Placed on caretaker status after the Civil War, Fort McRee's reservation was used during the Endicott Program for two concrete batteries. To the west of the old fort, a battery of two 8-inch guns on disappearing carriages and a battery of four 3-inch guns on masking pedestal mounts were installed in 1901. Fort McRee's served as a sub-post of Fort Barrancas, which was the headquarters of the 1st Coast Artillery District. The fort's batteries were disarmed and the fort abandoned after World War I. The importance of the naval yard and the naval air station caused the 1940 Program to build a #200 Series battery with two 6-inch barbette guns on top of the Endicott Program batteries. After World War II, Fort McRee was declared surplus. Transferred to the National Park Service in 1972 as part of the Gulf Islands National Seashore, no trace of the old fort remains, but the tops of the buried Endicott Program batteries and all of the #200 Series battery are visible. The site may be reached by boat or a long walk over the sand dunes.

PENSACOLA HARBOR
FORT McREE

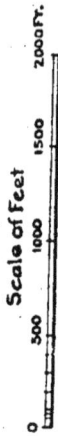

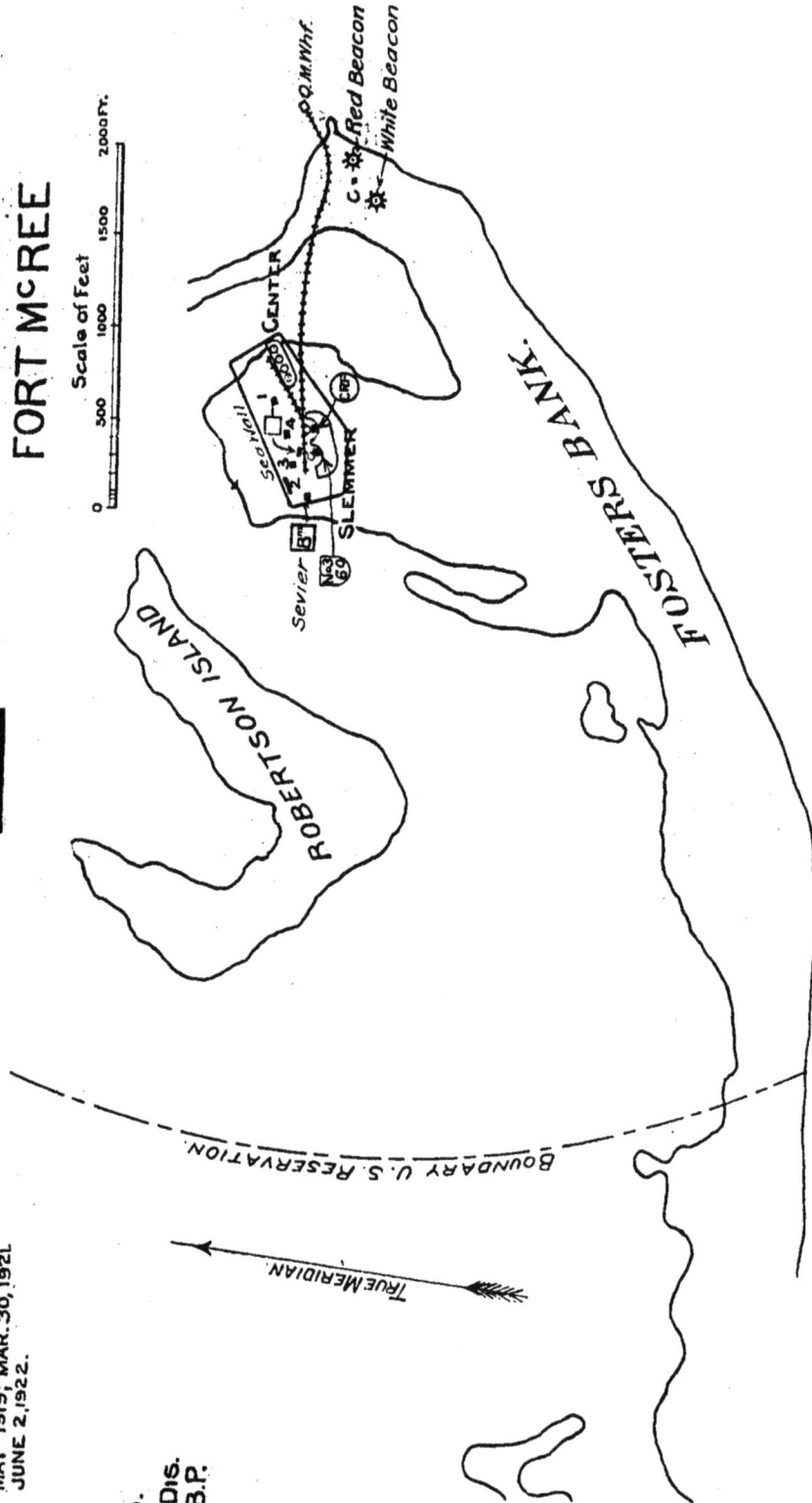

Scale of Feet

0   500   1000   1500   2000 FT.

SERIAL NUMBER 124

ROBERTSON ISLAND

Sevier

Slemmer

Seatail

Center

Q.M.Whf.

Red Beacon

White Beacon

FOSTERS BANK.

BOUNDARY U.S. RESERVATION

TRUE MERIDIAN

LEGEND.

1 QUARTERS.
2 BARRACKS.
3 ENGR. STORE HOUSE.
4 Q.M.COAL SHED.

BATTERIES.

* 1. SLEMMER..2-8"Dis.
  2. CENTER....4-3"B.P.

* *Dismounted and Guns removed*

EDITION OF APR. 23, 1915.
REVISIONS: DEC. 7, 1915;
MAY 1919; MAR. 30, 1921
JUNE 2, 1922.

PENSACOLA BAY

PENSACOLA BAY

VICINITY MAP
Scale — 1 : 80,000

N

△McRae

△BC Cons 233

△McRae

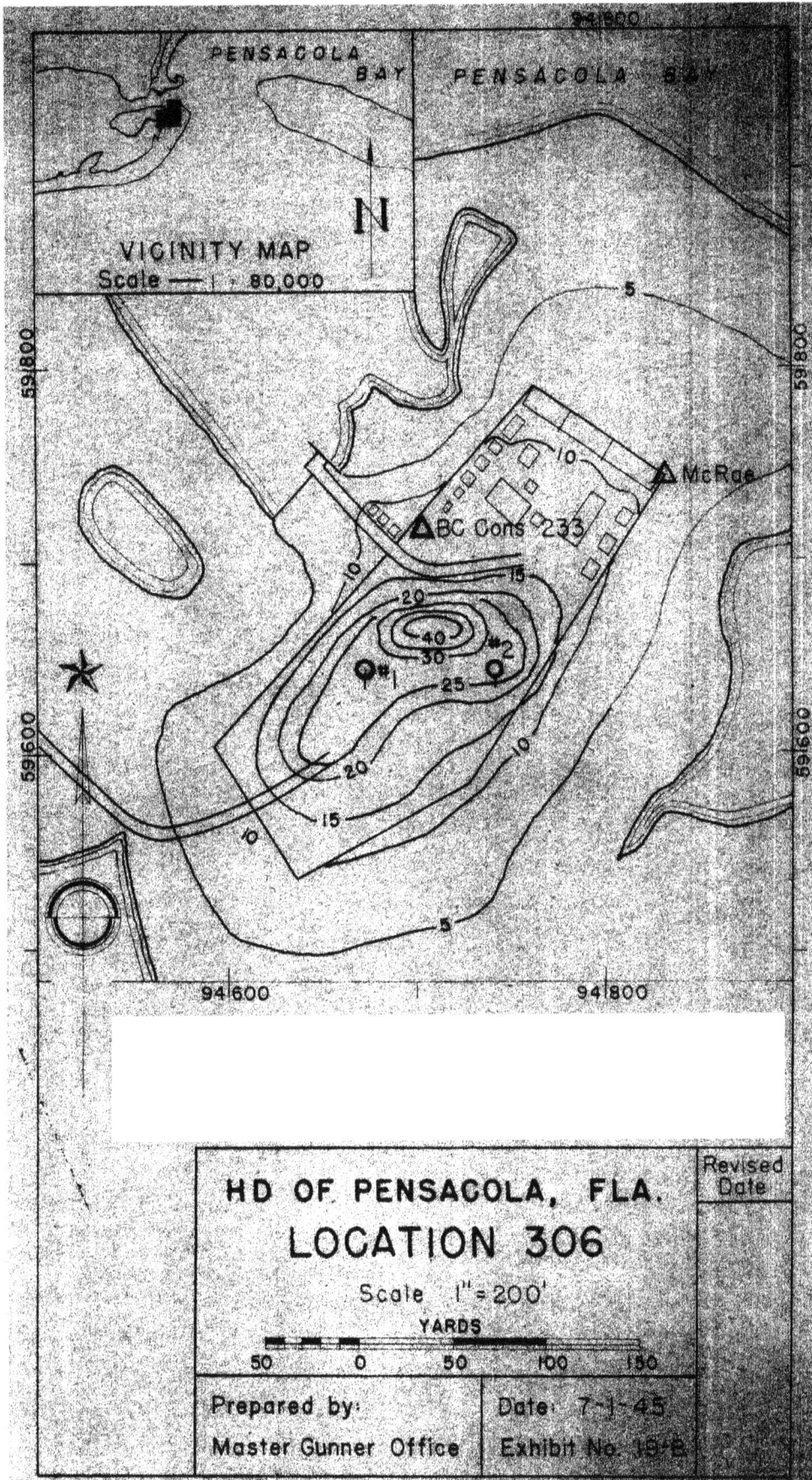

HD OF PENSACOLA, FLA.
LOCATION 306

Scale 1" = 200'

YARDS

50   0   50   100   150

Prepared by:	Date: 7-1-45
Master Gunner Office	Exhibit No. 16-B

Revised
Date

Fort McRee Gun Batteries

- **SLEMMER:** A battery for two 8-inch disappearing guns emplaced as the first modern battery on the western side of the Pensacola channel. Plans were submitted on March 21, 1898, using funds from the emergency National Defense Act passed in the same month. It was located at the Perdido Key reservation of Fort McRee, somewhat to the west of the older works that were rapidly being eroded away. The position fired to the southeast, covering the entrance to the bay all the way around to the Navy Yard in Pensacola Bay. The plan had some adjustments due to the location. In the rear was placed a concrete wall to protect the battery from fire and storm damage. The central traverse between the two platforms (gun separation of 131-feet) was made particularly heavy in protection. Also, the traverse was carried back further than required to better protect the No. 2 emplacement from flank fire. Magazines were on the lower right, ammunition service by hoists. Work was started promptly in April 1898 and the platforms reported done by June 30th. Armament received in November 1899. One gun was temporarily lost when the trestle it was being raised on collapsed, tossing it and the engine pulling it. Considerable efforts and additional funds were needed to recover it. The battery was transferred on March 21, 1900 for a cost of $113,806.47. It was armed with the gun tubes moved here from the temporary battery at St. John's Bluff, Florida. It was named in General Orders No. 16 of February 24, 1902 for Lt. Colonel Adam J. Slemmer, 4th U.S. Infantry who commanded Fort McRee at the start of the Civil War. It was armed with two 8-inch guns Model 1888MII guns on Model LF 1896 disappearing carriages (Watervliet tube #41/carriage #14 and Bethlehem tube #19/carriage #15). In 1902 a new communication gallery was added connecting the loading platforms, costing $700. The guns served until authorized for removal in August 1917, that being done in February 1918. In 1923 emplacement No. 2 was converted into a fire control station. The entire battery was filled in during 1942 to allow the construction of new Battery #233 immediately adjacent to it. In this state it still exists at the Perdido Key Area of the Gulf Islands National Seashore. The top of the battery site is open to the public.

- **CENTER:** A battery for four 3-inch masking parapet guns erected at the Perdido Key reservation of Fort McRee. Plans were submitted on December 18, 1898, though initially it was for just two emplacements. The battery was on the left flank of Battery Slemmer. The plan was for standard M1898 emplacements, with two internal platforms with a spacing of 29-feet. Magazines were on the lower right flank, service by hand. These were built using funds from the Act of July 7, 1898. In January 1899 an emplacement for two 3-inch guns was cancelled for its intended spot on the left flank of Battery Cullum at Fort Pickens and was relocated here to become the third and fourth emplacements of this battery. Initial work took place in 1898-1899, transfer being made on May 12, 1901 for a total cost of $20,603.84. It was armed with four 3-inch Model 1898 Driggs-Seabury guns on balanced pillar mounts (#10/#10, #11/#11, #20/#20 and #25/#25). The battery was named in General Orders No. 16 of February 14, 1900 for Lieutenant J. O. Center, 6th U.S. Infantry who was killed in the Seminole Wars in 1837 at Okeechobee. The battery was seriously damaged by storm action on September 26, 1906. Sand washed away and platform No. 3 was undermined and No. 4 completely broken up. In 1910 a substantial reconstruction of the entire battery was authorized, and the emplacements were rebuilt with a wider spacing. The rebuilt No. 3 and No. 4 were built to a different design, having rounded, flank emplacements and spaced at a more commodious 62-foot spacing with a magazine on either flank at a lower level. The original armament was re-emplaced, but some work lingered until about 1912. The carriages and emplacements were modified to the M1898M1 pedestal status about 1916. Armament was removed in June 1920 when all of these types of guns and mounts were declared obsolete. Parts of the emplacement still exists, though some sections are broken up and other sections are filled in at the Perdido Key Area of the Gulf Islands National Seashore. The battery site is open to the public.

- **Battery #233**: A 1940 Program dual 6-inch barbette battery emplaced at Fort McRee. It was required to fill-in most of old Battery Slemmer and build up the area behind the battery to create space for this emplacement. It was funded with FY-1943 Budget money. The plan was consistent to standard designs for 200-series emplacements. Concrete work was done from April 1, 1943 to October 15, 1943. It was conditionally transferred (pending arrival and mounting of armament) on June 5, 1944 at a cost of $284,498.83. The battery was never named, just known as Battery Construction No. 233 during work. Apparently, carriage M4 #55 and #56 were delivered to the site but never emplaced. With the shortage of available gun tubes, those were never sent to the fort. Work was suspended late in the war and never completed. In 1947 the site was abandoned, Battery #233 still incomplete. The emplacement still exists, though of difficult access, at the Perdido Key Area of the Gulf Islands National Seashore. The battery is open to the public, but access requires a boat or a long beach walk.

Barrancas Barracks/Fort Barrancas (1820-1947) was the location of the old Spanish fort and the rebuilt Third System works. It was the main garrison reservation are for the troops assigned to the Pensacola harbor defenses from 1900 to 1945.

6-inch gun in a shielded barbette carriage at Battery #234,
Fort Pickens Gulf Shores National Seashore (Mark Berhow)

6-inch gun in a disappearing carriage at Battery Cooper
Fort Pickens Gulf Shores National Seashore (Mark Berhow)

PENSACOLA HARBOR

FORT BARRANCAS

GENERAL MAP
Scale of Feet

1000   500   0   1000

SERIAL NUMBER 124

Woven Wire Fence

Ball Grd.

Country   Road

Open   Ditch

Q.M.Whf.

N⁴
60 (Portable)

F²¹ Aux.

True Meridian

Range   Lights

Pensacola Lt.

EDITION OF MAY 1919.
REVISIONS: MAR.30,1921.
JUNE 2,1922.

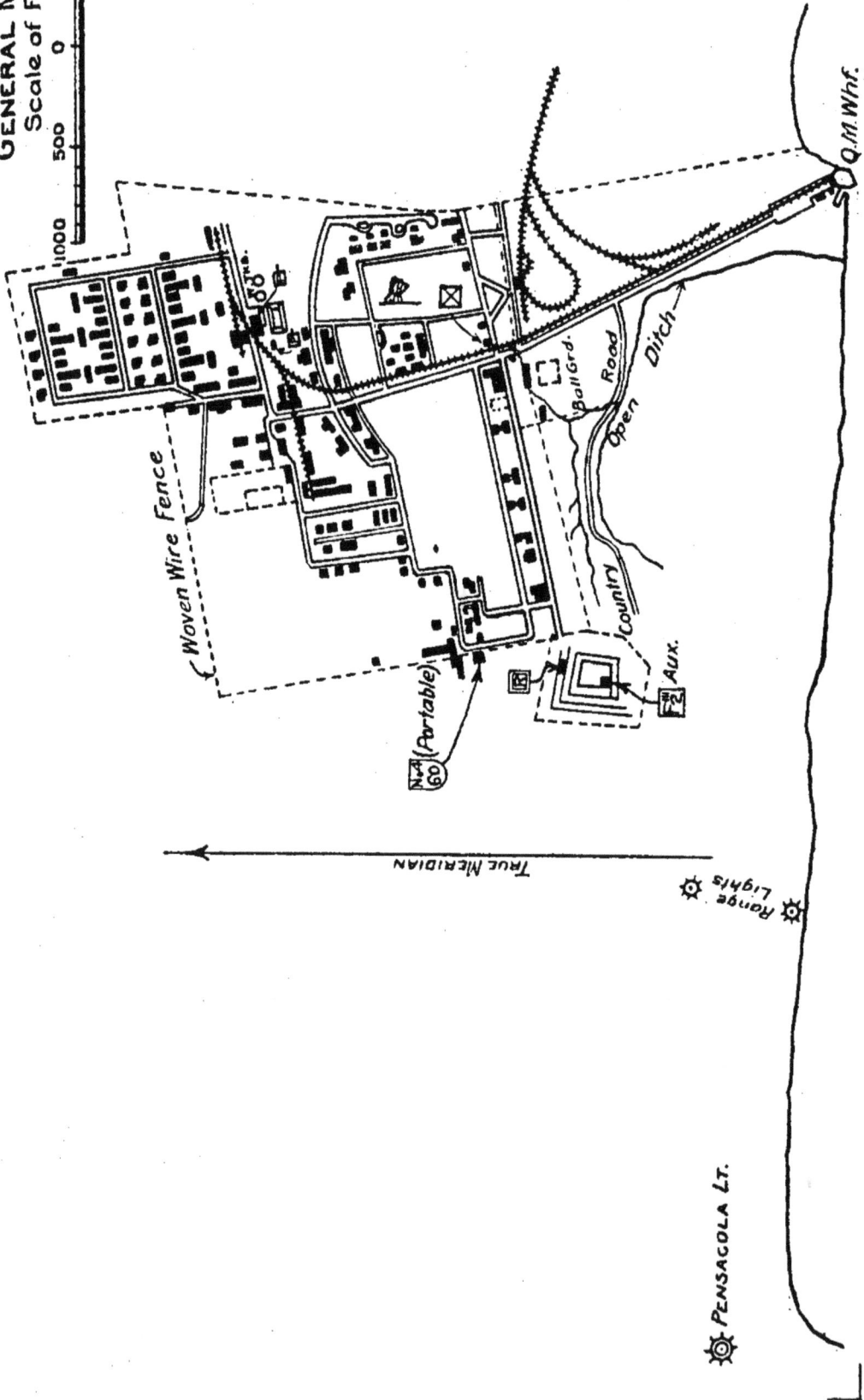

FORT BARRANCAS D-1.

PENSACOLA HARBOR

True Meridian

Pensa. Elec. Ry.

Q.M.Whf.

Ditch

Road

Open

Country

Ball Grd.

Ten. Ct.

PARADE

GROUND

(Portable)

Aux.

SERIAL NUMBER 124

LEGEND

1. ADMINISTRATION BLDG.
2. COMM. OFF. QRS.
3. OFFICER'S QRS.
4. HOSPITAL.
5. HOSPITAL STWD'S. QRS.
7. BARRACKS.
9. Q.M. SHED.
11. WHARFINGER'S QRS.
12. CONST'ING. Q.M. OFFICE.
13. SUB. STATION.
14. GRAND STAND.
15. OPEN AIR PAV.
16. POST OFFICE.
17. PENSA. ELEC. RY. STA.
18. BAND STAND.

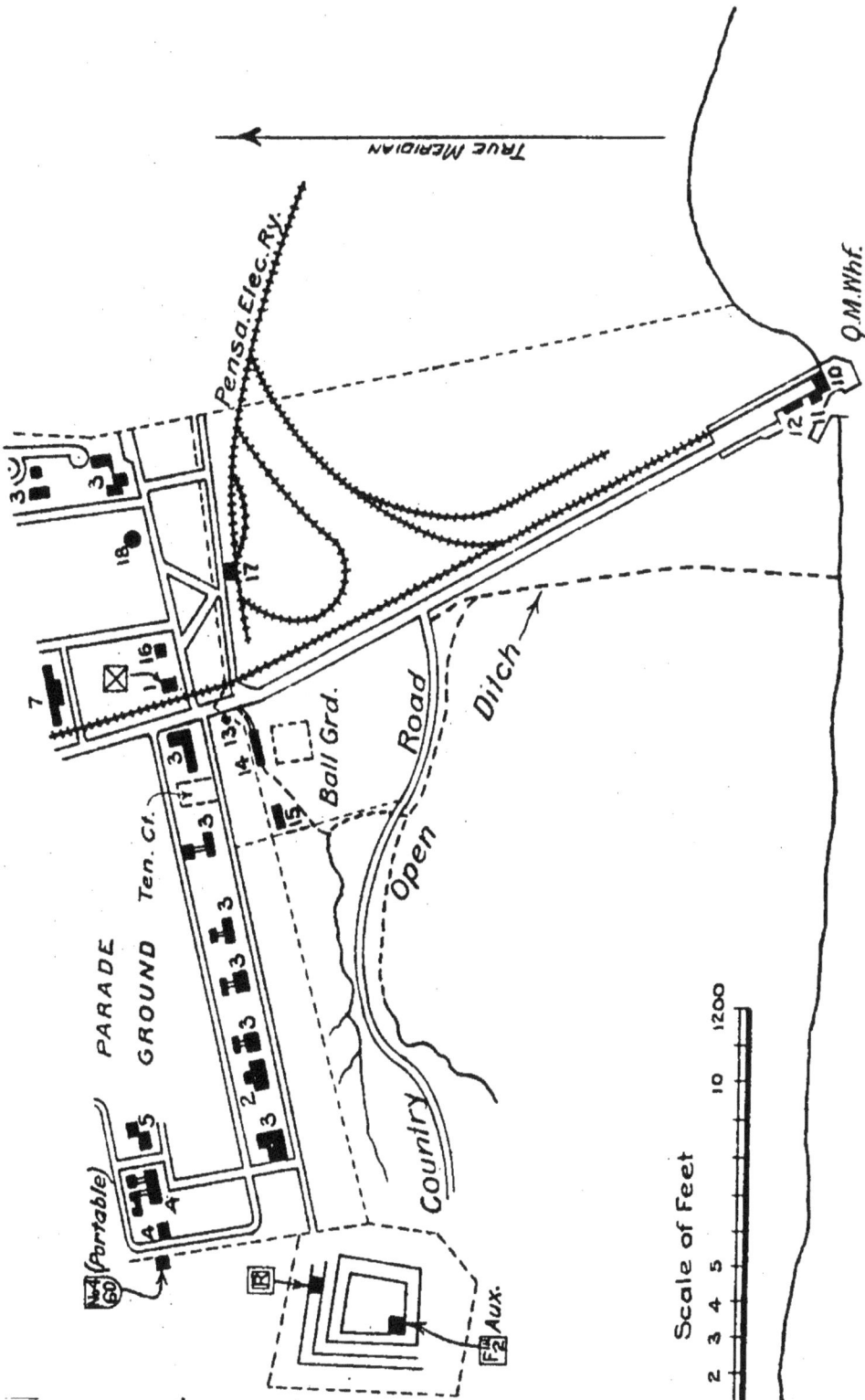

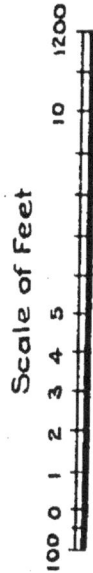

EDITION OF MAY 1919.
REVISIONS: MAR. 30, 1921.
JUNE 2, 1922.

Scale of Feet

100   0   1   2   3   4   5   10   1200

PENSACOLA HARBOR

FORT BARRANCAS D-2.

Scale of Feet

EDITION OF MAR.30, 1921.

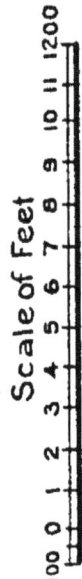

SERIAL NUMBER *124.*

TRUE MERIDIAN

PARADE GROUND

LEGEND

3 OFFICER'S QRS.
4 HOSPITAL.
5 HOSPITAL STWD'S QRS.
6 N.C. OFFICER'S QRS.
7 BARRACKS.
8 GUARD HOUSE.
9 POST EXCHANGE.
19 COMPANY OFFICE.
20 KITCHEN.
21 MESS HALL.
22 Q.M. SUB. STORE HO.
23 ORDNANCE ST. HO.
24 DISINFECTING HO.
25 FIRE ENGINE HO.
26 BAKERY.
27 ART. ENGR. ST. HO.
28 Q.M. BUILDING.
29 Q.M. STORE HOUSE.
30 BOWLING ALLEY.
31 WORK SHOP.
32 WAGON SHED.
33 Q.M. STABLE.
34 RECLAM. STORE HO
35 Q.M. SHOE REPAIR.
36 COAL SHED.
37 STORE HOUSE.
38 CARPENTER SHOP.
39 HOSTLER'S QRS.
40 Y.M.C.A. BUILDING.
41 PAINT SHOP.
42 GARAGE.
43 COMPANY ST. HO.
44 QUARTERS.
45 CREMATORY.
46 SOLDIER'S QRS.
47 MORGUE.

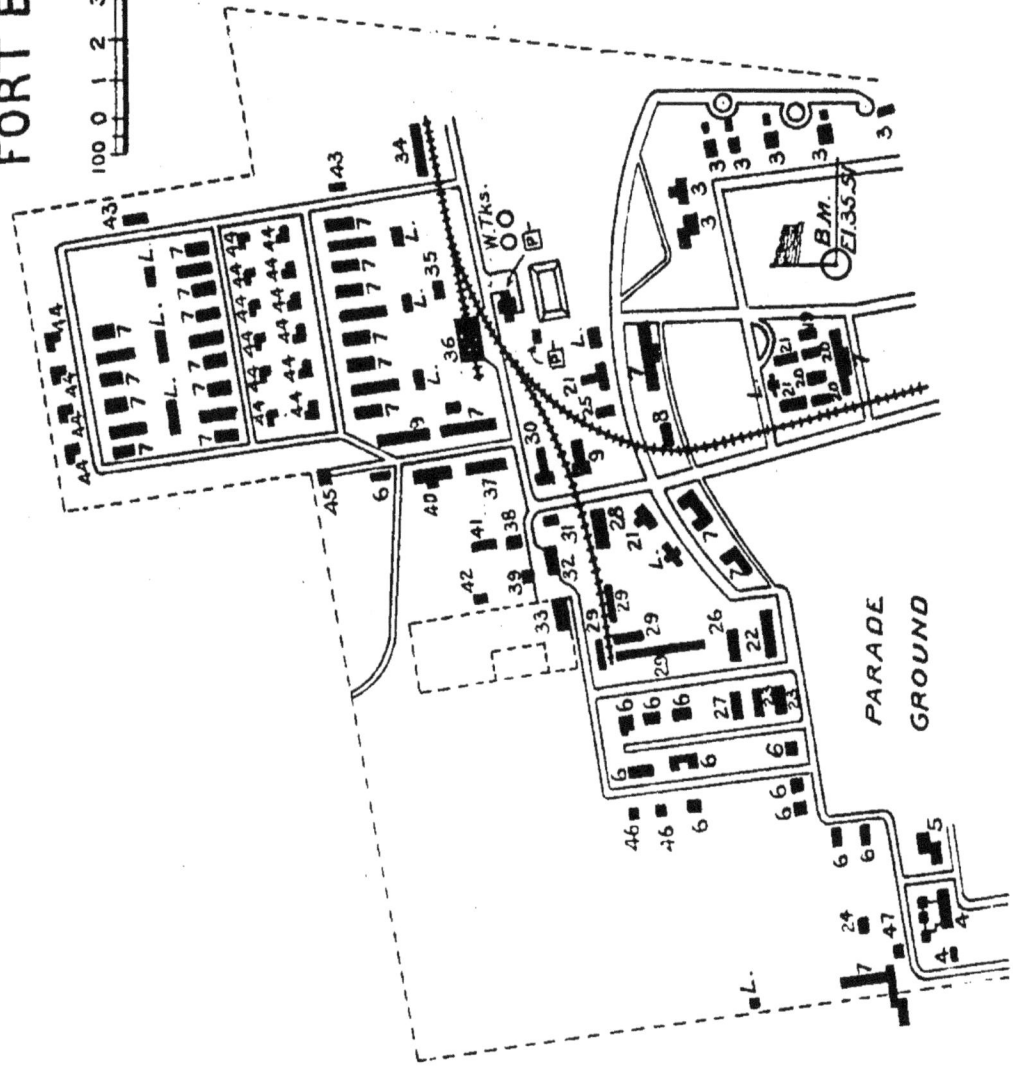

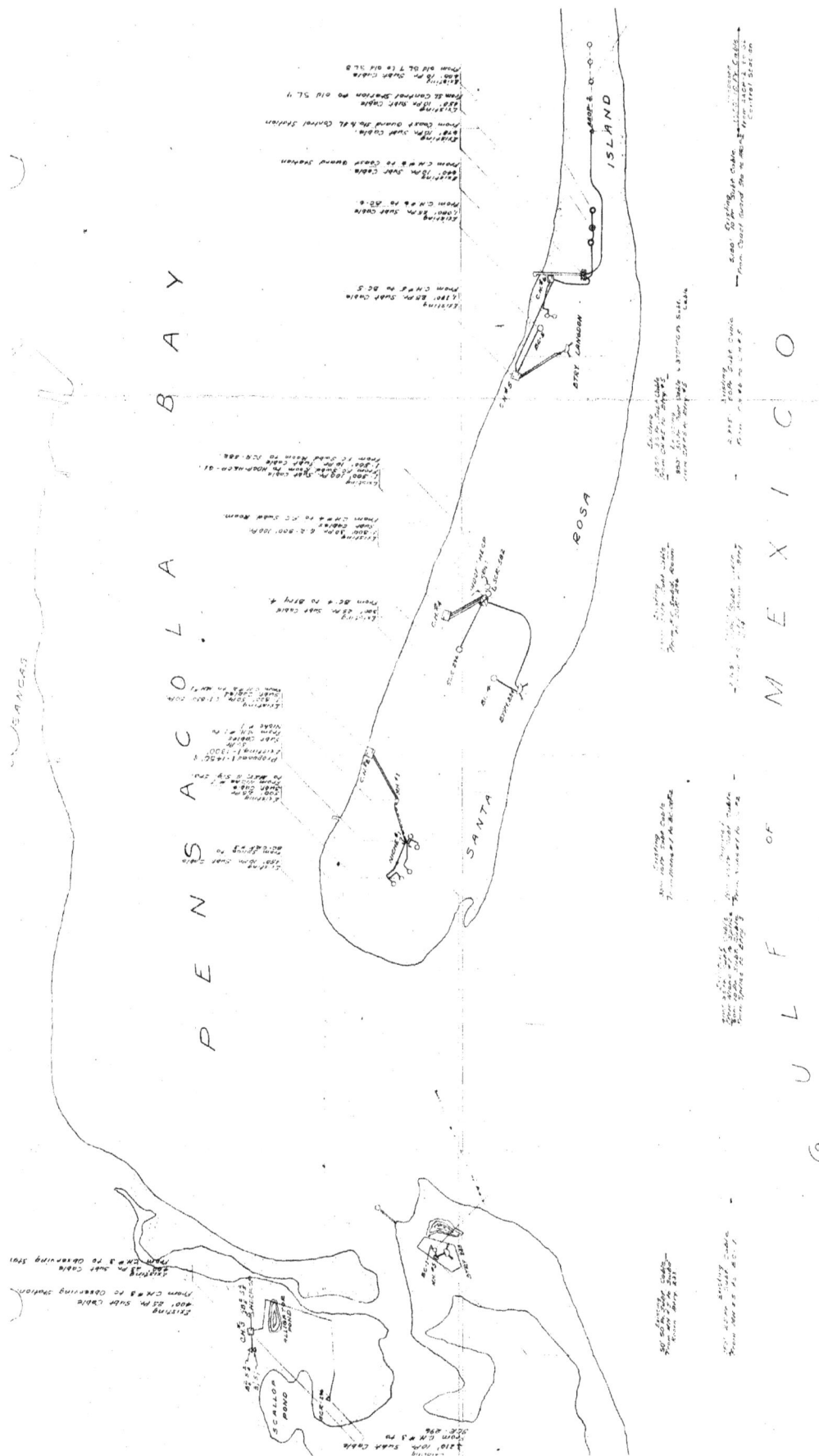

HARBOR DEFENSES OF
PENSACOLA, FLA.
SCHEMATIC LAYOUT, MAIN CABLES

Pensacola World War II-era Site Locations. Stations housed in a single structure are connected by dashes (-)

location	Loc#	Purpose
Gulf Beach	301	searchlight
Red Bluff	303	BS4/Langdon
Fort Barrancas	305	BS2/234, B1/Payne, BS2/Langdon
Fort McRee	306	Batt. Tact. #1 BCN 233, BC/233
Fort Pickens	308	Batt. Tact. #2 Payne, Batt. Tact. #3 Trueman, Batt. Tact. #3A AMTB, Bn2, BC/Paynje. BC/Trueman, SL 4,5
Fort Pickens	309	Batt. Tact. #4 BCN 234, BC/234, SCR296-234
Fort Pickens	310	SCR582, HECP-HDCP
Fort Barrancas	311	SBR
Naval Air Station	312	searchlight
Fort Pickens	313	BS1/Langdon, HDOP2, BS2/233, BS1/234
Fort Pickens	314	Batt. Tact. #5 Langdon, BC/Langdon
Fort Pickens	315	SCR296-Langdon
Pensacola MR	318	BS3/Langdon-BS3/234
Pensacola Beach	321	searchlight
Bald Point	322	BS5/Langdon

Battery Worth, Fort Pickens Gulf Shores National Seashore (Terry McGovern)

Coleman, James C. and Irene S Coleman. *Guardians of the Gulf, Pensacola Fortifications, 1608-1980.* The Pensacola Historical Society. Pensacola, FL, 1982.

Historic Resource Study, Pensacola Harbor Defense Project, 1890-1947, Gulf Islands National Seashore, Denver Service Center, Historic Preservation Division, US Dept. Interior, NPS, Denver, CO, 1982.

Muir, Thomas, Jr. and David P. Ogden. *The Fort Pickens Story.* The Pensacola Historical Society. Pensacola, FL, 1989.

Ogden, David P. *Frontline on the Home Front, the 13th Coast Artillery at Pensacola 1930-1947.* Gulf Islands National Seashore, US Dept. of the Interior, NPS, 1991.

THE HARBOR DEFENSES OF MOBILE BAY — ALABAMA

A large, initially shallow bay about 60 miles to the west of Pensacola. Spanish presence in the area had been sporadic from the 1500s, the initial settlement was by the French in 1702 on Dauphin Island and later at Mobile on the Mobile River, the later location was the initial capital of French Louisiana. Seceded to the British in 1763, taken by the Spanish in 1781, then annexed by the United States in 1813, the harbor had become an import commercial port. The U.S. Army built new defenses at the entrance to the Bay during the Third System construction program, started new defensive works in 1870, and modernized the defenses with seven concrete batteries and mine defenses during 1895-1905. Placed in caretaker status after World War I, the defenses were dropped from the defenses priority in 1923 and abandoned as harbor defenses by 1924 and transferred to the State of Alabama. A temporary battery of 155mm GPF guns were located at Fort Morgan in 1942 on Panama mounts, but the property was returned to the state.

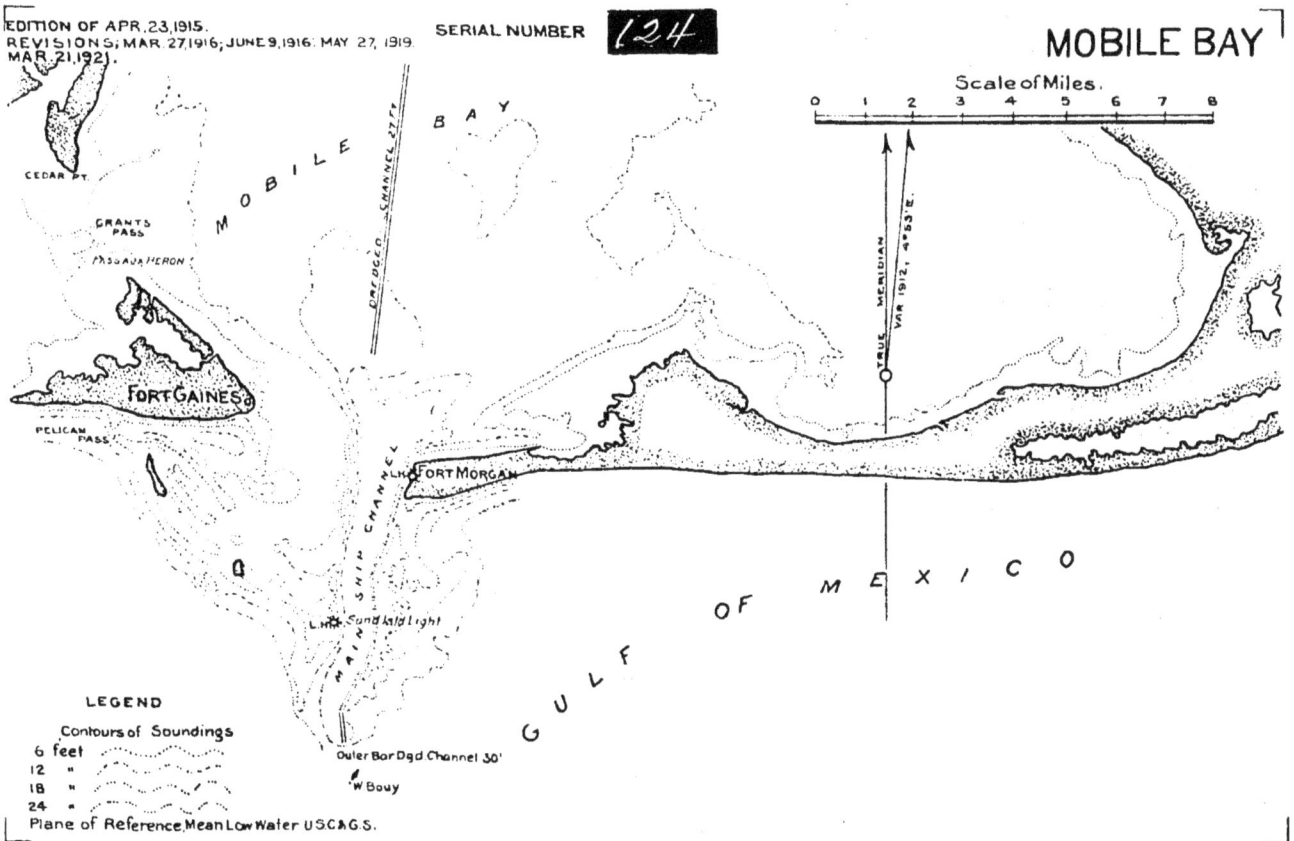

Fort Morgan (1833-1928) is located on Mobile Point at the entrance to Mobile Bay, about twenty miles west of Gulf Shores, Alabama. A Second System work, Fort Bowyer, first occupied this location during the War of 1812 until a large Third System fort was built from 1819 to 1833. It was named in General Order 38 of 1833 for Brig. Gen. Daniel Morgan of the Continental Army. The five-sided bastioned masonry fort served as the primary defense of Mobile Bay, along with its companion fort, Fort Gaines, located on Dauphin Island. Fort Morgan was occupied by Confederate forces at the beginning of the Civil War and only fell to Union forces after the Union Navy forced its way into Mobile Bay in August, 1864 and Union troops laid siege the fort until it surrendered. Placed on caretaker status after the Civil War, Fort Morgan's reservation was used during the Endicott Program for five concrete batteries and a controlled mine facility. Installed inside the walls of the old fort was the fort's primary battery of two 12-inch guns on disappearing carriages. To the east of the old fort, a battery of four 8-inch disappearing guns and a battery of eight 12-

inch mortars was also installed. The fort was used as training center during World War I, but in 1924 the fort was declared surplus. The State of Alabama took the site over as a state park in 1927. During World War II, the federal government took back the state park and installed a 155mm GPF battery on Panama mounts to defend Mobile Bay. Today Fort Morgan State Historic site features large day use facilities and a visitor center. The Third System is open daily with an entrance fee. All but a few of the garrison buildings have been destroyed. The fort is connected by a seasonal ferry to Dauphin Island and Fort Gaines. At the park entrance, to the south in the dunes, is the remains of a test battery that was built in 1914 for evaluating the effect of naval gun fire on seacoast defense batteries in 1916.

Fort Morgan Gun Batteries

- **THOMAS**: A war emergency battery for two 4.7-inch Armstrong guns assigned to the defenses of Mobile Bay. Local engineers were assigned the project on March 25, 1898 and provided funds from the recent National Defense Act. After consideration it was decided to emplace them behind the 12-inch battery assigned for the forward position in a spot in the outer works, in front of the Northwest bastion on top of the mine casemate there. Both were flank emplacements, separated by 39.6-feet, at a trunnion height of 37-feet. There were two magazines between the platforms under the center traverse. Work was started immediately, and the guns were mounted and reported ready by May 23, 1898 for an interim platform cost of $6,000. After completion of other featured it was turned over on February 27, 1899 for $15,000. It was named in General Orders No. 78 of May 25, 1903 for Captain Evan Thomas, 4th U.S. Artillery, killed in 1873 in battle with the Modoc Indians. The batterywas armed with two 4.7-inch Armstrong guns and pedestals (#10812/#10838 and #11853/#10844). These were then removed early in 1918, remounted for a short time in 1919, and finally removed for scrapping on July 22, 1919. One of the emplacements was subsequently converted in a CRF station in the early 1920s. The emplacement still exists at the Fort Morgan State Historic Site. The battery is open to the public.

- **SCHENCK**: A battery for three 3-inch guns assigned to Fort Morgan for coverage of the mine fields. Plans were submitted on January 31, 1899 for set of two masking parapet guns. They were sited to the north of Thomas, but on the slope below that battery at a relatively low height (trunnion height of just 20-feet) so that Thomas could fire over it. It fired to the northwest. It had two platforms separated at 29-feet, with an internal shape on the left and a flank emplacement on the right. Each platform had a magazine on its lower left, ammunition service was by hand. A plan for a third gun was submitted on May 2, 1903. It was for a pedestal type mount and emplaced on the right flank. The emplacement was set back slightly to the rear, and both a magazine and storeroom were located on the lower left so that a gun spacing of 56-feet was possible between this new No. 1 and the slightly older No. 2 emplacement. Transfer of the first two emplacement was made on June 8, 1900 for $6,000. They were armed with two 3-inch Model 1898 Driggs-Seabury guns on balanced pillar mounts (#70/#70 and #72/#72). The final emplacement had a 3-inch Model 1902 Bethlehem Steel gun and pedesta mount (#1/#1). It was transferred on April 28, 1904 for an additional cost of $7,963.97. All three were named on General Orders No. 78 of May 25, 1903 for 1st. Lieutenant William T. Schenck who was killed in 1900 in the Philippines. The two M1898 guns were modified to the M1898M1 pedestal configuration in 1915 but then dismounted and scrapped in mid-1920. The pedestal gun was removed in 1923. The emplacement still exists at the Fort Morgan State Historic Site. The battery is open to the public.

MOBILE BAY, ALA.

FORT MORGAN

MOBILE POINT

SERIAL NUMBER 124

EDITION OF APR. 23, 1915.
REVISIONS: DEC. 7, 1915; MAR. 27, 1916; MAY 27, 1919;
MAR. 21, 1921.

MOBILE BAY

GULF OF MEXICO

VAR. 1812, 4°32'E

TRUE MERIDIAN

Quarantine Whf.

Engr Whf.

SEA WALL

BOWVER

DEARBORN

SCHENCK

THOMAS

DUPORTAIL

10° Experimental Emplacement

LEGEND.

1. ADMINISTRATION BLDG.
2. COM'T. OFFICER'S QRS.
3. OFFICER'S QRS.
4. HOSPITAL.
4a. MORGUE.
4b. HOSPITAL STOREHOUSE.
5. HOSPITAL STEWARD'S QRS.
6. N.C. OFFICERS' QRS.
7. BARRACKS.
8. GUARD HOUSE.
9. POST EXCHANGE.
10. GYMNASIUM.
11. MESS HALLS AND LAVATORIES.
12. BAKERY.
13. ELECTRIC LIGHT AND ICE PLANT.
14. FIRE APPARATUS HO.
15. RAMP FOR FIRE ENGINE.
16. FIREMAN'S QRS.
17. COAL SHED.
18. STABLE.
19. WAGON SHED.
100. GROINS.
101. TELEGRAPH STATION.
20. Q.M. SUPT. OF CONSTRUCTION OFFICE.
21. Q.M. AND SUB. ST. HO.
22. Q.M. OILHOUSE.
23. Q.M. CARPENTER SHOP.
24. POST OFFICE.
25. ARTILLERY ENGR.
31. ORDNANCE ST. HO.
40. ENGINEER OFFICE.
41. ENGR. CARPENTER SHOP.
42. ENGR. BLACKSMITH SHOP.
43. ENGR. TOOL ROOM.
44. ENGR. CEMENT SHED.
45. ENGR. OILHOUSE.
46. ENGR. BUILDINGS.
80. LIGHT-KEEPER'S HO.
70. CHURCH
71. Y.M.C.A.

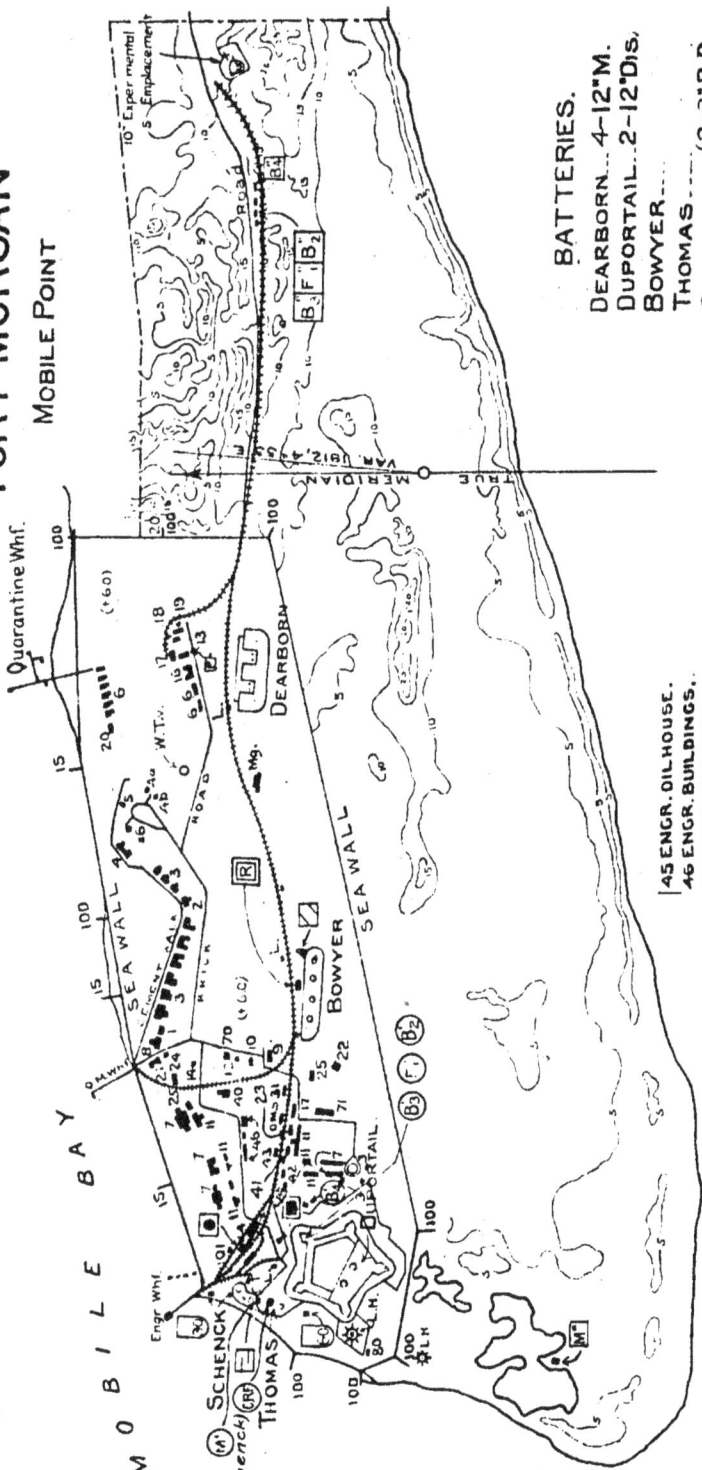

BATTERIES.

DEARBORN	4–12" M.
DUPORTAIL	2–12" D.'s.
BOWVER	
THOMAS	
SCHENCK	{2–3" B.P. {1–3" P.

- **DUPORTAIL**: A battery for two 12-inch disappearing guns erected inside the perimeter of old Fort Morgan. Very early Endicott plans projected a 12-inch lift battery here, replaced later with a scheme for a three-gun 12-inch battery. Finally, a pair of 12-inch disappearing guns was authorized. Plans were submitted on March 22, 1898 with an initial appropriation from the National Defense Act passed earlier that month. The site cut across the interior parade of the battery, with firing to the southwest. Construction required the removal of top layer of the three seaward fort faces to allow clearance for fire. Both positions were flank emplacements. The constricted space demanded a variation from type plans, with closer gun centers (just 106-feet) and the magazines on each outside flank, not under the center traverse. Ammunition service was by hoists. Additional authorization was given in February 1899 to remove the rest of the parapet around the old masonry work. Work was done from May to November 1898, the guns were mounted in 1899 and the battery turned over on June 8, 1900 for a total cost of $172,647. It was named in General Orders No. 78 of May 25, 1903 for Major General Louis L. Duportail, Chief of Engineers in the Continental Army. It was armed with two 12-inch Model 1888MII Bethlehem guns on Model LF 1896 disappearing carriages (#6/#16 and #7/#17). In 1916 new Taylor-Raymond back delivery hoists were installed, and the gun carriages were modified for higher elevation a little later. The battery served until the defenses were declared obsolete after World War I. Even then it seems the armament remained unused, until eventually being turned over to the State of Alabama in 1932. It was probably donated during the scrap drives of World War II. The emplacement still exists at the Fort Morgan State Historic Site. The battery is open to the public.

- **BOWYER**: A battery for four 8-inch disappearing guns, this was the first modern Endicott work for the Mobile defenses. The site was on ground 1000-feet away to the immediate east of the old fort and approved on November 10, 1894. It fired to the south. Plans were soon submitted for the first pair of guns (emplacements No. 1 and 2) for the LF M1894 carriage. A third emplacement (for M1896) was submitted on July 9, 1896, and the final fourth (also a model M1896) on January 4, 1897. It followed fairly typical early design types, with multi-level shot and shell galleries, magazines on the right flanks and initially lifts later replaced by hoists (in 1909). Gun spacing was 106-feet, there were three internal and one flank emplacement (No. 4). Because the construction authorization was spread over several years, construction took place between 1895 and 1898. The first emplacement was completed by January 1897, two more in May and the fourth in June 1898. By June 30, 1898 all was finished except some fill on the parapet and mounting of the guns. The battery was named in General Orders No. 43 of April 4, 1900 for Colonel John Bowyer, for whom an earlier fort was named in Mobile Bay during the War of 1812. The first two emplacement were for Model 1894 carriages that arrived in September 1897 and mounted complete by the following February. The last two received Model 1896 carriages and were mounted on March 11, 1898. The battery was transferred on September 26, 1898 for a total cost of $187,811. It was armed with four 8-inch Model 1888MI and 1888MII Watervliet guns on disappearing carriages as noted (#29/M1894 #9, #32/M1894 #10, #45/M1896 #3, and #48/M1896 #4). By authority of July 3, 1901 $2,000 was supplied to construct a gallery behind the traverses to allow ammunition transfer. Modifications were also made to the loading platforms, for electrification, and new hoists in 1909. All four guns were removed in 1917-1918 for use on railway carriage, the carriages were scrapped in place in 1920 and the hoists removed for use elsewhere in 1923. The emplacement was not subsequently used for armament. It does still exist at the Fort Morgan State Historic Site. The battery is open to the public.

- **DEARBORN:** The mortar battery for Fort Morgan. It was placed to the east of the main part of the fort, firing to the south. It was one of the later Endicott mortar batteries, and might be the final mortar battery of the initial series built in the U.S. It was of type design, with the wide pits and magazines on adjacent flanks and under the central traverse. Funding came from the Act of May 7, 1898 with the concrete plant and framing ready on June 30, 1899. Work was delayed due to the loss of two boatloads of cement sent to Mobile from the Atlas Cement Company of New York. Nonetheless it was structurally completed by June 30, 1900. Four carriages were received on January 10th and 23rd of 1902 and mounted by year's end. Problems were encountered with the sides of the traverses falling into the pits. Finally, transfer was made on May 20, 1902 for a construction cost of $150,000. It was armed with eight 12-inch Model M1890M1 Watervliet mortars on Model 1896 carriages (#80/#112, #123/#145, #78/#144, #126/#147, #87/#143, #72/#146, #113/#194, #73/#168). It was named on General Orders No. 43 of April 4, 1900 for General Henry Dearborn, an officer in the Revolutionary War and former Secretary of War 1801-1809. On October 27, 1915 it was proposed to ship four guns and carriages San Diego, but this was rescinded on June 27, 1916. Four guns were removed (#78, #80, #87, and #113 from the rear No. 1 and No. 3 positions of each pit) under authority of May 3, 1918 completed by July 23, 1918. The carriages for these were scrapped in 1921. The battery stayed armed for a number of years, though it was abandoned in the late 1920s with the ending of the Mobile defenses. Even as late as 1936 the fort's training 9.2-inch howitzers were stored in the pits of the mortar battery, along with the final four, abandoned mortars. The latter were removed for scrap between 1938 and 1942. The emplacement still exists at the Fort Morgan State Historic Site. The battery is open to the public.

- **Experimental Battery:** An experimental 10-inch emplacement battery was built at Fort Morgan in 1915. It was emplaced east of the reservation proper, on newly acquired property. It was part of a specific program to test the endurance of "modern" disappearing batteries from live-fire naval bombardment. Work was done from June 1915 to February 1916, for transfer on June 2, 1916 at a cost of $49,381.84. Its design closely approximated that of a current type 14-inch disappearing battery plan, but without refinement like hoists, power, or even magazines. It was armed for the tests with one 10-inch Model 1888 Watervliet gun on Model 1896 disappearing carriage (#56/#27). This was not mounted for fire, but rather to examine the effect of fire upon it. The actual test firing came from battleships USS *New York* and USS *Arkansas*. The test was successful in finding that the emplacement was both difficult to spot and the damage inflicted was minimal. Authority for removal of the gun came on July 12, 1918 and the carriage in 1922. It does appear that at least some major parts of the carriage may still be left in the pit, but they are currently covered with sand. The emplacement, at times filled with sand and brush, still exists on the eastern part of the Fort Morgan State Historic Site. The battery is open to the public.

Test Battery Fort Morgan State Park (Terry McGovern)

Battery Bowyer Fort Morgan State Park (Terry McGovern)

Battery Dearborn, Fort Morgan State Park (Terry McGovern)

Fort Morgan State Park (Terry McGovern)

Fort Gaines (1821-1928) is located on the east end of Dauphin Island at the entrance to Mobile Bay, about twenty-five miles south of Mobile, Alabama. The Third System fort was built from 1846 to 1859. It was named in General Order 38 of 1833 for Brig. Gen. Edmond P. Gaines, U.S. Army. The five-sided bastioned masonry fort served the defenses of Mobile Bay, along with its companion fort, Fort Morgan, located on Mobile Point. It was under Confederate control until the garrison surrendered in August 1864. Placed on caretaker status after the Civil War, Fort Morgan's reservation was used during the Endicott Program for two concrete batteries. Installed inside the walls of the old fort was the fort's primary battery of three 6-inch guns on disappearing carriages. To the north of the old fort, a battery of two 3-inch gun on masking pedestal mounts was also installed in 1901. The fort was used as training center during World War I, but in 1924 the fort was declared surplus. The county government took the site over as a park in 1928. During World War II, the federal government took back the county park. Most buildings have disappeared, but the old fort and the Endicott Program batteries remain. Today the county park fort is open during daylight houses with an admission fee charged. Significant masonry damage had been sustained during hurricanes and tropical storms during its lifetime. Though this damage has been largely repaired, the fort continues to be under threat from erosion.

Fort Gaines Gun Batteries

- A temporary battery for two 8-inch guns on reinforced Rodman carriages allocated to the Dauphin Island reservation of Fort Gaines. These were two of the twenty-one special ordnance combinations ordered emplaced at the start of the Spanish American War in temporary emplacements. Local engineers prepared plans and submitted them on May 20, 1898. A new concrete platform to hold iron traversing rings of the strengthened 15-inch Rodmans spaced at 50-feet was planned, along

MOBILE BAY, ALA.
FORT GAINES
DAUPHIN ISLAND.

SERIAL NUMBER 124.

EDITION OF APR. 23, 1915.
REVISIONS: DEC. 7, 1915: MAR. 27, 1916: MAY 27, 1919
MAR. 21, 1921.

LEGEND.
1
2
3 OFFICERS QUARTERS.
3a OFFICERS QRS. (QBS.)
4
5
6 N.C.OFFICERS' QUARTERS.
7 BARRACKS.
8 GUARD HOUSE.
9
10 MESS HALL.
11 ORIGINAL STOREHOUSE.
12 ARTESIAN WELL.
21 Q.M. STOREHOUSE.
41 U.S.ENGR.BUILDING

BATTERIES.
STANTON ---- 2-6" DIS
TERRETT ---- { 2-3" B.P.
{ 1-3" P.

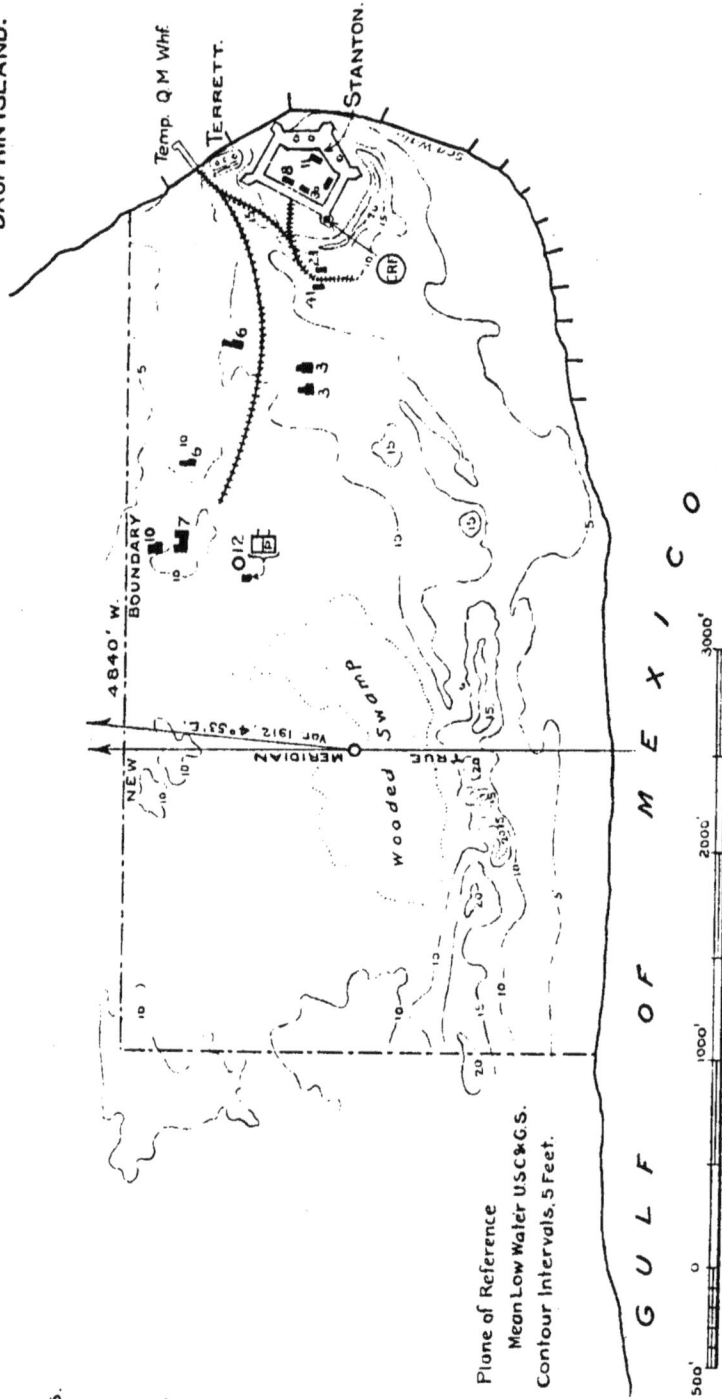

Plane of Reference
Mean Low Water U.S.C.&G.S.
Contour Intervals, 5 Feet.

GULF OF MEXICO

Wooded Swamp

TRUE MERIDIAN
NEW MERIDIAN
Var. 1912. 4° 53' E.
4840' W.
BOUNDARY

Temp. Q.M. Whf.
TERRETT.
STANTON.

500'   0   1000'   2000'   3000'

with adjacent magazines withdrawn on either flank. It was funded from the March 1898 National Defense Act. Work was done that summer. The two tubes (Bethlehem 8-inch Model 1888MII #4 and #7) were shipped here on August 15, 1898, but a delay due to the lack of an appropriate wharf prevented them from being delivered until January 1899. One gun was finally mounted in June of 1899, the other being indefinitely delayed to required alterations to its carriage which were never actually accomplished. The one gun served only a short time, probably being dismounted in 1902. The emplacement was mostly destroyed in 1903 for the construction of the last emplacement of 6-inch Battery Stanton, though the two magazines were retained as bombproofs for the new structure. Little else remains of this emplacement at Fort Gaines Historic Site

- **STANTON**: A battery for ultimately three 6-inch disappearing guns emplaced inside the walls of old Fort Gaines. Plans were submitted on March 15, 1899 for a pair of 6-inch Model 1897 disappearing guns in a new emplacement. They were placed inside the old work, parallel to the east face, firing to the east. Plans followed mimeograph recommendations, with guns spaced at 77-feet and magazines in the outside flanks. Hoists were used for shell handling. Work was funded from the Fortification Act of July 7, 1898; the concrete was done in the latter half of 1899. The guns were received on January 20, 1901 and mounted in August. Transfer was made on May 20, 1901 for a cost of $67,250. It was armed with two 6-inch Model 1897M1 guns on Model M1898 disappearing carriages (#22/#18 and #27/#19). It was named in General Orders No. 78 of May 25, 1903 for Captain Henry W. Stanton, 1st US Dragoons killed in 1855 in action with the Apache Indians in New Mexico. On April 9, 1903 plans were submitted for an additional emplacement to the battery for a later Model 1903 disappearing 6-inch gun. The new emplacement replaced the temporary 8-inch battery built during the Spanish American War along the southeast face of the fort. It was close to but entirely separate from the earlier pair of guns. A standard 1903 type platform with magazine on its lower left was planned, though the old 8-inch magazines on the lower level were kept in the plan as storerooms or bombproofs. Concrete work was done by January 1904, transfer coming on May 31, 1904 for an additional cost of $24,134.89. Its armament was mounted on June 24, 1903: one 6-inch Model 1903 gun on Model 1903 disappearing carriage (#67/#61). The Model 1903 gun was taken out in 1918 as a unit intended to be put on a wheeled field mount. The two M1897MI guns were removed in the late 1920s. The emplacement still exists at the Fort Gaines Historic Site. The battery is open to the public.

Battery Terrett, Fort Gaines County Park (Mark Berhow)

- **TERRETT**: A battery for ultimately three 3-inch rapid-fire guns built at Fort Gaines. Original plans for a two-gun, masking parapet type emplacement was submitted on July 27, 1900. It was emplaced outside the old fort to the northeast, and fired to the east, northeast. This section was fairly standard in plan, with two platforms separated by 29-feet and magazines on the lower right. The No. 2, left emplacement was built as a flank position, the right-hand one as an internal position. Work started in May 1900 and finished the following January. It was transferred on May 20, 1901 for only $6,500. It carried two 3-inch, 15-pounder Model 1898 guns on balanced pillar mounts (#118/#118 and #119/#119). On January 12, 1903 a plan for a third emplacement was submitted and approved for the more modern 3-inch Model 1903 gun and pedestal. While still in line and on the right, it was spaced at 47.5-feet from the No. 2 platform and was also built as an internal platform. Its magazine was on the lower left (all were dependent on hand delivery of ammunition). This unit was finished for transfer on April 28, 1904 for a cost of another $6,965.11. It was armed with one altered Model 1898 gun tube to fit in a standard Model 1903 pedestal mount (gun #3/pedestal #63). This was a unique combination, not reproduced anywhere else. The battery was named in General Orders No. 78 of May 25, 1903 for 1st Lieutenant John C. Terrett, 1st US Infantry who was killed at Monterey, Mexico in 1846. The two Model 1898 guns and pillars (never modified to the M1898M1 standard) were removed in 1921, the Model 1898/1903 combination was removed in 1923 and sent to arm a battery in Charleston Harbor. The emplacement still exists at the Fort Gaines Historic Site. The battery is open to the public.

Fort Gaines County Park (Terry McGovern)

Gaines, William C. "The Modern Coast Defenses of Mobile Bay 1865-1945" *Coast Defense Journal* Vol. 24, No. 1, Feb. 2010, p. 61; Vol. 24, No. 2, May 2010, p. 57; Vol. 24, No. 3, Aug. 2010, p. 4.

THE HARBOR DEFENSES OF THE MISSISSIPPI RIVER – LOUISIANA

The mouth of the Mississippi River has been an important trade and commercial passage since the times of French exploration in the 1680s. New Orleans was established as a French settlement on the first high ground inland along river in 1718. Seceded to the Spanish in 1763, who began a significant fortification program. Briefly returned to the French in 1803, who then sold to the United States as part of the Louisiana Purchase. The old Spanish fortifications were re-occupied and renovated by the American forces and defended during the British attack in 1814-15. The locations of all the British probes and landings were fortified by United States in the early stages of the Third System construction program and saw significant action during the Civil War. The defenses were modernized with temporary batteries during the Spanish American War and eight new concrete batteries during the Endicott construction program. However, the defenses were placed in caretaker status after World War I and ordered deactivated and abandoned in 1924. During World War II, the only defenses installed were 155mm GPF guns at the mouth of Mississippi River.

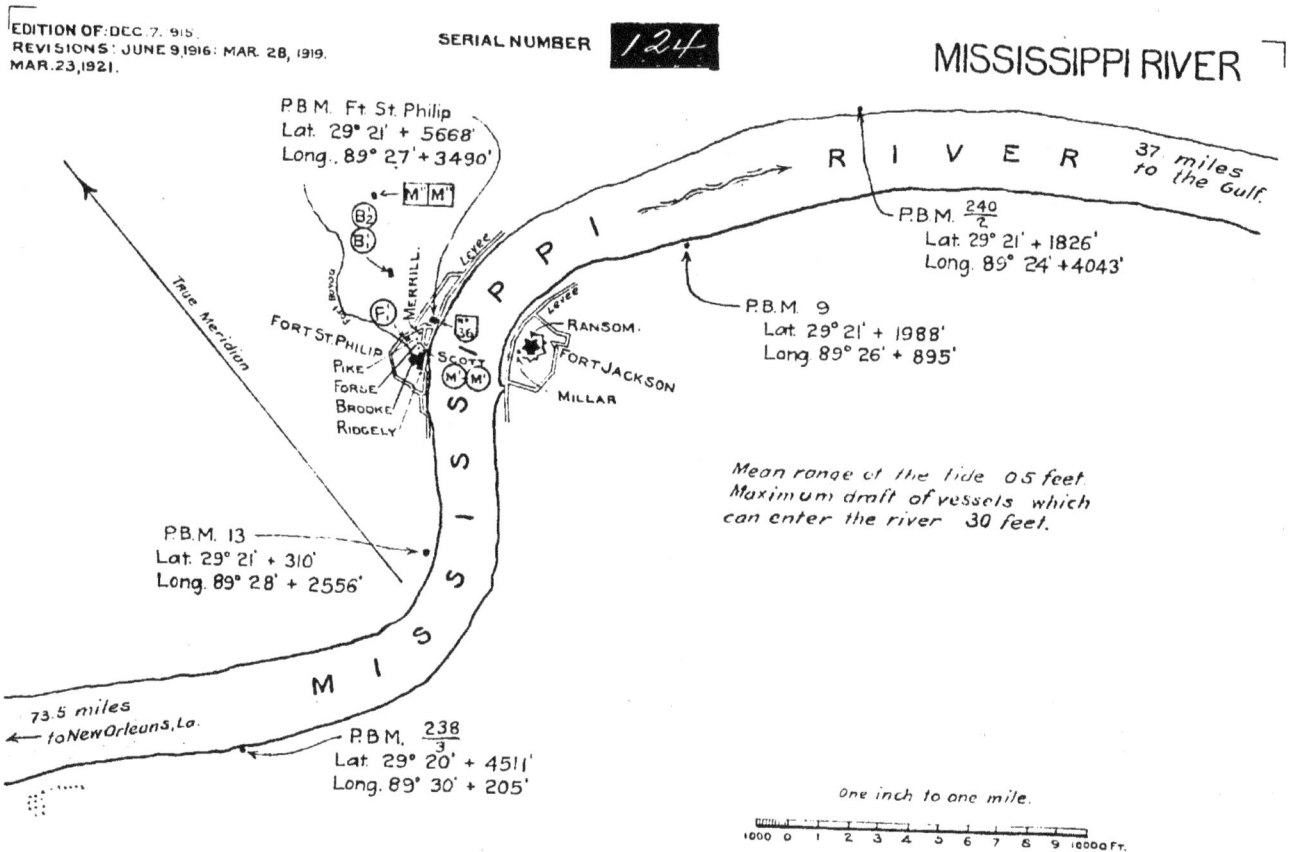

Fort St. Philip (1814-1924) is located on east bank of the Mississippi River at Plaquemines Bend, about sixty miles downriver from New Orleans, Louisiana. This fort was original constructed in 1793 by the Spanish as Castillo San Felipe. During the War of 1812, the American's restored this fort and were bombarded by English warships. The fort deteriorated over the years when the existing fort was rebuilt as a Third System fort from 1840 to 1858. The fort's long history is responsible for its strange parallelogram shape. It was finally officially named in General Order 1 of 1937. The masonry fort served as the primary defense of the Mississippi River, along with its companion defensive work, Fort Jackson, located on the west bank. Fort St. Philip was occupied by Confederate forces at the beginning of the Civil War and only fell to Union forces after the Union Navy forced its way by the forts and captured New Orleans. Placed on caretaker status after the Civil War, Fort St. Philip's reservation was used during the Endicott Program for six concrete batteries and a controlled submarine mine facility. The fort's primary batteries were two 10-inch guns and two 8-inch guns, on disappearing carriages. The fort was used as training center during World War I, but in 1924 the fort was declared surplus and sold to private interests in 1926. The area round the fort has been used for farming and cattle grazing, as well as spiritual commune in the former officer quarters. Hurricanes have destroyed all buildings and damaged the berm around the fort, this has allowed the fort to be filled with many feet of silt and flooding during high waters. The old fort and Endicott Program batteries remain today. Privately owned, restricted access is only available by boat or heliocopter.

Fort St. Phillip Gun Batteries

- **PIKE:** The first modern Endicott battery at the Mississippi River reservation near old Fort St. Philip. First plans were submitted on February 8, 1893 for the foundation of a battery of two 10-inch disappearing guns. Fort St. Philip was a very difficult place to build heavy structures, essentially just river silt, and extensive surveys and then deep piles and grillage were required for each emplacement. Costs and time to build the New Orleans defenses were just about the greatest experienced during the Endicott Program. On a site located about 700-feet east of the old masonry work, only in 1895 was authority granted to actually start placing the foundation piles. On April 25, 1896 new plans were submitted using the Model 1896 disappearing carriage. Up to this point all work had been done just for the north emplacement, in October proposals were made for the south emplacement. Expenses exceeded estimates due to the cost of labor, interruptions for yellow fever outbreaks, and the cost of delivering materials by boat as no road existed on this side of the river. It was finally reported ready for armament on May 18, 1898. The battery was fairly conventional in layout, with a flank platform on the right, and an internally shaped on the left. Magazines were on the lower level on the flanks, gun centers were just 96-feet. To aid as a reduction in weight, extensive use was made of steel roofs and floors instead of concrete. Most structural work was done in 1896-1897. It was transferred on June 2, 1898 for $160,000. The original armament was two 10-inch Model 1888MII Watervliet guns on Model LF 1896 disappearing carriages (#54/#2 and #51/#3). Due to settling the guns were dismounted for releveling on August 22, 1901, when new anchoring bolts were emplaced. It was then found that the carriages had defects and were returned to Watertown for repairs on October 25, 1901. Eventually they were repaired and were sent to Fort Wetherill, while Battery Pike received replacement Model 1896 carriages #73 and #74, being mounted in February 1902. The battery was named in General Orders No. 78 of May 25, 1903 for Brigadier General Zebulon Pike who was killed in action at York, Canada during the War of 1812. Beginning in 1904 modifications were made at enlarging the loading platforms and adding new hoists and communications stairways and walks. The siting of this battery (along with Forse and Ransom) was later heavily criticized as being unable to command all but a small portion of the river approach. It served until listed on May 25, 1918 for removal, dismounted on June 22nd of that year. Apparently, the tubes did not leave the reservation, as they were remounted for a short time in 1919. The battery was reclassified to a low category "C" of readiness in early 1920. The guns were removed permanently shortly thereafter. By memorandum of March 11, 1920 these defenses were discontinued, and the reservation disposed of in 1921. The emplacement was subsequently abandoned but still exists in poor condition and of very difficult access on private property. The battery is open, but fort is closed to the public.

- **FORSE:** A battery for two 8-inch disappearing guns on the Fort St. Philip reservation. Plans were submitted, at first for just a single emplacement, on April 13, 1897 at this site in a gun line to the southwest of 10-inch Battery Pike. Funding came from the Act of March 3, 1897. It fired along

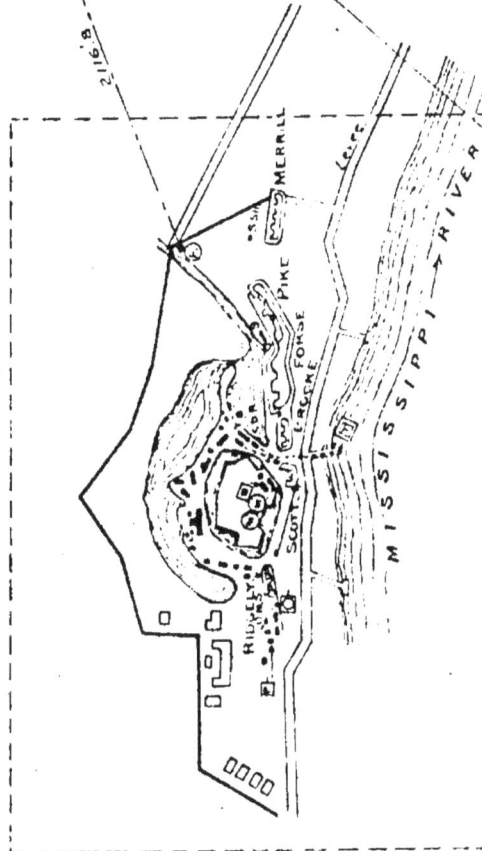

EDITION OF: DEC. 7, 1915.
REVISIONS: JUNE 9, 1916: MAR. 28, 1919.
MAR. 23, 1921.

SERIAL NUMBER

DEFENSES OF NEW ORLEANS

FORT ST. PHILIP

LOUISIANA.
GENERAL MAP.

BATTERIES

PIKE
FORGE
MERRILL... 2 - 6" PED.
RIDGELY
SCOTT
BROOKE

FOR STRUCTURES WITHIN THIS AREA
SEE PLAN FORT ST. PHILIP D-1
REVISION OF MARCH 23, 1921.

Scale of Feet.

500   0   500   1000   1500   2000   2500   3000

MISSISSIPPI RIVER

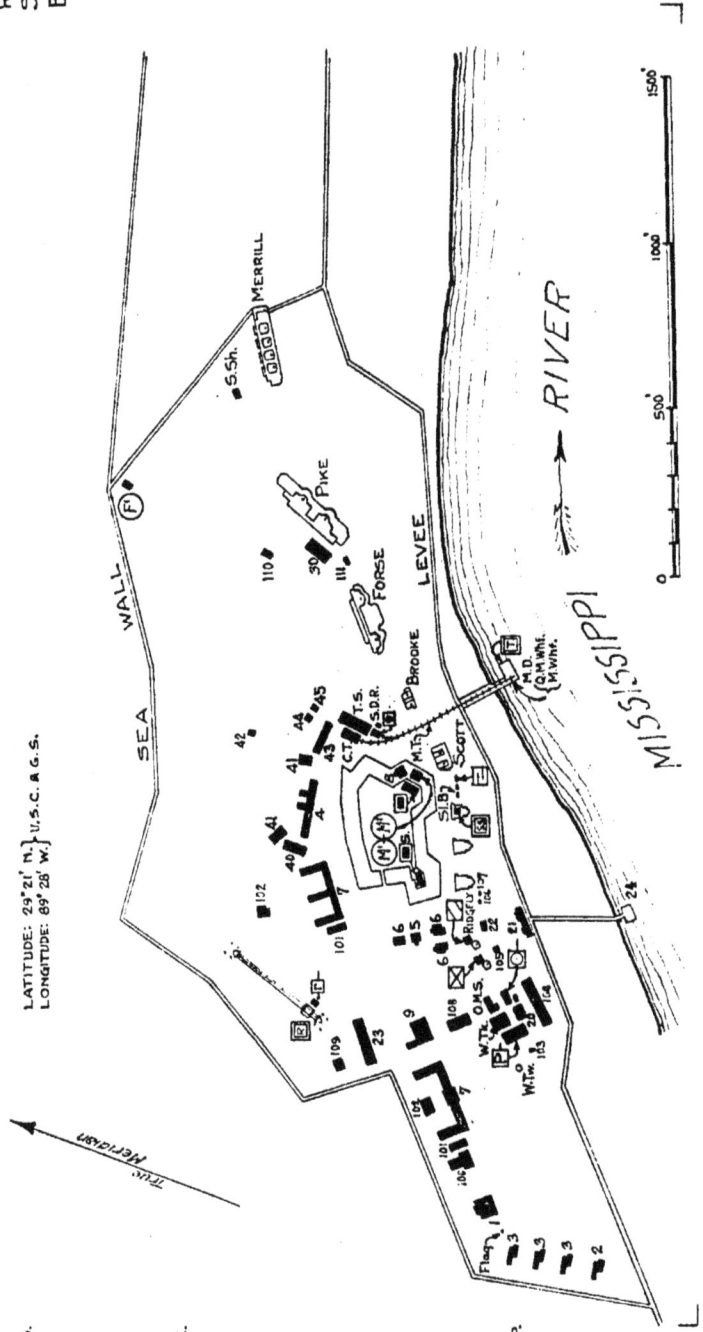

DEFENSES OF NEW ORLEANS

FORT ST. PHILIP-DI.

LOUISIANA

SERIAL NUMBER 1-24

CONFIDENTIAL.

LATITUDE: 29°21' N.
LONGITUDE: 89°28' W. } U.S.C. & G.S.

EDITION OF MARCH 26,1919
REVISIONS: MAR. 23,1921.

TRUE MERIDIAN

SEA WALL

LEVEE

MISSISSIPPI RIVER

0　　500　　1000　　1500

BATTERIES

PIKE -----
FORSE -----
MERRILL --- 2-6" PED.
RIDGELY ---
SCOTT -----
BROOKE -----

LEGEND

1 ADMINISTRATION BUILDING.
2 COMMANDING OFFICER'S QRS.
3 OFFICERS QUARTERS.
4 HOSPITAL.
5 HOSPITAL STEWARD'S QRS.
6 NON-COM. OFFICERS QRS.
7 BARRACKS.
8 GUARD HOUSE.
9 POST EXCHANGE.

100 MESS HALL.
101 DAY ROOM.
102 LAVATORY.
103 SCALE.
104 ARTILLERY STORE HSE.
105 FIRE PATROL.
106 TOOL HOUSE.
107 PAINT HOUSE.
108 RECREATION HALL.
109 BAKERY.
110 STABLE.
111 OIL HOUSE.

20 COAL BIN.
21 Q.M. MESS & QRS.
22 Q.M. CARPENTER SHOP.
23 Q.M. STORE HOUSE.
24 Q.M. BOATHOUSE.

30 ORDNANCE STORE HSE.

40 ENGR. OFFICE.
41 ENGR. STORE HOUSE.
42 ENGR. OIL HOUSE.
43 ENGR. MESS HALL.
44 ENGR. CARP. SHOP.
45 ENGR. BLACKSMITH SHOP.

the river approach like most of the other batteries, to the southeast. Detailed plans were finished on August 10, 1897. Building required destruction of several old 1870s platforms and magazines of the old water battery. It had to be built on sheet pilings for stability in the loose silt foundation. The plan was fairly conventional; gun separation was 129-feet. It had a single magazine complex in the center traverse, shell service was by hoists. Both platforms were standard internal plans. Concrete work was done from December 27. 1897 to May 1898, the battery being practically complete and the guns mounted by June 30, 1898. Transfer was made on October 25, 1899 for a total cost of $100,000. The battery was named in General Orders No. 78 of May 25, 1903 for Major Albert G. Forse, 1st U.S. Cavalry, killed in 1898 during the Spanish American War. It was armed with two 8-inch Model 1888MII guns on disappearing carriages Model LF 1896 (Bethlehem tube #18/carriage #6 and Watervliet tube #37/carriage #7). Hoists were added for transfer in May of 1898 and the platforms widened as early as December 1903. On August 24, 1917, Battery Forse was listed for armament removal, accomplished in early 1918. It was considered officially eliminated on October 8, 1918, though the disappearing carriages remained in place until scrapped in 1920. The site was subsequently sold into private ownership, but the emplacement still exists though partially flooded. The battery is open, but the fort is closed to the public.

- One of the Spanish American War emergency batteries emplaced in the south. This was an emplacement intended for two modern 8-inch guns on reinforced older 15-inch Rodman traversing carriages. An excess of finished tubes for this type of carriage as an expediency for emergency defenses during this conflict. Twenty-one such gun combination in new emplacements were authorized; this was an emplacement for two such guns. Orders to prepare a plan were given on April 27, 1898. Plans were submitted on May 31st. In an attempt to accelerate service availability and reduce cost, it was decided to convert two existing platforms from the 1870s water battery that ran between the fort and the river into this new emplacement. Two emplacements from this older work, specifically platforms No. 52 and No. 53, with a traverse magazine between were selected. This was located just to the southwest of the 8-inch disappearing gun battery already under construction. Two 8-inch Model 1888MII guns (Bethlehem #8 and #9) were shipped to Fort Jackson on July 23, 1898 and then transported across the river to Fort St. Philip shortly thereafter. The battery served only a short time, the tubes being dismounted and sent back to Fort Jackson on June 17, 1899 for use in Battery Ransom. Parts of the platform still exist on the private property with access being difficult due to soft soils and only through the use of a boat. The battery is open, but the fort is closed to the public.

- **MERRILL**: A battery for four 6-inch, rapid-fire pedestal guns emplaced at Fort St. Philip. The site chosen was on the far eastern portion of the reservation, requiring purchase of the adjacent property. It was sited to fire south, southeast. Plans were submitted on May 16, 1901 using funds from the Act of March 1, 1901. It was placed on piles connected to grillage under the foundation. It was designed for a single gun per platform, with 43.5-foot gun center spacing. Shell magazines were on the lower left of each platform, with ammunition service using hand-operated hoists. Work started on June 10, 1901 and was finished by September 1, 1902. Transfer was made on November 21, 1907 for a total cost of $95,052.15. Repairs made from 1907-1910 added another $2276 to expenditures. It was named in General Orders No. 78 of May 25, 1903 for Captain Moses E. Merrill, 5th U.S. Infantry, who was killed in 1847 during the Mexican War. Originally the battery received Model 1900 6-inch gun tubes (#29, #30, #33, and #34) on Model 1900 pedestal mounts (#32, #37, #38, and #39) in 1904. By authority of a letter of February 24, 1905 the tubes were returned to Watervliet Arsenal for modifications. Shortly thereafter a new set was returned to Merrill, arriving in mid-1906 (this time #6, #35, #39 and #41). New hoists were added in 1914. Emplacements No. 3 and No.

4 were disarmed in May of 1918, the guns being relocated to temporary batteries at the Delaware Capes. The last two emplacements were finally disarmed in 1920. A memorandum for reservation closing was issued on March 11, 1920. The emplacement, now in private hands and difficult to access (requiring watercraft), still exists. The battery is open, but the fort is closed to the public.

- **RIDGELY**: A battery for two recently purchased Armstrong 4.7-inch guns provided to the fort at the start of the Spanish American War. It was located south of the masonry work, firing to the southeast. The submission plan of April 14, 1898 advocated using existing 1870s platforms as foundation, and old Rodman magazines re-purposed for the 4.7-inch ammunition. Work was initially done with NDA funds from May 23 to October 1898. Transfer was done on October 25, 1899 for a cost to date of only $7,000. It was armed with two 4.7-inch Armstrong guns on pedestal mounts (#11009/#11019 and #11933/#11021). It was named in General Orders No. 78 of May 25, 1903 for 1st Lieutenant Henderson Ridgely, who was killed in the Mexican War in 1847. Plans for a permanent emplacement were approved in 1903 to use new concrete magazines, one each behind and to the left flank of each platform. On account of ground moisture, they were placed 6-feet above the floors of the old magazines and then given extensive overhead earth cover. Work was submitted on February 9, 1903 for an estimated cost of $11,100. The armament served only a short while, being removed on July 24, 1913 and sent to Hawaii. The battery itself was subsequently abandoned but was used for a post telephone and protected switchboard room. It is currently on private property at the old fort reservation and is difficult to access but still exists. The battery is open, but the fort is closed to the public.

- **SCOTT**: An emplacement for two 3-inch, 15-pounder guns submitted (simultaneously to sister-battery Millar at Fort Jackson) on December 9, 1898. It was placed south of the old fort, firing to the southeast. The plan followed type mimeographs, with 29-foot gun centers, two rounded flank platforms, and lower-level magazine on the left flank of each platform. Work was done grom December 28, 1898 to July 1, 1899. Transfer was made on January 17, 1901 for a construction cost of $24,278.67. It was named on General Orders No. 78 of May 25, 1903 for 1st Lieutenant Richard W. Scott, 7th U.S. Infantry who was killed in 1817 during the Seminole Wars. It carried two 3-inch Model 1898 Driggs-Seabury guns and masking parapet mounts (#57/#57 and #58/#58). These were modified in the 1914-1916 period to the M1898M1 pedestal configuration. The battery served until removed in 1920 with the obsolescence of this type of gun. The emplacement still exists on private property but difficult to access. The battery is open, but the fort is closed to the public.

- **BROOKE**: An emplacement for two 3-inch, 15-pounder masking parapet guns emplaced along the rail to the mine wharf on the eastern side of the old fort. Plans were submitted on May 25, 1900, details coming the following month. Construction demanded the destruction of old Rodman-type emplacements from the 1870s water battery. The plan was of conventional type design, with guns at a 29-foot spacing, and platforms square-shaped as internal emplacements. Each platform had a single magazine on its lower left flank, ammunition service was by hand. Work began on June 25, 1900 and continued through September into the fall. Transfer was then made on February 23, 1904 for a cost of $9,816.81. It was armed with two 3-inch Model 1898 Driggs-Seabury guns on balanced pillar mounts (#76/#76 and #77/#77). The battery had been named on General Orders No. 78 of May 25, 1903 for 1st Lieutenant Francis J. Brooke, 6th US Infantry killed in action against the Seminole Indians on December 25, 1837. The armament was modified about 1916 to the M1898M1 pedestal standard. It served until being removed in mid-1920 with the obsolescence of this model of weapon. The emplacement still exists on private property but difficult to access. The battery is open, but the fort is closed to the public.

Fort Jackson (1832-1924) is located on west bank of the Mississippi River at Plaquemines Bend, about sixty miles downriver from New Orleans, Louisiana. The Third System fort was built from 1822 to 1832. In 1826 it was officially named for Gen. Andrew Jackson, the eighth President of the United States in General Order No.108. The five-sided masonry fort served as the primary defense of the Mississippi River, along with its companion defensive work, Fort St. Philip, located on the east bank. Fort Jackson was occupied by Confederate forces at the beginning of the Civil War and only fell to Union forces after the Union Navy forced its way by the forts and captured New Orleans. Placed on caretaker status after the Civil War, Fort Jackson was used during the Endicott Program for two concrete batteries. Installed inside the walls of the old fort was the fort's primary battery of two 10-inch guns on disappearing carriages. Directly outside the old fort's moat a 3-inch rapid fire battery was constructed. The fort was used as training center during World War I, but in 1924 the fort was declared surplus. Fort Jackson was sold to private interests in 1927. The fort was donated to the Plaquemines Parish in 1959 and restored as a parish park. The Fort Jackson Museum and Welcome Center was opened in 2010 about ½ mile north of the historic fort. The fort was damaged by Hurricane Katrina (2005) and remains closed to the public since that time.

Fort Jackson Gun Batteries

- **RANSOM:** An Endicott battery for two 8-inch disappearing guns at Fort Jackson on the banks of the Mississippi River. It was emplaced within the structure of the old fort, as a self-standing emplacement within the parade ground. It faced the northeast curtain wall, and fired to the north, northeast. Plans were submitted on April 28, 1898. The emplacement was of conventional design, following type mimeographs. It had two separate platforms, with gun centers spaced at 122-feet. Both were flank emplacements, with magazines shared together under the center traverse and served with shell and shot hoists. Work was done from April of 1898 to June of 1899, Transfer was made on October 25, 1899 for a cost of $150,000. Additional work was authorized on December 26, 1899 to remove some of the old fort to clear fields of fire and involved removing thirteen old guns of the former parapet level. It was named in General Orders No. 78 of May 25, 1903 for Colonel Truman B. Ransom, 9th US Infantry who was killed in 1847 at the Battle of Chapultepec. It was armed with two 8-inch Model 1888MII guns (Bethlehem #8 and #9 previously at the Fort St. Philip temporary battery just across the river) on Model LF 1896 disappearing carriages (#11 and #9 respectively). The battery served until listed for removal on August 24, 1917. After the guns were removed, the carriages were scrapped in place. The emplacement still exists at the Fort Jackson Plaquemines Parish Park. Access to the battery is limited due to the closure of the 3rd System fort since Hurricane Katrinia (2005). The battery is open, but the fort is closed to the public.

- **MILLAR:** A battery for two 3-inch, 15-pounder rapid-fire guns on masking parapet pillars mounted at Fort Jackson. It was emplaced to the west of the old fort, close to the levee along the river, firing to the north to cover the mine fields. Plans were submitted on December 9th, and $24,500 was allocated on December 20, 1898, the plan followed mimeograph designs, with two platforms and gun spacing of 29-feet. The No. 1 emplacement was of internal layout, No. 2 was a flank position. Work was rapidly accomplished. Transfer was made on January 17, 1901 for a completion cost of $22,867.73. It was named in General Orders No. 78 of May 25, 1903 for Captain James F. Millar, 14th US Infantry, who was killed in 1877 in action against the Apache Indians in Arizona. The battery was armed with two 3-inch Model 1898 Driggs-Seabury guns on balanced pillar mounts (#59/#59 and #61/#61). The mounts and emplacement were modified by 1916 to the 1898M1 fixed pedestal configuration. The battery was disarmed in July 1921 but still exists at the Fort Jackson Plaquemines Parish Park with public access as Mississippi River viewing platform. The battery is open to the public.

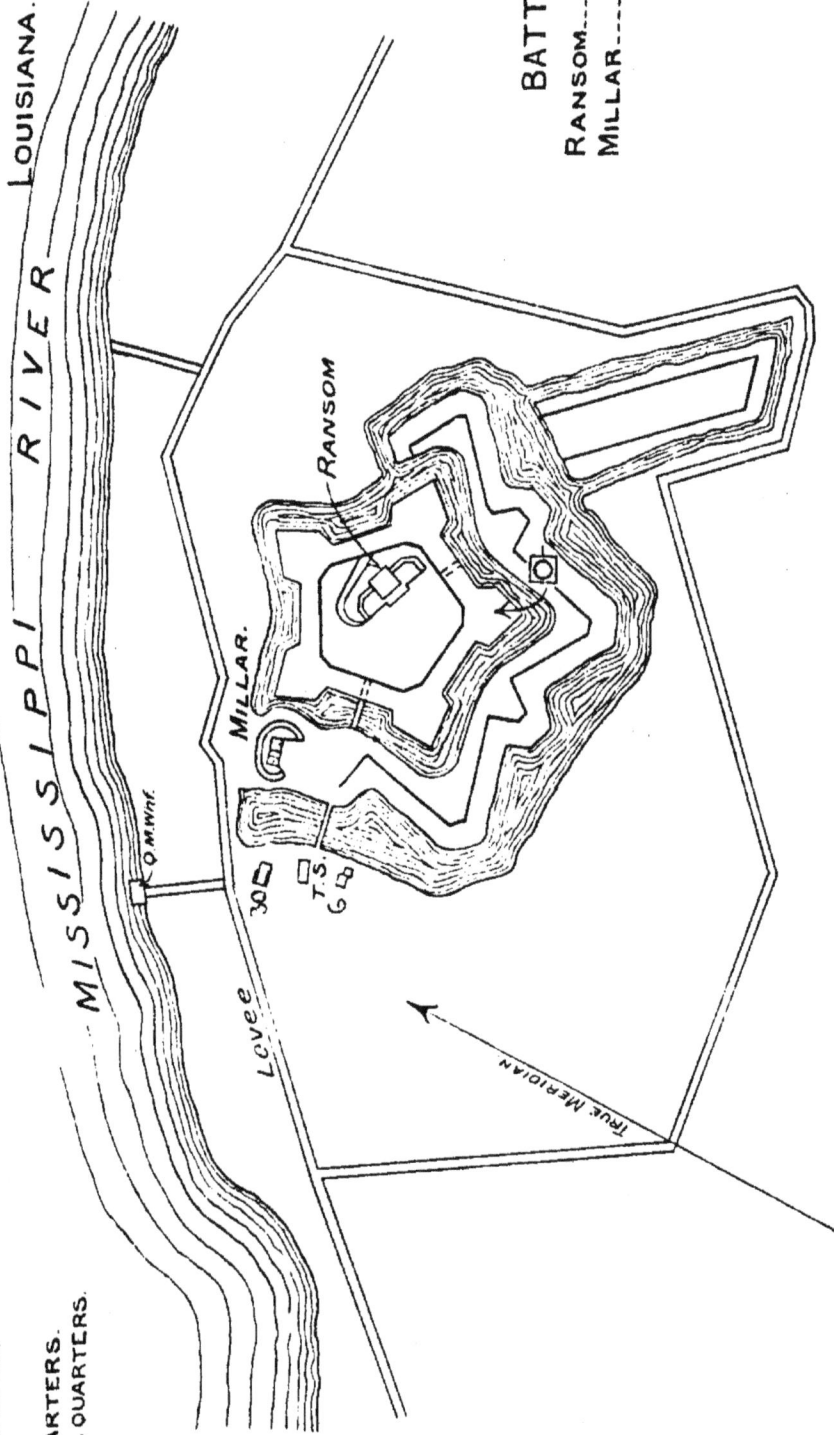

SERIAL NUMBER 124

DEFENSES OF NEW ORLEANS
FORT JACKSON
LOUISIANA.

MISSISSIPPI   RIVER

Levee

O.M.Wnf.

RANSOM

MILLAR.

30
T.S.
6.B.

TRUE MERIDIAN

EDITION OF DEC. 7, 1915;
REVISIONS: JUNE 9, 1916; MAR. 28, 1919.
MAR. 23, 1921.

LEGEND.

6. N.C.O. QUARTERS.
30. ORD. SGT. QUARTERS.

BATTERIES.
RANSOM-------
MILLAR-------

0   500   1000   1500 Ft.

Fort St. Phillip (Glen Williford)

Fort Jackson Parish Park (Terry McGovern)

Jackson Barracks (1834-current) was the main garrison reservation for the troops stationed at the Mississippi River defenses. Caretakers looked after the two forts downriver and soldiers were shuttled down river and back for drills.

Sabine Pass Military Reservation (1898-1945) is located at the Sabine Pass which is the natural outlet of Sabine Lake into the Gulf of Mexico. It borders Jefferson County, Texas, and Cameron Parish, Louisiana. Located on Texas Point were three unnamed coastal defense batteries from 1898-1899 (one M1888 8-inch BL gun on modified 15-inch Rodman carriage, two 5-inch guns, and two 7-inch guns) which are all destroyed or buried. During World War I, a two-gun 5-inch gun battery was installed and removed in 1918. During World War II (1942-1944), a two-gun 155mm battery on Panama mounts, which replaced a temporary field-positioned 105mm howitzer battery. A Harbor Entrance Control Post (HECP) was built at the U.S. Navy Section near the present-day Coast Guard Station to control ship traffic into the Pass. Other 155mm guns were emplaced at Calcasieu Pass and Freeport, LA, and Port Aransas, TX. The two Panama mounts and four concrete ammunition magazines from World War II still exist. The magazines are located at the Sabine Pass Battleground State Park.

Sabine Pass Gun Battery

- The Sabine River was given temporary defenses with funding from the March 1898 National Defense Act at the start of the Spanish American War. These were composed of an earthen battery for four siege guns and a semi-permanent battery for one modern 8-inch gun on strengthened 15-inch Rodman carriage. There were twenty-one such gun combinations made available at the start of the war for such emergency emplacements. Permission to use the land was secured (only lease or permission was necessary for these emplacements, not outright purchase). The 8-inch battery consisted of a simple concrete platform with traversing rings for the carriage and an adjacent magazine. Work was done in the summer of 1898 with a $6000 engineer allocation, and it was reported ready for arming on August 23, 1898. One Model 1888 8-inch tube (West Point Foundry #6) was shipped here on July 2, and the carriage transferred from New Orleans' Fort St. Philip. Apparently serving for just a short while, the 8-inch gun was authorized for removal in late 1899. The 8-inch tube was shipped for use at Ft. Ward, Washington, on November 11, 1899. The battery site was then abandoned. Subsequently, on August 28, 1913 an explosion of powder occurred in one of the magazines and a young boy was killed; the emplacement was also damaged. According to relatively recent magazine article, there are no remains left of the original emplacement.

Panama mount at Sabine Pass (Terry McGovern)

Smith, Bolling W. "Fire Control Puzzle at New Orleans." *Coast Defense Journal*, Vol. 23, No. 3, 2009, p. 4.

DEFENSES OF NEW ORLEANS

JACKSON BARRACKS

New Orleans, La.

Scale of feet

100' 0 100' 200' 300' 400' 500' 600' 700' 800'

SERIAL NUMBER

EDITION OF MARCH 26, 1919.

N

BURGUNDY

DELERY ST.

DAUPHINE

ROYAL

CHARTRES

DOUGLAS

BIENVENUE

PONTALBA

N. PETERS

STREET RAILWAY

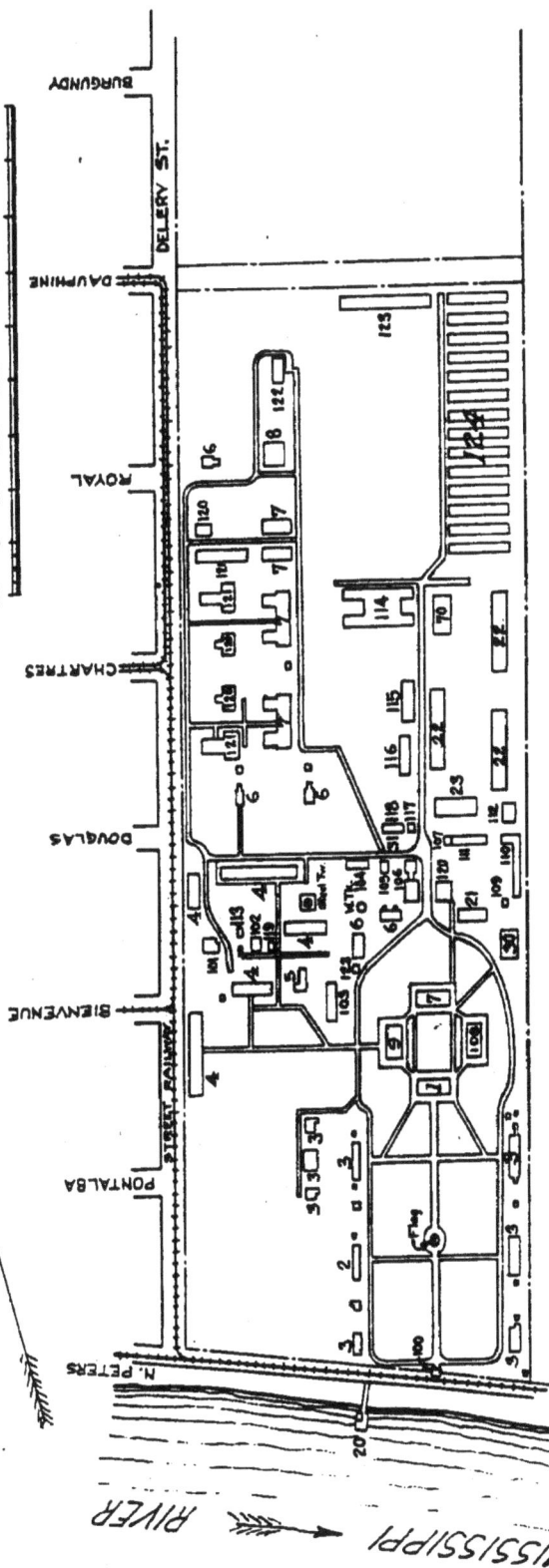

LATITUDE: 29°57' N. } U.S.C.&G.S.
LONGITUDE: 90° W.

MISSISSIPPI RIVER

LEGEND

1 ADMINISTRATION BUILDING.
2 COMMANDING OFF. QRS.
3 OFFICER'S QUARTERS.
4 HOSPITAL.
5 HOSPITAL STEWARD'S QRS.
6 NON-COM. OFFICERS QRS.
7 BARRACKS.
8 GUARD HOUSE.
9 POST EXCHANGE.

100 GUARD STATION.
101 QUARTERS.
102 RECRUIT ROOM.
103 NURSE'S BUILDING.
104 COAL SHED.
105 ELEC. SUB. STATION.
106 FINANCE OFFICE.
107 OLD BAKERY.
108 G.H. & BAND ROOM.
109 GARAGE.
110 STABLES.
111 WAGON SHED.
112 CORRAL MASTER'S QRS.
113 DEAD HOUSE.
114 RECRUIT MESS HALL.
115 CARPENTER SHOP.
116 REGIMENTAL ST. HO.
117 OIL HOUSE.
118 GARAGE.
119 DETENTION WARD.
120 LAVATORY.
121 ___ESS HALL.
1__ ___KERY.
125 RECRUITS LAVATORY.
124 CANTONMENTS.

20 Q. M. WHARF.
21 Q. M. OFFICE.
22 Q.M. STORE HOUSE.
23 COMMISSARY ST. HO.

30 ORDNANCE ST. HO.
31 GUN SHED.

70 Y. M. C. A. BUILDING.

The Harbor Defenses of Galveston – Texas

Galveston is located on a barrier island at the entrance to Galveston Bay one of the largest estuaries along the Gulf coast of Texas. Initially settled by pirates in years following the War of 1812, the area was cleared by the U.S. Navy in 1820s, and a port of entry was established by the Spanish in 1825. A key city and port during the years of the Texas Republic and during the Civil War, the bay was the center of economic growth for the State of Texas. By 1890 it was an important commercial port and received nine new concrete batteries at three locations after 1898. The hurricane of 1900 destroyed or damaged all of the new defenses, and they had to be rebuilt, along with new seawalls completed in 1905 and 1912. Two new long-range 12-inch batteries were added in the 1920s, and five new batteries were added during the World War II years. The defenses were closed in 1947 and properties dispersed to various public and private entities.

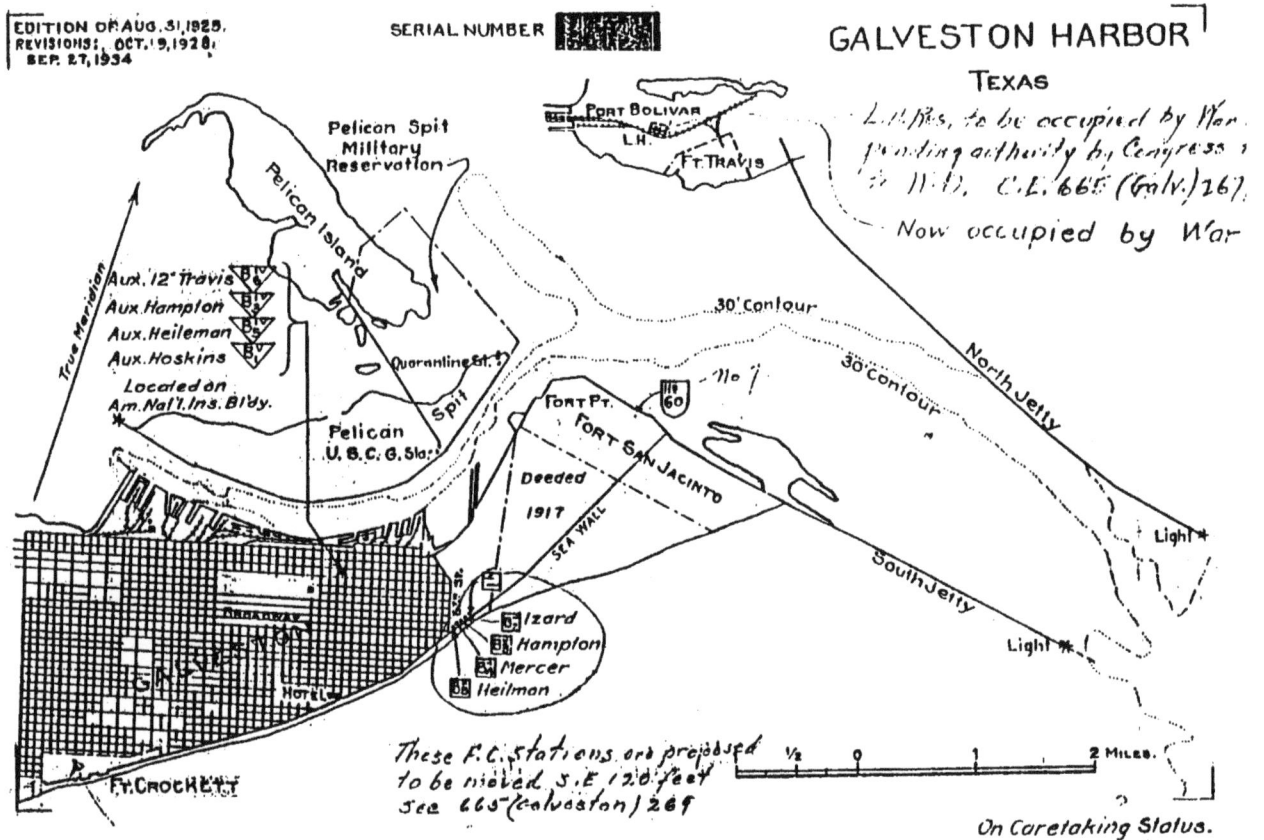

Fort Travis (1898-1947) is located on Bolivar Point at the entrance to Galveston Bay, just south of Port Bolivar, Texas. Site of Fort Greene, a Confederate earthwork during the Civil War, Fort Travis became the site for two Endicott Program batteries in 1900. It was named in General Order 43 of 1900 for William B. Travis, commander of the forces at the Alamo who was killed with his command on Mar. 6, 1836. Battery Kimble, a battery of two 12-inch guns on long-range barbette carriages was also built there between 1917 and 1922. During the Word War II, a 6-inch shielded battery was built there, and Battery Kimble was disarmed, and its guns were moved to Fort Moultrie, South Carolina to become Battery #520. After World War II, the fort was declared surplus and is currently a Galveston County Park, Fort Travis Seashore Park. Battery Kimble, as it is one of the three long-range 12-inch batteries in the US that was not casemated during World War II. The former reservation has been cleared of all shrubs and trees. The county park is open during daylight hours. All the batteries are surviving and of special interest is Battery Kimble, as it is one of the few long-range 12-inch batteries that was not casemate during World War II.

Fort Travis Gun Batteries

- **DAVIS:** An Endicott battery for two 8-inch disappearing guns emplaced at the Bolivar Point reservation of Fort Davis. Its plans were submitted on March 31, 1898 using authorization from the recently passed National Defense Act. It was located on the central portion of the fort, along the shoreline firing to the southeast. The plan was fairly conventional in design, with guns at 130-foot centers spacing. It had a large magazine under the central traverse with access to both platforms and hoists for ammunition service. Piles had to be used to support the foundation. Work for many years had been delayed for lack of a completed road into the portion of the peninsula. Work was done during 1898; the platforms being reported ready by June 30th. The guns were delivered for mounting by late 1899. It carried two 8-inch Model M1888MII Bethlehem guns on Model LF 1896 disappearing carriages (#3/#20 and #16/#19). Repairs were needed almost immediately due to cracking and settling, but it was ready to be turned over on October 25, 1899 for $111,000. The hurricane of September 1900 damaged the emplacement badly, causing the No. 2 platform to collapse. Repairs were made and the battery was eventually re-transferred on August 21, 1911 for a new cost of $100,829.68. It was named in General Orders No. 194 of December 27, 1904 for 2nd Lieutenant Thomas Davis mortally wounded at Cerro Gordo Mexico in 1847. It served as one of the primary defensive works for Galveston, until the guns were removed in 1918 for railway mounts. The emplacement, though somewhat damaged, still exists along the current seawall line at the Fort Travis Seashore Park. The battery site is open, but the interior is closed to the public.

- **ERNST:** A battery for three 3-inch masking parapet guns emplaced on the western end of the reservation near the shore at the Fort Travis reservation. It fired to the southwest. Plans were submitted on December 24, 1898, the same date as the other two 3-inch batteries in the defenses. Funds came from the July 7, 1898 Act. The plan had two internal platforms and one flank position on the right side (No. 1 emplacement). Also, the left-side end wall was extended to the rear to provide better protection from enfilade fire. There were two shared magazines, one on the lower level between each two-gun positions. Work began on January 28, 1899 and the structural part was completed by June 30th. It was turned over on March 31, 1900 for a cost of $28,287.64. The hurricane of September 1900 damaged the battery and interrupted the delivery of the prospective armament. The battery was repaired over the next couple of years, but it was rebuilt for pedestal mounts. It was eventually transferred on August 21, 1911 for a cost of $37,709.14. It had been named in General Orders No. 194 of December 27, 1904 for 2nd Lieutenant Rudolph F. Ernst, 6th U.S. Infantry who was killed at Molino del Rey Mexico in 1847. It was finally armed with three 3-inch Model 1903 guns on Model 1903 pedestal mounts (#78/#50, #79/#51, and #73/#49). The battery continued to serve for a considerable period of time, not being deleted until after World War II in 1946. The emplacement still exists, and forms part of the seawall structure at the Fort Travis Seashore Park. The battery site is open to the public.

- **KIMBLE:** A 1915 Program battery for two 12-inch, long-range barbette guns emplaced on the north side of the harbor entry at Fort Travis. During much of planning in 1915-1916 it was listed as just a single-gun emplacement, but a standard dual type was finally adopted before construction began. It was submitted on March 25, 1917, at the same time with the same authority as sister Battery Hoskins at Fort Galveston. It was placed at the northwest corner of the reservation, firing to the southeast. It was strictly of standard dual gun type, with an open back to the central traverse and completely exposed gun platforms on either side. Work was done from August 1917 to April of 1921 for transfer on May 12, 1922 at a cost of $310,237.63. When completed it was armed with-

GALVESTON HARBOR
FORT TRAVIS
BOLIVAR POINT.

SERIAL NUMBER 124

EDITION OF DEC. 7, 1915;
REVISIONS: MAR.14,1916; MAY 13, 1919.
APR. 5,1921; DEC. 30,1921.

Retaining Wall

U.S. ENGR.
RESERVATION

40ᵐ

2-12"G.

SEA WALL

Retaining Wall

CRF

RUDOLPH ERNST.

11

10   7

8

3

6

SEA WALL

THOMAS DAVIS.   B4

B5

TRUE MERIDIAN

ABANDONED

ENGR. Whf.

0   500'   1000'   2000'   3000'

LEGEND
3. OFFICERS' QUARTERS.
6. N.C. OFFICERS' QUARTERS.
7. BARRACKS.
10. MESS HALL.
11. BATH HOUSE.
12. BARBER SHOP.
40. ENGR. OFFICE.

BATTERIES.
THOMAS DAVIS:
RUDOLPH ERNST 3-3"Pca.
2-12" B.

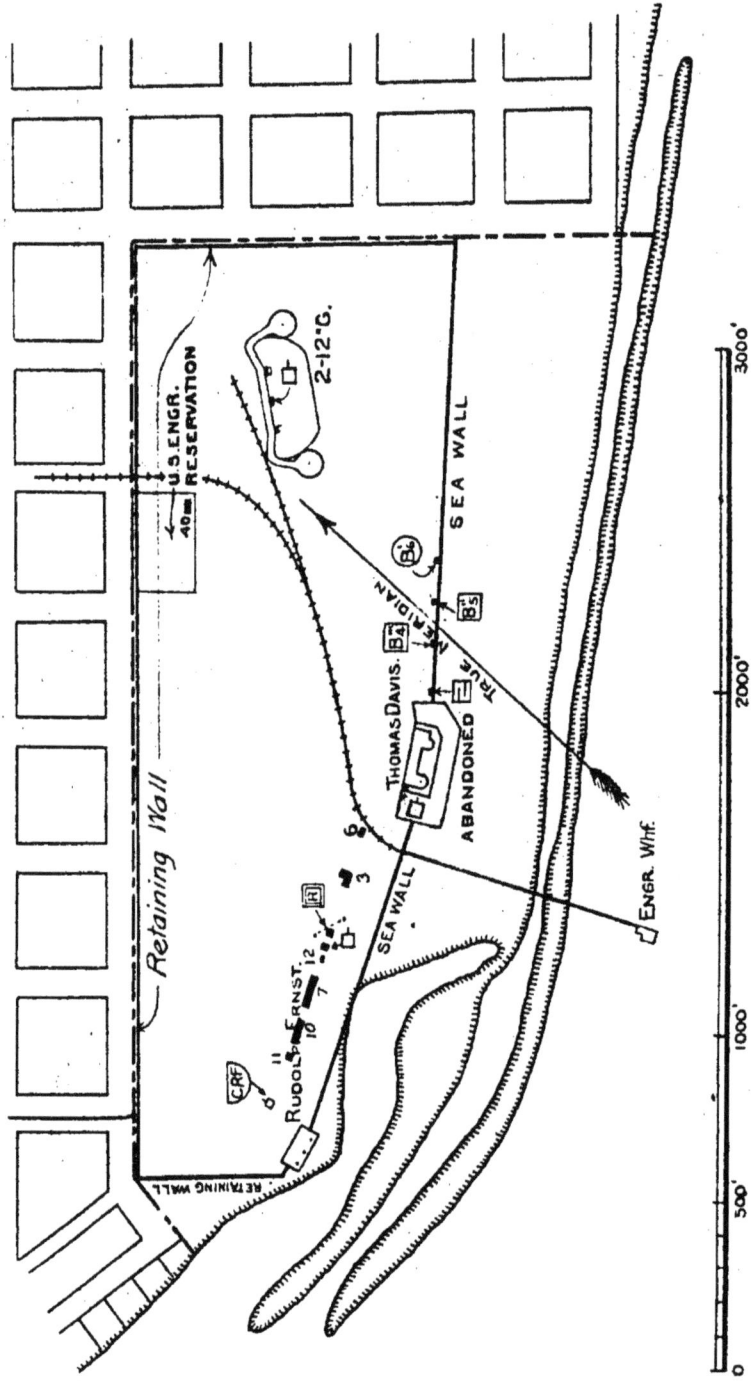

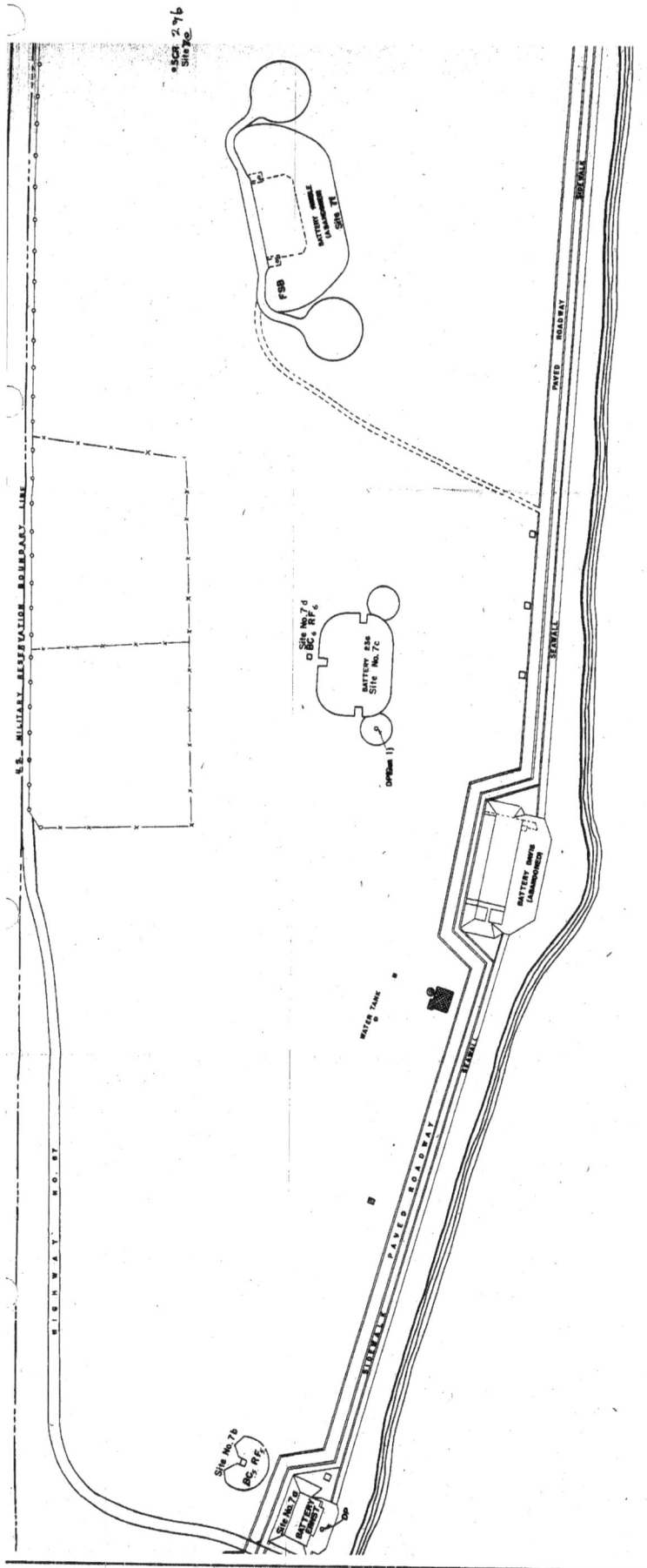

HARBOR DEFENSES
OF
GALVESTON

LOCATION FIRE CONTROL
INSTALLATIONS SITE NO.7

| Prepared | By | Exhibit No. 198 |
| Office of the Art'y Engineer | | 11th January 1945 |

SCR 296
Site 7e

BATTERY 236
Site 7c

Site No.7d
BC₄ RF₆

BATTERY (A.A.MUNITION)
Site 7f

FSB

U.S. MILITARY RESERVATION BOUNDARY LINE

HIGHWAY NO. 87

WATER TANK

PAVED ROADWAY

SEAWALL

PAVED ROADWAY

BATTERY DAVIS (A.A.MUNITION)

Site No.7b
BC₅ RF₅

BATTERY SMITH
Site No.7a

two 12-inch Model 1895M1A4 guns on Model 1917 barbette carriages (#48/#21 and #50/#22). It was named in General Orders No. 13 of March 27, 1922 for Major Edwin R. Kimble, Corps of Engineers who died in 1918. Battery Kimble was one of only three 12-inch long-range batteries not given overhead protection during World War II. The armament was removed in early 1943 and used to arm Battery #520 completing at the Marshall Reservation in Charleston, South Carolina. The emplacement still exists in good condition at the Fort Travis Seashore Park. The battery site is open, but the interior is closed to the public.

- **Battery #236**: A 1940 Program dual 6-inch barbette battery assigned to Fort Travis. It was emplaced to the southwest of Kimble, firing to the southeast. In plan it followed the standard layout and design for the 200-series batteries. It was not initially assigned a high priority and not funded until the FY-1943 Budget. Work was done from October 17, 1942 until July 31, 1943. Transfer was made on August 9, 1945 at a total cost of $305,310 completed except for the gun tubes. It was to be armed with two new 6-inch M1(T2) guns on Model M4 barbette carriages. The gun tubes were never received, though the M4 carriages #47 and #48 were apparently delivered to the site. The battery was not named, just being known as Battery Construction No. 236 during construction. The battery was structurally complete but was abandoned in 1946. The emplacement still exists at the Fort Travis Seashore Park. The battery site is open, but the interior is closed to the public.

Fort Travis County Park (Terry McGovern)

Fort San Jacinto (1898-1947) is located on Fort Point at the entrance to Galveston Bay, just east of Galveston, Texas. Site of a Confederate fort during the Civil War, Fort San Jacinto became the site for four Endicott Program batteries and a controlled submarine mine facility in 1898. It was named in General Order 78 of 1899 to commemorate the Battle of San Jacinto which occurred on Apr. 21, 1836. In 1900, the fort experienced the impact of a massive hurricane that basically destroyed its concrete batteries. These batteries were restored by 1911, and a massive seawall was completed by 1925. The 1940 Program added a 6-inch shielded battery and an AMTB battery. After World War II, the fort was deactivated and became a spoil dumping area for material being removed from the main shipping channel by the U.S. Army Corps of Engineers. Batteries Hogan and Heileman were leveled in 1953, while the mortar battery and Battery #235 are now mostly covered by dredge spoil, leaving only Battery Crogan and its mine casemate abandoned but exposed.

Fort San Jacinto Gun Batteries

- **HEILEMAN**: A battery for two 10-inch disappearing guns authorized to be emplaced at the Fort Point reservation of Fort San Jacinto. Original plans were submitted on September 14, 1896, using funding from the Act of June 6, 1896. It was northwest of mortar Battery Mercer, on the central part of the reservation firing to the east. To deal with ships that might run into the bay behind the coast defense forts, the left emplacement was to be designed for all-round fire, using one of just three ARF Model 1896 disappearing carriages produced. However, unlike Battery Mishler at Fort Stevens, the expense of creating a completely enclosed loading platform was avoided. The platform was thus fairly conventional despite the special carriage; while the gun could indeed fire at any angle, it had to be moved back into its forward-facing position to be reloaded. Otherwise, it was a compact design, the guns separated by only 134-feet, and sharing a single magazine in the lower level of the central traverse. Ammunition service was by lifts, later replaced with hoists. Work was done from early 1897 to early 1898. One gun had been mounted in June of 1898, while the ARF carriage was being waited on. It was to be armed with two 10-inch Model 1888 Watervliet guns; one on Model 1896 LF and one Model 1896 ARF carriages (#13/#37 and #37/#1). Just after completion severe damage was done to the emplacement by the hurricane of September 1900. Repairs could not be made, and a new appropriation of $175,000 was approved on March 1, 1902 for dynamiting out the old structure (using the broken concrete for a new riprap on the shore) and completely rebuilding the battery. The ordnance material, including guns and carriages, were saved and re-installed. New railroad tracks to the construction site was completed by May 14, 1902 and work done by the end of 1904. A second transfer was made on August 11, 1911 at a new cost of $181,632.82. It retained the design features of the original structure. It was named in General Orders No. 194 of December 27, 1904 for Brevet Lt. Colonel Julius F. Heileman of War of 1812 and Seminole Indian War service. The rebuilt, rearmed battery was kept throughout the World War 1 and by the 1932 Reviews, not being deleted until authority given in January 1943. The emplacement was subsequently leveled in 1953, no remains exist today.

- **MERCER**: The mortar battery emplaced at Fort San Jacinto. Original plans were submitted on November 6, 1896, revised plans (among other things now being an electric plant) were submitted on April 7, 1897. Funding came from the Act of June 6, 1896. It was located on the central part of the reservation, north of the initial reservation line firing to the east. As originally built it had small pits with magazines under the forward parapets. It was finished and transferred on October 25, 1899 for a cost of $117,700. It was severely damaged and undermined by the Galveston hurricane of September 1900, the walls and platforms being destroyed. It could not be repaired and had to be destroyed and replaced. Under the emergency authorization of March 1, 1902, $290,000

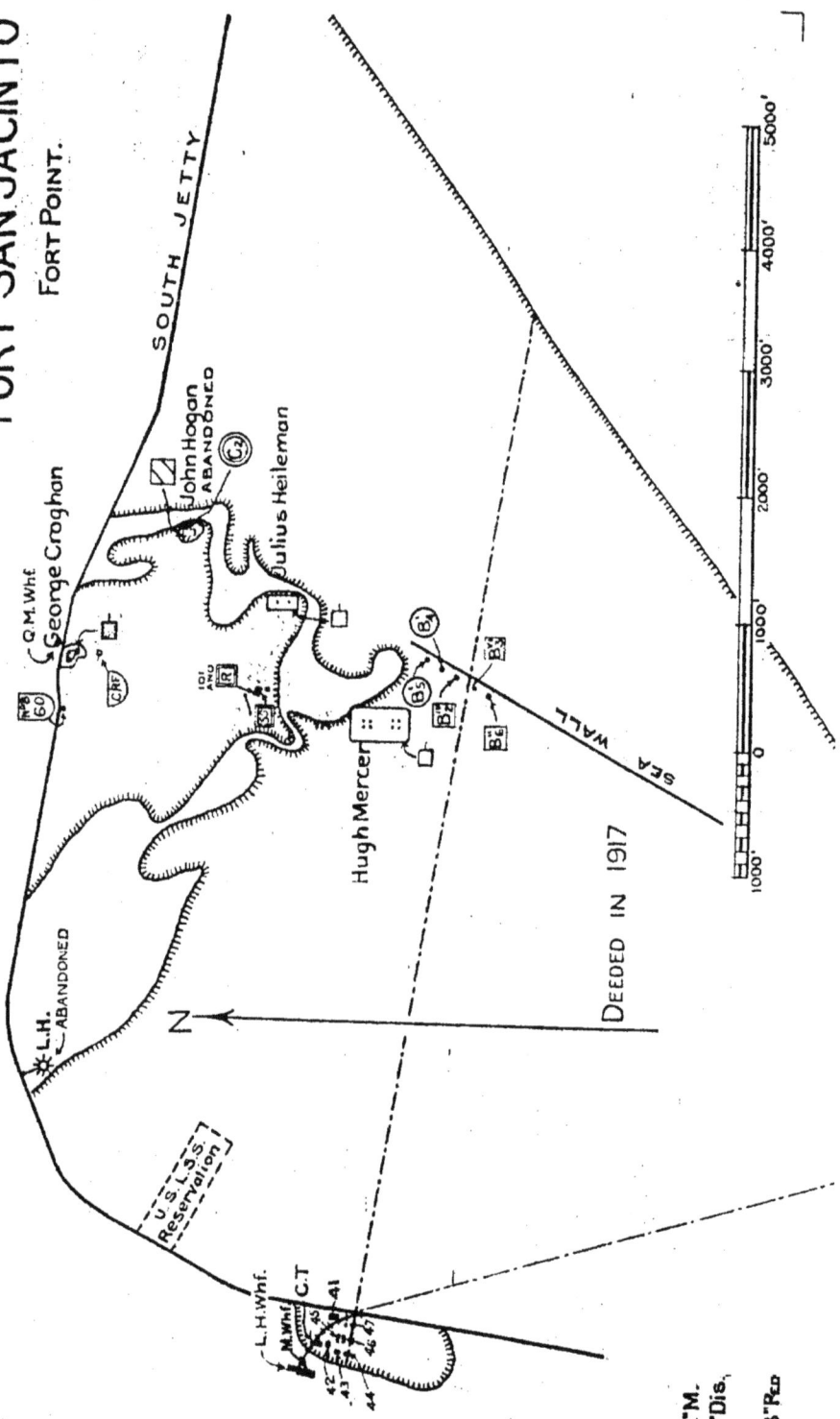

GALVESTON HARBOR

FORT SAN JACINTO

FORT POINT.

SERIAL NUMBER **124**

EDITION OF: DEC. 7, 1915;
REVISIONS: MAR. 14, 1916: MAY 13, 1919.
APR. 5, 1921. DEC. 30, 1921.

SOUTH JETTY

SEA WALL

DEEDED IN 1917

U.S.L.S.S. Reservation

L.H. ABANDONED

John Hogan ABANDONED

George Croghan

Q.M. Whf.

Julius Heileman

Hugh Mercer

L.H. Whf.

N

1000'   0   1000   2000   3000   4000   5000'

LEGEND.

H ENGINEER STOREHOUSE.

I 2 OFFICE (ENG'R)
3 BLACKSMITH SHOP (ENG'R)
4 CRAB HOUSE
5 CARPENTER SHOP
6 MACHINE
7 PAINT. ST. HOUSE
21 COMPANY QRS.

BATTERIES.

HUGH MERCER....8-12"M.
JULIUS HEILEMAN..2-10"Dis.

GEORGE CROGHAN 2-3"R.F.

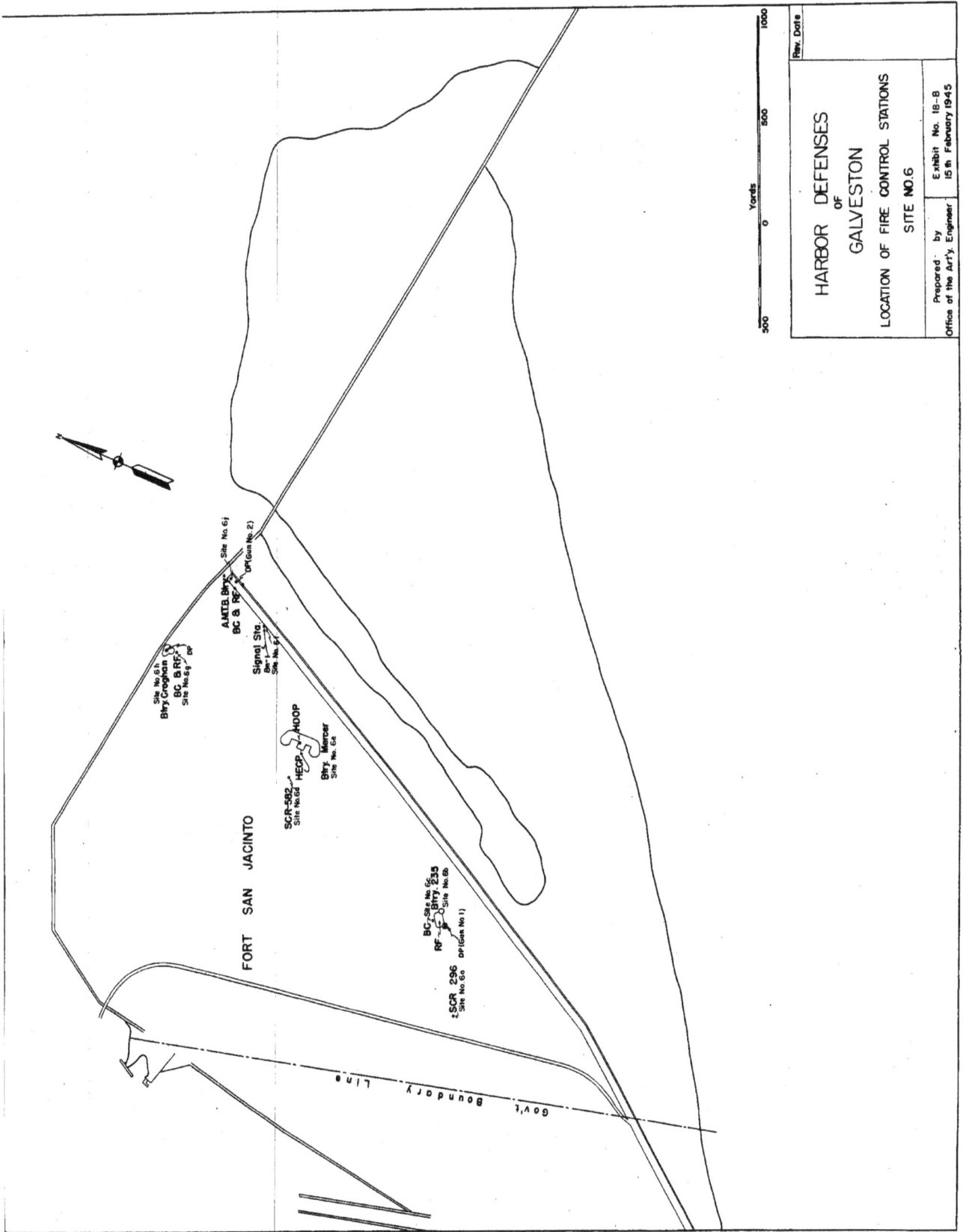

FORT SAN JACINTO

Site No.6h
Btry Croghan
BC & RF
Site No.6g   DP

AMTB Bry
BC & RF
Site No.6j

DP(Gun No.2)

Signal Sta.
Bn-1
Site No.6f

SCR-582
Site No.6d   HECP   ADOP

Bry. Mercer
Site No.6e

BC Site No.6c
RF   Bry 235
Site No.6b

DP(Gun No.1)

SCR 296
Site No.6a

Gov't Boundary Line

Yards

500   0   500   1000

HARBOR DEFENSES
OF
GALVESTON

LOCATION OF FIRE CONTROL STATIONS

SITE NO.6

Prepared   by
Office of the Art'y. Engineer

Exhibit No. 18-B
15th February 1945

Rev. Date

The site of Battery Mercer at Fort San Jacinto (Terry McGovern)

Battery Croghan at Fort San Jacinto (Terry McGovern)

was allocated for this project. The old work was completely removed, but the armament was saved and reused. Work of the rebuilt battery was completed by early 1904. New transfer was delayed however until August 11, 1911 for a new cost of $291,309.29. It was named in General Orders No. 194 of December 27, 1904 for Brigadier General Hugh Mercer of the Continental Army. It was armed with eight 12-inch Model 1890M1 Bethlehem mortars on Model 1896 mortar carriages (#13/#75, #16/#74, #8/#71, #19/#70, #9/#82, #15/#35, #11/#68 and #14/#81). Two mortars from each pit were removed and sent to Battery Izard at Fort Crockett in April of 1922. The final four guns continued to serve until authority to remove in January of 1943. The emplacement was subsequently modified to become the combined HECP/HDCP for Galveston and still exists through in poor condition and almost fully filled with dredging spoil. The battery site is open, but access is very difficult.

- **HOGAN:** A battery for two war emergency, Armstrong 4.7-inch guns emplaced in 1898 at Fort San Jacinto. The design plans were submitted for approval on April 23, 1898. It was northeast of the 10-inch battery, parallel to the shore and main gun line, firing to the northeast. The plan had each gun on its own platform, with a small traverse to protect from enfilade fire of ships that might actually be in the channel. Work was done in the late spring of 1898, using the National Defense Act funds from right at the start of the Spanish American War. Itwas armed with two 4.7-inch Armstrong guns on pedestal mounts in July of 1898 (#12126/#11007 and #12125/#11008). It was originally turned over on October 25, 1899. This battery, like all of those at Fort San Jacinto, was severely damaged by the Galveston hurricane of September 1900. However, this position was repaired rather than entirely reconstructed. As the foundation (except the pilings) had been swept away, new sheet pilings and concrete were added from a $50,000 appropriation. Work was done by early 1903, and a new transfer made on August 21, 1911 for $47,318.29. The battery was named in General Orders No. 194 of December 27, 1904 for Brigadier General John B. Hogan, Alabama volunteers, who saw service in the War of 1812. The battery was disarmed in 1917. In 1921 a new switchboard room was built in the structure and transferred for a coat of $2744. Also, later it was used for a while as the fort C-2 station. The emplacement itself was leveled in 1953.

- **CROGHAN**: A battery for two 3-inch masking parapet guns allocated to the Fort Point reservation in 1898. Plans were submitted on December 24, 1898 for the battery. It placed the battery on the northeast shore of the point, in contact with the extension of the south jetty. It fired to the northeast. Soft ground meant it had to be placed on essentially an island of wood grillage on piles. A special dredge had to be purchased to fill in the location and the $5682 cost was charged against the battery account. Two platforms were spaced at 37-feet, a long protective wall on the eastern flank protected it from enfilade fire. Work was done during 1899 and finished without armament installed by June 30, 1900. It was turned over to troops on March 3, 1900 for $32,988.75. The two 3-inch Model 1898 guns and pillars were shipped here just nine days prior to the Galveston hurricane. The storm severely damaged the battery, sweeping away the fill under the battery. It did not collapse, and new sheet pilings and fill were able to be used to repair rather than requiring reconstruction. A 1902 allotment for an additional $35,000 for this work was made, the battery being repaired during 1903. During the rebuilding on the right flank of the battery a mining casemate was included at the same time. New transfer came on September 21, 1911 at a cost of $33,673.57. Opportunity was taken to convert the battery during repair to pedestal mounts rather than balanced pillars, the original two guns were shipped away from storage, and it was finally armed with two 3-inch Model 1903 guns and pedestal mounts (#6/#29 and #48/#30). It was named in General Orders No. 194 of December 27, 1904 for Colonel George Croghan, Inspector General during the

War of 1812. The battery served until final deletion in 1946. Still in surprisingly good condition, the emplacement still exists. The battery is open to the public.

- **Battery #235**: A 1940 Program battery for two 6-inch barbette guns allocated for the Galveston defenses. It was to be emplaced southwest of Battery Mercer, near the southern limit of the reservation, firing to the east. The plan was entirely consistent with the type for other 200-series, 6-inch barbette batteries. It was funded under the FY-1942 budget as national priority #16. Construction was done from February 15, 1942 to January 28, 1943 for transfer being made on November 28, 1944 at a total cost of $310,091. The battery was never named, just known as Battery Construction No. 235 during construction and service. It was armed with two 6-inch guns Model 1905A2 and Model M1 barbette carriages (#19/#60 and #29/#61). It had three base-end station assignments in towers assigned to the defenses. The battery then served until disarmed at the end of the local harbor defenses, probably in 1946 or 1947. The emplacement still exists, though in poor condition and surrounded and partly filled with dredging spoil. The battery site is open, but access is very difficult.

- **AMTB #4a**: A 1943 Program AMTB battery consisting of two 90mm fixed and two 90mm mobile guns. It was assigned to a position on the northern coastline of the fort, between batteries Croghan and Hogan. Concrete gun blocks were built from December 2, 1942 to April 2, 1943 for transfer on August 4, 1943 at a cost of $17,039. The battery was then abandoned soon after the end of the war, probably in 1946. The blocks were covered by the reconstruction of the seawall in the early 1960s, so only one emplacement remains visible at the end the seawall. The battery site is open to the public.

Fort Crockett (1898-1953) is located at City Beach, to the southwest of downtown Galveston, Texas. Fort Crockett was established as an Endicott Program fort of three concrete batteries built from 1898 to 1902. It was named in General Order 43 of 1900 for David Crockett, U.S. Congressman, who died at the Alamo on Mar. 6, 1836. This was the main garrison post for the Galveston defenses. Battery Hoskins with two 12-inch guns on long-range barbette carriages was built in 1924 on the southern edge of the reservation. During the 1920s and early 1930s, Fort Crockett housed the United States Army Air Corps' (USAAC) 3rd Attack Group using an airfield located about a mile SW of the fort. During World War II, Battery Hoskins was casemated. The post also served as a German POW camp until 1946. After World War II, the fort received some additional buildings for military use. Parts of the reservation along the beachfront were declared surplus and sold to private interests. In 1950s, the buildings and remaining federal property were transferred to the National Marine Fisheries Service (NOAA Fisheries). NOAA has refurbished of most the remaining garrison and post World War II Army buildings. Several buildings were also used by the Texas A&M Maritime Academy. The mortar battery and the 10-inch disappearing battery were destroyed for the beachfront roadway, the partial remains of Battery Laval are imbedded in the seawall, and Battery Hoskins serves as the foundation for the San Luis Hotel and Conference Center and is closed to the public. The casemates can be seen from the seawall highway and the rear entrances to each casemate can be viewed from each side.

Fort Crockett Gun Batteries

- **HAMPTON**: A battery for two 10-inch disappearing guns emplaced as the first new battery for the Galveston City Beach reservation of Fort Crockett. Plans were submitted on March 31, 1898 using funds from the new National Defense Act. Its first design was conventional; however, it did have ammunition for both guns in a shared magazine area under the central traverse of the two guns. Ammunition service was by hoists. Work was done in 1898-1899, completing on August 31st of that year. It was transferred initially on October 25, 1899 for $109,000. The battery was armed with two 10-inch Model 1895 Watervliet guns on Model LF 1896 carriages (#2/#50 and #3/#49). The battery was damaged by the hurricane of September 1900, but not so severely that it couldn't be repaired. It was re-transferred on August 21, 1911 for a new cost of $87,929.28. It had been named in General Orders No. 194 of December 27, 1904 for Brigadier General Wade Hampton, who served during the Revolutionary War and War of 1812. The guns were designated for removal during World War I for potential use on railroad mounts but this was not carried out. The battery then served until authorized for removal in late 1942, the guns and carriages being scrapped in January 1943. The emplacement was mostly destroyed during the construction of the new seawall in 1961-1962, but the outline of the structure and some evidence of remains along the main street that surmounts the current protective wall.

- **IZARD**: The mortar battery for Fort Crockett, located about in the center of the reservation, approximately 200-feet from the shoreline. Initial plans were submitted on July 1, 1898 using funds from the Act of May 7, 1898. It fired to the southeast. Work began on September 17, 1898 and was substantially advanced by June 30, 1899. Most other work was done the following year, coming for a total expenditure of $106,139.76. It was damaged by the hurricane in September 1900, before transfer could be made, and the mortars were not yet mounted. Repairs could be made, though new wing walls had to be built around the pits. By 1903 the reconstruction was done, but the mortars were still not mounted. Finally on August 21, 1911 retransfer was made for a cost of $150,204.10. It had been named in General Orders No. 194 of December 27, 1904 for Major General George Izard, U.S. Army who served in the War of 1812. It was armed with eight 12-inch mortars Model 1890M1 on carriages Model 1896 (Watervliet tubes #57/#153, #63/#166, #54/#167, #106/#156, #65/#154, #104/#152, #58/#155 and Niles tube #23/#157). In 1918 all eight mortars were removed, but the carriages remained behind. After the war, in 1922, four mortars were removed from Battery Mercer at Fort San Jacinto and placed on four of the remaining carriages of Battery Izard (the battery winding up with Bethlehem tubes #15/#166, #16/#167, #14/#157, and #19/#153). The four spare carriages were then scrapped. This final armament continued to serve until finally being deleted in December of 1943. The emplacement itself was totally destroyed for the seawall expansion in the early 1960s.

- **LAVAL**: A battery for two 3-inch, 15-pounder masking parapet guns erected on the right flank of Battery Izard, close to the shoreline at Fort Crockett. Plans were submitted on November 7, 1898, using funding from the Deficiency Act of July 7, 1898. It was built between January 9 and June 30, 1899. By a year later most of the installation was complete. Damage from the 1900 hurricane necessitated the reconstruction and repair to the yet incomplete emplacement. This was done to the end of 1902. As rebuilt after the storm damage it had two irregularly shaped platforms, and just a single large, shared magazine between the two guns with an exit door leading to each platform. Final transfer was delayed until August 21, 1911 for a cost of $27,321.88. It was named on General Orders No. 194 of December 27, 1904 for Colonel Jacint Laval, 1st U.S. Dragoons of War of 1812 service. It was armed with two Model 1898 Driggs-Seabury guns on balanced pillar mounts (#37/#37 and #38/#38). While this was the original armament on hand since 1900, the original base rings had been lost in the hurricane and obtaining new ones was one of the reasons for the long delay in mounting. These carriages were modified in 1916 in place to the M1898M1 fixed parapet standard and then removed in June of 1920 in common with all such armament. A CRF station was added to the rear of the battery in 1921. The emplacement was mostly filled-in with concrete for the new seawall built in 1961-1962, but the outline and some structural elements can still be seen today just off the side of the street that runs along the protective seawall.

GALVESTON HARBOR

FORT CROCKETT

CITY BEACH.

SERIAL NUMBER 124

EDITION OF: DEC. 7, 1915;
REVISIONS: MAR. 14, 1916; MAY 13, 1919.
APR. 5, 1921; DEC. 30, 1921.

Riprap retaining wall.

SEAWALL

Wade Hampton

George Izard

Jacint Laval

SEAWALL

TRUE MERIDIAN

Leonard Hoskins

U.S. ENGR. RES'N.

LEGEND

1 ADMINISTRATION BLDG.
2 COMMANDING OFFICER'S QRS.
3 OFFICER'S QUARTERS.
4 HOSPITAL.
5 HOSPITAL STEWARD'S QRS.
6 N.C. OFFICERS' QUARTERS.
7 BARRACKS.
8 GUARD HOUSE.
9 POST EXCHANGE.
10 MESS HALL.
11 ARTY. ENGINEER
12 BAKERY.
13 GRANARY.
14 STABLE.
15 CORRAL.
20 Q.M. OFFICE.
21 Q.M. AND COMMISSARY ST.HO.
70 AMUSEMENT HALL.
16 WAGON SHED.
17 OIL STORAGE HOUSE.
18 BARBER SHOP.
19 GARAGE.
23 Q.M. WORK SHOP.
24 COAL SHED.
30 ORDNANCE SHOP.
50 SIGNAL OFFICERS OFFICES
100 FIRE STATION.
40 ENGR. OFFICE.
101 SHELTER FOR
      ANTI-AIRCRAFT GUNS

BATTERIES.

GEORGE IZARD... 8-12"M.
WADE HAMPTON... 2-10"DIS.
JACINT LAVAL... 2-3"B.P.
LEONARD HOSKINS... 2-12" B.

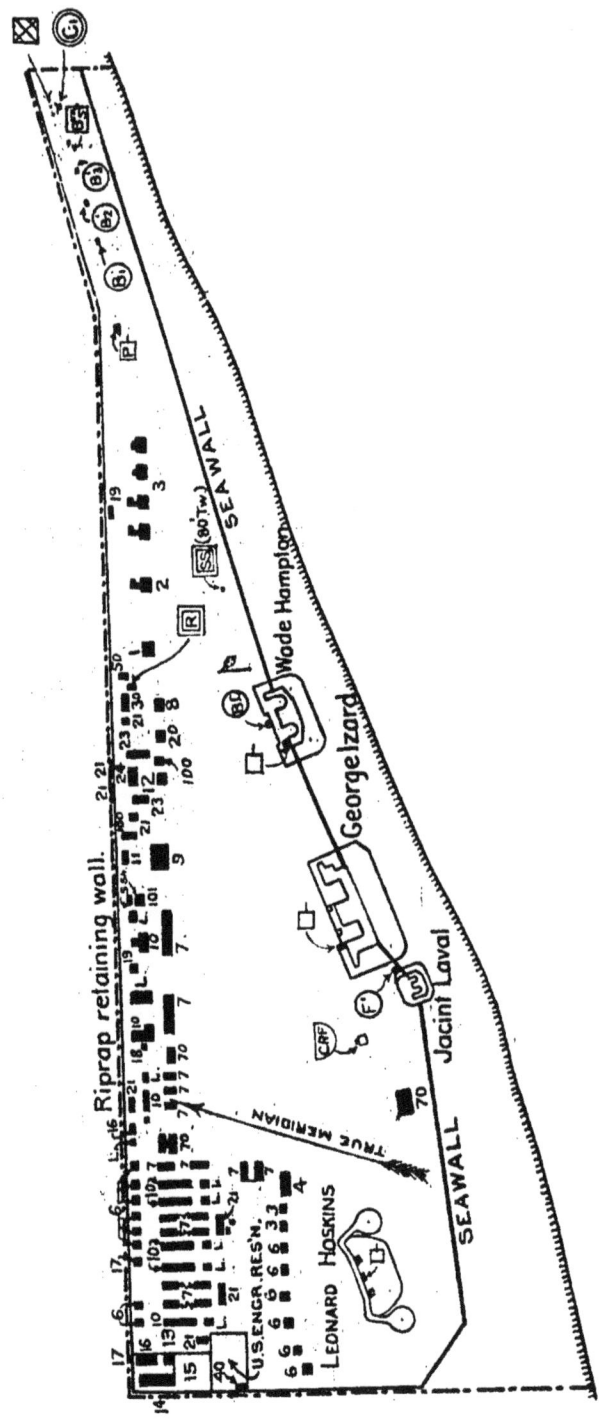

0   500'   1000'   1500'   2000'

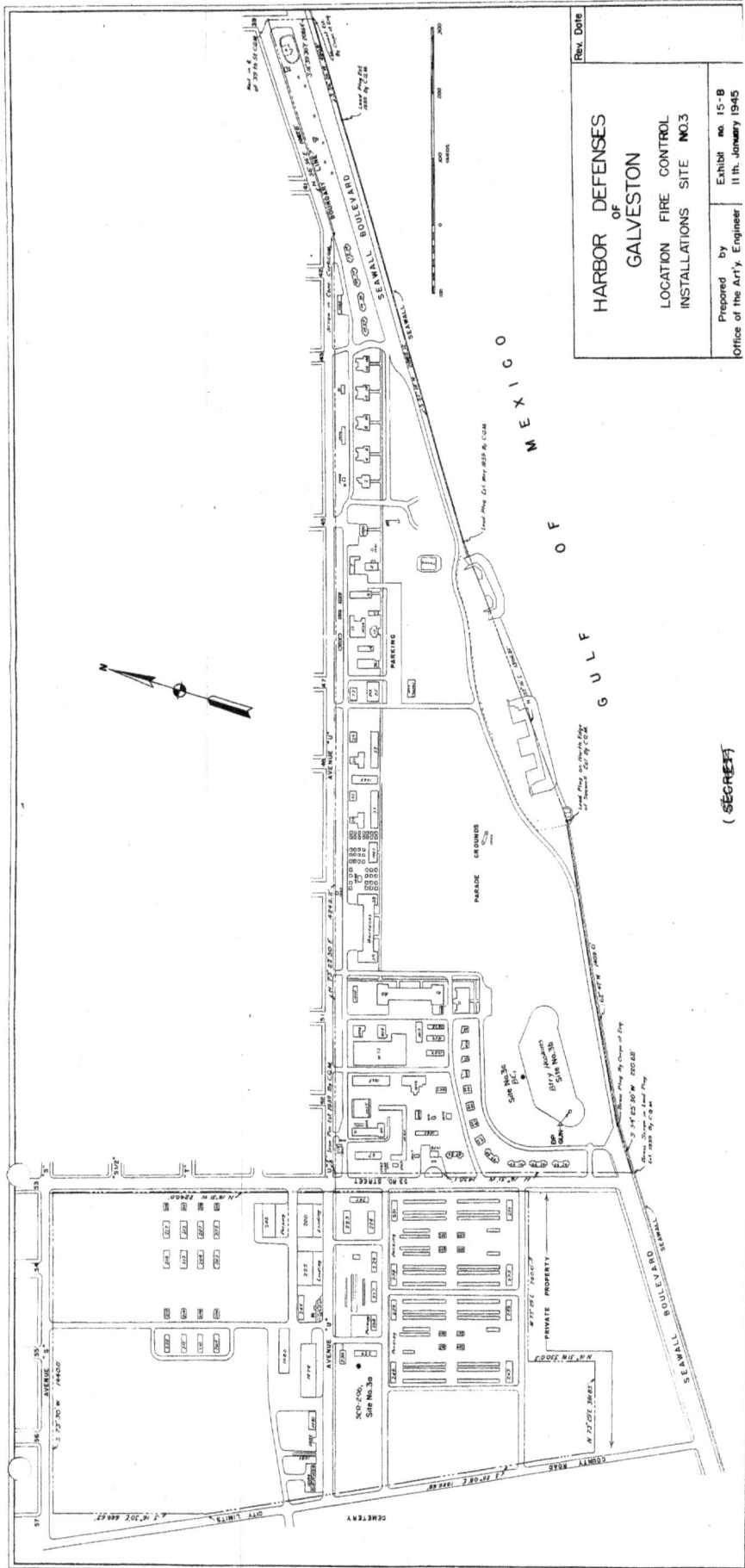

HARBOR DEFENSES
OF
GALVESTON

LOCATION FIRE CONTROL
INSTALLATIONS SITE NO.3

Exhibit no. 15-B
11th. January 1945

Prepared by
Office of the Art'y. Engineer

Rev. Date

GULF OF MEXICO

(SECRET)

Fort Crockett 1942 (NARA)

Fort Crockett 1932 (NARA)

- **HOSKINS**: A 1915 Program battery for two 12-inch guns on new long-range barbette carriages emplaced on the far western edge of the Fort Crockett reservation. It was placed to fire in a southeastern direction. It closely followed the recommended type plans and had the "open-back" type of construction. Plans were submitted (along with sister Galveston Battery Kimble) on January 20, 1916. Final approval of revised plans came on March 28, 1917. Work was started in August 1917 and the concrete portion was finished by year's end. Transfer was made on May 27, 1921 at a cost of $304,237.63. It was armed with two 12-inch Model 1895M1A4 Watervliet guns on the new Model 1917 long-range barbette carriage (#45/#11 and #56/#10). It was named in General Orders No. 12 of March 15, 1921 for 2nd Lieutenant Leonard C. Hoskins, killed in action in Europe during World War I in June of 1918. It served as an important element of the Galveston defenses for the next twenty-five years. It was provided with heavy overhead protection with new reinforced concrete gun houses over its two-gun positions and a complete enclosed gallery behind the magazine in the traverse between the guns. This work was done from September 1942 to March 1943. The battery was finally deleted in 1947. The emplacement still exists, though on the property of a hotel which has been partially built over the top of it. The hotel has used the battery's service gallery and rooms for utilities and storage. The battery can be seen from the exterior, but the interior is closed to the public.

The San Luis Resort atop Battery Hoskins (Terry McGovern)

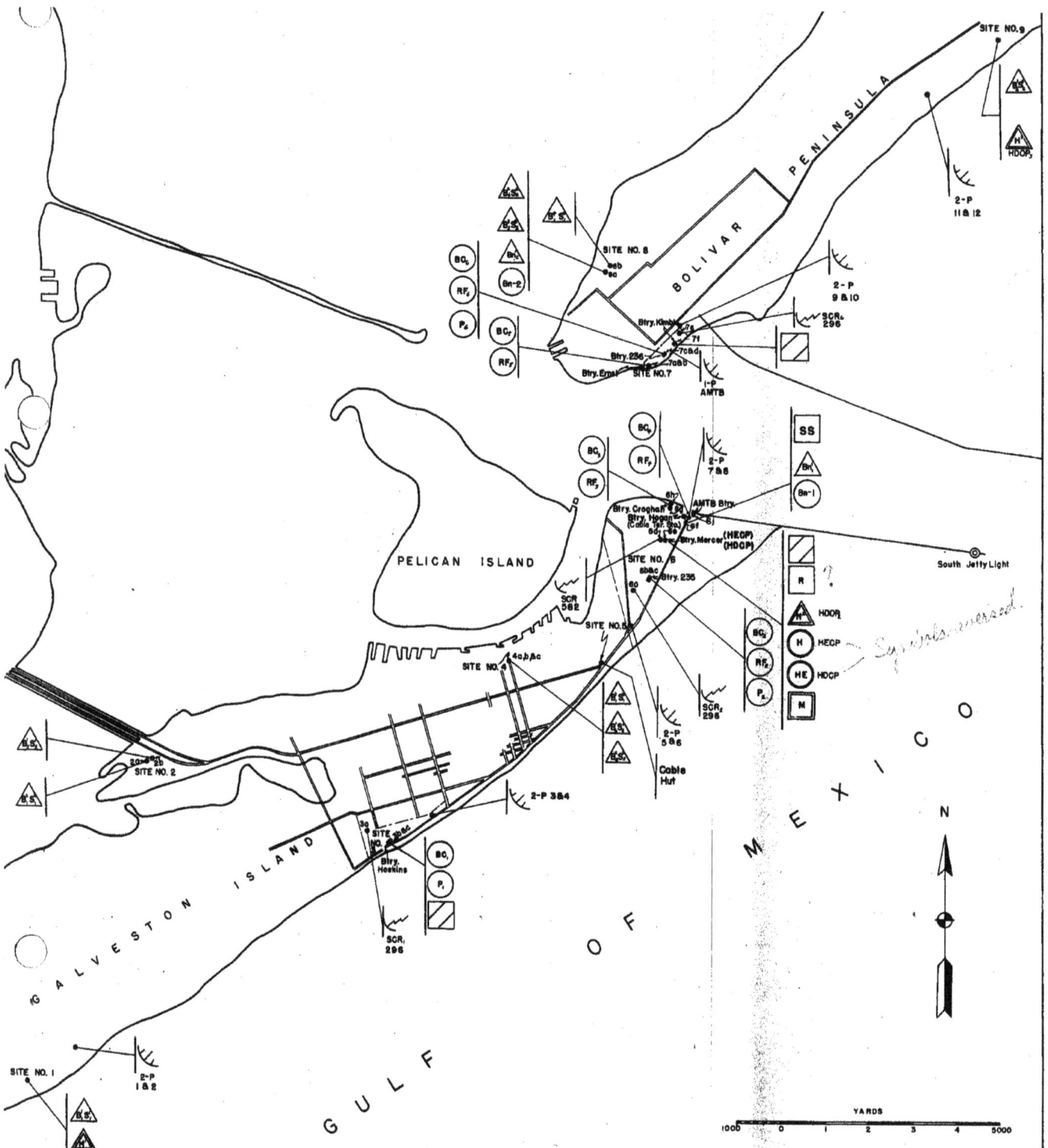

Site No.	Location
1	Gal. West Beach.
2	Gal. Causeway
3	Ft. Crockett
4	Roof Am. Nat'l. Ins. Bldg.
5	6th. St. & Blvd.
6	Ft. San Jacinto
7	Ft. Travis
8	Bol. Township
9	Bol. East Beach

HARBOR DEFENSES
of
GALVESTON
LOCATION CHART OF FIRE CONTROL
DEFENSE ELEMENTS

| Prepared by Office of The Art'y. Engineer | Exhibit No. 1-A 15 October 1944 |

Rev. Date 25 May 1945

Galveston World War II-era Site Locations. Stations housed in a single structure are connected by dashes (-)

location	Loc#	Purpose
West Beach	1	BS1/Hoskins-HDOP1
Causeway	2	BS1/235, BS2/Hoskins
Fort Crockett	3	Batt. Tact. #1 Hoskins, BC/Hoskins, SBR, SCR296-Hoskins, SL 3,4
ANICO Building	4	B2/235, B1/236, B3/Hoskins
6th Street & Blvd	5	SL 5,6
Fort San Jacinto	6	Batt. Tact. #2 BCN 235, Batt. Tact. #3 Crogan, Batt. Tact. #4 AMTB, BC/Croghan, BC/AMTB, BC-B1/235, SBR, HECP-HDCP-HDOP2, SCR296-235, SL 7,8
Fort Travis	7	Batt. Tact. #5 Ernst, Batt. Tact. #6 BCN 236, BC/Ernst, BC/236, SBR, SCR296-236, SL 9,10
Bolivar Townsite	8	B3/235-BS2/236-Bn1/2, B4/Hoskins
Bolivar East Beach	9	SL11/12, B3/236-HDOP3
Bolivar Lighthouse Res		not used

Barracks, Fort Crockett, NOAH (Mark Berhow)

Rinks, J.C., *Galveston Coast Artillery Forts 1895 thru the Second World War,* privately published 2025

Gaines, William C. "The Seacoast Defenses of Galveston, Texas." *Coast Defense Journal* Vol. 21, No. 4, Nov. 2007, p. 4.

Battery Cullum-Sevier Fort Pickens Unit Gulf Shores National Seashore (Mark Berhow)

90 mm M1 gun on M3 mount at Battery Parrott Fort Monroe National Monument (Mark Berhow)

www.ingramcontent.com/pod-product-compliance
Lightning Source LLC
Chambersburg PA
CBHW040259100426
42811CB00011B/1313